REIMAGINING COMMUNICATION: MEANING

Reimagining Communication: Meaning surveys the foundational theoretical and methodological approaches that continue to shape communication studies, synthesizing the complex relationship of communication to meaning making in a uniquely accessible and engaging way.

The *Reimagining Communication* series develops a new information architecture for the field of communications studies, grounded in its interdisciplinary origins and looking ahead to emerging trends as researchers take into account new media technologies and their impacts on society and culture. *Reimagining Communication: Meaning* brings together international authors to provide contemporary perspectives on semiotics, hermeneutics, paralanguage, corpus analysis, critical theory, intercultural communication, global culture, cultural hybridity, postcolonialism, feminism, political economy, propaganda, cultural capital, media literacy, media ecology and media psychology. The volume is designed as a reader for scholars and a textbook for students, offering a new approach for comprehending the vast diversity of communications topics in today's globally networked world.

This will be an essential introductory text for advanced undergraduate and graduate students and scholars of communication, broadcast media, and interactive technologies, with an interdisciplinary focus and an emphasis on the integration of new technologies.

Michael Filimowicz, PhD, is a Senior Lecturer in the School of Interactive Arts and Technology at Simon Fraser University. His research is in the area of computer-mediated communication, with a focus on new media poetics applied in the development of new immersive audiovisual displays for simulations, exhibition, games and telepresence as well as research creation.

Veronika Tzankova is a PhD candidate in the School of Interactive Arts and Technology, Simon Fraser University and a Communications Instructor at Columbia College – both in Vancouver, Canada. Her background is in human–computer interaction and communication. Sports shape the essence of her research which explores the potential of interactive technologies to enhance bodily awareness in high-risk sports activities.

REIMAGINING COMMUNICATION: MEANING

Edited by Michael Filimowicz and Veronika Tzankova

NEW YORK AND LONDON

First published 2020
by Routledge
52 Vanderbilt Avenue, New York, NY 10017

and by Routledge
2 Park Square, Milton Park, Abingdon, Oxon OX14 4RN

Routledge is an imprint of the Taylor & Francis Group, an informa business

British Library Cataloguing-in-Publication Data
A catalogue record for this book is available from the British Library

Library of Congress Cataloging-in-Publication Data
Names: Filimowicz, Michael, editor. | Tzankova, Veronika, editor.
Title: Reimagining Communication : Meaning / Edited by Michael Filimowicz and Veronika Tzankova.
Other titles: Meaning
Description: First edition. | Abingdon, Oxon ; New York, NY : Routledge, 2020. | Includes bibliographical references and index. | Summary: "Reimagining Communication: Meaning surveys the foundational theoretical and methodological approaches that continue to shape Communication Studies, synthesizing the complex relationship of communication to meaning making in a uniquely accessible and engaging way"– Provided by publisher.
Identifiers: LCCN 2019056098 | ISBN 9781138542860 (hardback) | ISBN 9781138542884 (paperback) | ISBN 9781351007924 (ebook) | ISBN 9781351007917 (adobe pdf) | ISBN 9781351007900 (epub) | ISBN 9781351007894 (mobi)
Subjects: LCSH: Communication and technology. | Media literacy. | Meaning (Philosophy)
Classification: LCC P96.T42 R449 2020 | DDC 302.23–dc23
LC record available at https://lccn.loc.gov/2019056098

ISBN: 978-1-138-54286-0 (hbk)
ISBN: 978-1-138-54288-4 (pbk)
ISBN: 978-1-351-00792-4 (ebk)

Typeset in Bembo
by Swales & Willis, Exeter, Devon, UK

CONTENTS

SERIES INTRODUCTION

In an age of information overflow, the word "communication" seems to have reached a buzz-status. From persuasive communication, to the importance of good communication, to communication skills for bridging gaps, the term and domain of "communication" have become increasingly obscure and even confusing. We see this confusion reflected in students in introductory communication courses. Their initial understanding of communication often tends to focus on linguistic exchanges between individuals, not accounting for the wide range of personal, social and contextual dynamics that influence not only the transfer of information, but also the complex processes of meaning making at individual and collective levels. Where initial conceptualizations of communication have focused on information-transfers between senders and receivers, contemporary scholarship seems to apply a much broader and to an extent pragmatically-informed approach looking not only at how communication practices work, but also how these shape the ways in which we make sense of the reality that surrounds us.

Partially due to the wide range of perspectives and approaches, the field of communication studies still lacks distinctive disciplinary boundaries. When we think of traditional academic fields – such as biology, sociology, philosophy and mathematics, for example – we know with some degree of certainty what the object of study is. When it comes to the field of communication however, we realize that establishing a definitive scholastic focus is not easy, if not impossible. This is to an extent related to the challenges associated with answering a fundamental, and seemingly simple question: "What is communication?" As an all-encompassing answer to this question should be able to account for all the situations, exchanges, contexts, interpretations, channels and any possible mix of these, we can see that this can easily become a hopeless task. In an attempt to resolve this complexity, scholars have emphasized different sides of communication processes that have influenced the development of a multiplicity of material, functional and experiential definitions. Most definitions revolve around one of the following categorical characteristics of communication: (1) transfer of information, (2) symbolic culture and (3) a ritual that facilitates the inherent social essence of humanity (see Carey, 2009). These varying types of exegesis pose fundamental challenges in defining and delimiting the concept and disciplinary boundaries of communication. Within such expanding views, the purpose of the *Reimagining Communication* series is to capture the existing and prospective trends and perspectives on and within communications studies as a whole. To systematize our approach, we have provided a specific theme for each of our volumes: Meaning, Experience, Action and Mediation. Each of these volumes extends on the existing notions and scholarship within the field of communication, in order to capture – as much as possible – the main contexts, communication technologies, institutions and social practices.

These volumes construct a new information architecture that reimagines a new organization for traditional and emerging themes in communication studies. To describe our editorial project, we appropriate the concept of "information architecture" (hereafter IA) from the domain of human–information interaction. The notion of IA – as opposed to the commonly used "map" or "framework" – highlights the constructed character of the series as a work of design, and as one design solution out of many possible others. We have decided to redesign and reimagine the "field of fields and disciplines" of communication through a forward-looking multidisciplinary lens that has its grounding in both established academic scholarship and developing domains focusing on the latest technological developments.

> [A]s much as architects are expected to create structure and order in the world through planning and building, information architects were expected to draw lines and derive some kind of order in dataspace, their primary task being to make this information simpler, more direct, and ultimately more comprehensible.
>
> *(Resmini and Rosati, 2012, 4)*

Similarly, the purpose of this series is to try to capture the "dataspace" of communication studies today directly and as a comprehensible whole grounded in where it has been and inferring where it seems to be going. The chapter topics also facilitate "crosstalk" across the volumes – a desired aspect of our schema. Thus, while eschewing maps and frameworks, we are also not offering "an intellectual patchwork" for a field that "[f]rom the beginning ... was critiqued for having no core" (Zelizer, 2016, 223)

The institutional origins of communications in the academy – (1) whether one locates it in post-World War 2 interests in journalism, mass media and propaganda studies (Jensen and Neuman, 2013, 231) or in the development of speech programs c. 1915 (Wilson, 2015) and (2) regardless of professional, liberal arts or mixed-orientation (Troester and Wertheimer, 2015, 2) – seem very distant from the kinds of questions explored by many contemporary communications scholars. When we think of the early days of communication studies, the associations that often come to mind are the printing press and broadcast towers for radio and television. But when we think of the future of communication, we tend to drift to the new territories of brain–computer interfaces and the hybrid arrays of virtual-augmented-mixed-crossed realities where our living environments become increasingly networked and informatic in character. We may even say that the future of communication reaches to the nano-level of neurons.

The space between neurons and socio-cultural spaces – populated by the technologies of mediation and social practices of meaning-making humans – is what gives shape to our IA across the series' four volumes – Meaning, Experience, Action and Mediation. These themes carry an empirical core as they relate directly to the animal level of sensorimotor interactions (Experience, Action) taking place within the affordances of the environment (Mediation), where all of these are understood as a source of information (Meaning). As humanity becomes increasingly more cyborg, we hope that such theoretical trace-back to primordial animal conditions will future-proof these volumes in a way that other volumes have not been able to.

While our IA approach is empirically motivated, it is neither an ontology, nor periodization of communication studies. Much communication scholarship has traditionally relied on a "grand-narrative" *modus operandi* based on essential dichotomies. Some of these dichotomies are implicit, such as the oppositions underlying the modalities of the one-to-one (interpersonal communication), one-to-many (broadcast technologies) or even many-to-many (social media) communication practices. Other dichotomies are explicit, formally expressed and have become canonical in communication studies:

> By now, several generations of scholars have relied on such dichotomies, grounding their grand historical interpretations on the ways in which media interact with social life. The work of Harold Innis (1951) is commonly associated with the contrasting influences of time-biased and space-biased media; McLuhan (1964) classified media as hot and cold; Ong (1982) traced social evolution through the prism of orality or literacy; Carey (1969) in one of his earliest works classified media as centripetal and centrifugal; and Turow (1997) talked about society-making and segment-making media.
>
> The dichotomies on which they rely frequently lead to periodization, or the attempt to locate pivotal moments in which some new essential aspects of social development suddenly emerge while others vanish. The ultimate

> purpose of periodization is to establish compelling, often teleological or cyclically structured narratives relying on a sequence of communication eras defined through different technological paradigms.
>
> *(Balbi and Kittler, 2016, 1972)*

With respect to conceptual temptations toward periodization, cycles and telos, our IA looks forward as much as backward, considering the animal plane of sensorimotor interactions with affordances as ontologically flat with the plane of the cybernetic meaning maker. Our claim is that this distinctive IA approach is somewhat different from those grand dichotomous periodizations of the field that have come before. As our IA cuts across disciplinary boundaries, it is difficult not to acknowledge "that the politics of interdisciplinarity is always *a politics of the moment*" (Shome, 2006, 2). So, the image of the brain–computer interface, strategically placed as the final chapter in the fourth volume, is of our present time. But in parallel, we have also chosen the most primordial and earliest human moment – sensorimotor interactions with affordances in an environment – if only to give our best wishes to the historical chances at the bio-cybernetic frontier of communication media technologies today.

The reconsideration of binaries becomes complex when it comes to institutionalized dichotomies in the historical record – such as the debates related to the categorization of communications in the academy as a field or a discipline (Phillips, 2016, 691) where such categorizations are often suspended between institutional forces of "cohesion and fragmentation" (Nordenstreng, 2007, 212). In an attempt to escape the limits of institutionalized dualities, our IA tries to acknowledge and accommodate a multiplicity of perspectives within an egalitarian treatment of both the phenomena studied and the methodologies by which these are studied. In this series, we have placed a strong emphasis on technologies as the recent convergence of media, content, industries and audiences have presented "significant difficulty separating our ideas about communication from the technological advance of media" (Herbst, 2008, 604). These difficulties are still far from being overcome and are at the core of the problematics studied by communication researchers. Debates around disciplinarity often relate to issues regarding the status of communications in the academy, such as "the desires for monopolization, legitimization, and recognition" (Stalker, 2014, 172), "expansion, monopolization and protection" (173) and "professional identity" (First and Adoni, 2007, 252). Such debates are somewhat removed from our IA since we presume the relevance of communication to all disciplines and indeed all fields, as well its essence as a permanently open form of inquiry capable of utilizing the full array of methods including those that are increasingly computational in nature. Communication studies have always been defined by its undefinability because what does not count as communication? It has always been "poly" in its methodologies and scope of interests since the founding of its first units in the academy.

> [T]he field consolidated as the result of the aggregation of academic interests in communication broadly defined – interpersonal, organizational, mediated, media industries, cultural studies, information studies, language, rhetoric, intercultural, journalism, and media and information policies, among others.
>
> *(Waisbord, 2016, 869)*

In addition to this broad disciplinary integration (and fragmentation), communication is somewhat unique amongst the academic disciplines in the way it has historically bridged professionally applied and theoretically inclined orientations.

> The so-called "field" of communication is, of course, not a field by any academic definition. It has no boundaries. It is equally hospitable to the advertising agency seeking evidence to support the slogan "it pays to advertise," and to the semanticist seeking closer definitions of "meaning."
>
> *(Karin, 2003, 4)*

> Some academic fields such as history and philosophy are more central in the pursuit of liberal arts, while others such as business administration and engineering are more related to career development. The discipline of communication is fairly unique as it crosses these boundaries.
>
> *(Morreale et al., 2000, 1)*

It is in this spirit of crossing the boundaries between theory and practice, or professional practice and liberal arts, that we have conceived of our editorial project as an IA, as IA is a form of communication design. Our research backgrounds are grounded in computer-mediated communication (multimodal display and interactive sports technologies, for Filimowicz and Tzankova, respectively). In our lines of research, there are no real institutional, practical or conceptual barriers that prevent easy crossover between poetic hermeneutic mullings and concrete systems design – to name just two ends of the many fluid spectrums of transdisciplinary inquiry in our work. For more than a decade now, the field of human–computer interaction has initiated an increased interest in so-called "third wave HCI" where meaning-making has become exponentially central to the design of computational artifacts. The material character of interactive artifacts is far from being foreign to communication as its origin – the Latin word *communicatio* – refers to social functions organized around tangibles.

> "Communication" is a word with a rich history[, f]rom the Latin *communicare*, meaning to impart, share, or make common … The key root is *mun-* (not *uni-*), related to such words as "munificent," "community," "meaning" … The Latin *munus* has to do with gifts or duties offered publicly – including gladiatorial shows, tributes, and rites to honor the dead. In Latin, *communicatio* did not signify the general arts of human connection vis symbols, nor did it suggest the hope for some kind of mutual recognition. Its sense was not in the least mentalistic: *commuicatio* generally involved tangibles.
>
> *(Peters, 1999, 7)*

> The notion of communication in ancient Rome, as well as the previous notion of rhetoric from ancient Greece, did not refer to transfer, to transmission, to interaction or dialogue, but rather pointed to acknowledging and performing specific social functions and group memberships, or to knowing and utilizing concrete technical devices for conveying specific social functions and group memberships.
>
> *(Nastasia and Rakow, 2005, 4)*

Such a broad conception of communication – which gets even more extended when we put into consideration its social and ritualistic aspects – does run some risks of definition and scope, but these are familiar risks that have always been associated with the discipline's "ontological deficit and epistemological pitfalls" (Kane, 2016, 88). Communications in its academic origins "was stunningly interdisciplinary from the start" (Herbst, 2008, 604).

> [T]he short tradition as an academic discipline, the external influences coming from meda industry and the state, the legitimacy deficit, the diffuse research topic "communication," the scattering across the university and the heterogeneous scientific origins of its scholars … these characteristics lead to a "lack of consensus" within the field concerning its subject matters and to difficulties to shape a self-conception.
>
> *(Löblich and Scheu, 2011, 2–3)*

The concept and practice of Information Architecture originates from domains of human–computer interaction, defined early on as "the conceptual structure and functional behavior, as distinct from the organization of data flows and controls, logical design, and physical implementation" (Amdahl et al., 1964, 21). We discussed our rationale for the *conceptual* structure of these volumes – the desire to avoid dichotomies, ontologies, periodizations and grand narratives through our strategic use of guiding analogies that allow a simultaneous look forward to new cyborg variations of humanity but also backward to complex organisms and their sensorimotor engagements within informational environments.

We now want to emphasize the *functional* aspect of our IA by linking it to our professional context and the connection to pedagogy. As teaching faculty, we have encountered difficulties finding adequate texts that cover the gamut of interdisciplinary *and* international considerations about and within communication studies in a way that interests those who are the basis of our employment.

> Those of us who teach communication theory face unique challenges. Undergraduates … come for something comprehensible and we offer them fragments of a subject no one can comprehend, up to 249 theories and still counting.
>
> *(Craig, 1999, 153)*

Craig's mapping of the traditions of communication contains eight main domains, all of which find representation in this collection: social-psychological, cybernetic, rhetorical, semiotic, critical, sociocultural, phenomenological and pragmatic (Phillips, 2016, 701). The functional aspect of our IA is to "cultivate a sense of the 'whole' of communication studies, not a 'unified,' stable entity but a polyphonic, unstable whole that can be developed as a theory-rich body of work through dialogue across difference" (700). This cultivation of a sense for the whole is not only for the benefit of our students, but also for us as researchers and communication designers – as by definition we ourselves are lifelong learners.

We are particularly pleased to be able to offer such an international assemblage of authors in these volumes hailing from 19 countries, literally from A to Z: Australia, Canada, China, Denmark, Estonia, France, Germany, Greece, Italy, New Zealand, Norway, Spain, Sweden, Switzerland, Turkey, UK, UAE, USA and Zimbabwe. There is a somewhat contrarian transnational impulse at work in bringing these global voices together, as many communication books used in classroom environments tend to overemphasize the national (e.g., the various communication books with country names in their title). Our aim is to foster the connectivity of ideas, which can be as national, transnational, international or cosmopolitan as one wishes to make of this gathering of voices from across the globe.

> If internationalism means exchanging knowledge and understanding across borders, then we would probably all sign up to it, confident that national approaches or concerns could find their place within this larger forum.
>
> *(Livingstone, 2007, 274)*

Where in the past, communication's scholarly traditions have been character-ized as "isolated frog ponds – with no friendly croaking between the ponds, very little productive intercourse at all, few cases of successful cross-fertiliza-tion" (Rosengren, 1993, 6), this IA-inspired collection aims to bring the frog ponds closer together (to strain the metaphor perhaps!). While these books cannot promise cross-continental cross-fertilization, they get the international croaking underway. Maybe for the moment the chapters will be read by its audiences sans brain implants, but as we continue to undergo the "mediation of everything" (Livingstone, 2009), surely not for long.

Michael Filimowicz
Veronika Tzankova

References

Amdahl, G. M., Blaauw, G. A. and Brooks, F. P. (1964). Architecture of the IBM System/360. *IBM Journal for Research and Development*, 8(2), 21–36.

Balbi, G. and Kittler, J. (2016). One-to-One and One-to-Many Dichotomy: Grand The-ories, Periodization, and the Historical Narratives in Communication Studies. *International Journal of Communication*, 10(2016), 1971–1990.

Carey, J. W. (2009). *Communication as Culture: Essays on Media and Society*. New York/London: Routledge.

Craig, R. T. (1999). Communication Theory as a Field. *Communication Theory*, 9(2), 119–161.

First, A. and Adoni, H. (2007). The Never-Ending Story: Structural Dilemmas and Changing Solutions in the Communication Field. *Mass Communication and Society*, 10 (3), 251–273.

Herbst, S. (2008). Disciplines, Intersections, and the Future of Communication Research. *Journal of Communication*, 58(4), 603–614.

Jensen, K. B. and Neuman, W. R. (2013). Evolving Paradigms of Communication Research. *International Journal of Communication*, 7(2013), 230–238.

Kane, O. (2016). Communication Studies, Disciplination and the Ontological Stakes of Interdisciplinarity: A Critical Review. *Communication and Society*, 29(3), 87–102.

Karin, W.-J. (2003). Wilbur Schramm Was Not the Founder of Our Discipline: New Findings on the History of Communication Research. Conference Paper: International Communication Association 2003 Annual Meeting, San Diego, CA, pp. 1–9.

Livingstone, S. (2007). Internationalizing Media and Communication Studies: Reflec-tions on the International Communication Association. *Global Media and Communica-tion*, 3(3), 273–288.

Livingstone, S. (2009). On the Mediation of Everything: ICA Presidential Address 2008. *Journal of Communication*, 59(1), 1–18.

Löblich, M. and Scheu, A. M. (2011). Writing the History of Communication Studies: A Sociology of Science Approach. *Communication Theory*, 21(1), 1–22, https://doi-org.proxy.lib.sfu.ca/10.1111/j.1468-2885.2010.01373.x.

Morreale, S. P., Osborn, M. M. and Pearson, J. C. (2000). Why Communication is Important: A Rationale for the Centrality of the Study of Communication. *Journal of the Association for Communication Administration*, 21(2000), 1–25.

Nastasia, D. and Rakow, L. (2005). What is Communication? Unsettling a Priori and a Posteriori Approaches. *The International Communication Association, Philosophy of Communication Division*, Washington, DC. 1 November 2005. Retrieved online http://citation.allacademic.com/meta/p_mla_apa_research_citation/0/9/3/2/6/pages93260/p93260-1.php. Accessed online Apr 18th, 2019..

Nordenstreng, K. (2007). Discipline or Field? Soul-searching in Communication Research. *Nordicom Review*, Jubilee Issue 2007, 211–222.

Peters, J. D. (1999). *Speaking into the Air: A History of the Idea of Communication.* Chicago: University of Chicago Press.

Phillips, L. (2016). Epistemological (Im)possibilities and the Play of Power: Effects of the Fragmentation and Weak Institutionalization of Communication Studies in Europe. *International Journal of Communication*, 10(2016), 689–705.

Resmini, A. and Rosati, L. (2012). A Brief History of Information Architecture. *Journal of Information Architecture*, 3(2), 1–12. Retrieved at http://journalofia.org/volume3/issue2/03-resmini/. Accessed online Apr 22, 2019 (Originally published in Resmini, A. and Rosati L. (2011). *Pervasive Information Architecture.* Morgan Kauffman. (Edited by the authors)).

Rosengren, K. E. (1993). From Field to the Frog Ponds. *Journal of Communication*, 43(3), 6–7.

Shome, R. (2006). Interdisciplinary Research and Globalization. *The Communication Review*, 9(1), 1–36.

Stalker, J. (2014). Disciplining Communication Study at the University of Illinois at Chicago, 1973–2007. *The Review of Communication*, 14(2), 171–181.

Troester, R. and Wertheimer, M. (2015). The Blending of Traditional and Professional Approaches to Communication: Department Chairs Share Administrative Challenges, Opportunities, and Best Practices. *Journal of the Association for Communication Administration*, 34(1), 2–11.

Waisbord, S. (2016). Communication Studies Without Frontiers? Translation and Cosmopolitanism Across Academic Cultures. *International Journal of Communication*, 10 (2016), 868–886.

Wilson, K.H. (2015). The National and Cosmopolitan Dimensions of Disciplinarity: Reconsidering the Origins of Communication Studies. *Quarterly Journal of Speech*, 101 (1), 244–257.

Zelizer, B. (2016). Communication in the Fan of Disciplines. *Communication Theory*, 26 (2016), 213–235.

INTRODUCTION

From popular texts and public discourses to scholarship and research, the use of the word "communication" has been on the rise in academic, professional and everyday contexts. But what exactly is "communication"? Is it possible to not communicate? How does communication work? What are the disciplinary boundaries of communication studies and what is its object of investigation?

Reimagining Communication: Meaning addresses these questions by presenting a survey of the foundational theoretical and methodological approaches in communication studies. As the field has been continuously growing and reaching new horizons, this volume specifically synthesizes major trends within the trajectory of fundamental ideas that have served as its base and continue shaping its sub-branches, covering topics related to the production of meaning in communication. By presenting perspectives illustrated by concrete examples on the topics of semiotics, hermeneutics, paralanguage, corpus analysis, critical theory, intercultural communication, global culture, cultural hybridity, postcolonialism, feminism, political economy, propaganda, cultural capital, media literacy, media ecology and media psychology, this volume synthesizes the complex relationship of communication to meaning making in a uniquely accessible and engaging way.

Chapter 1 – Paul Cobley's "Reimagining Semiotics in Communication" begins the volume by presenting a new approach to sign relations within the light of suprasubjectivity, interpretation and continuously changing contexts. As contemporary communication practices reach beyond human-to-human interactions, the chapter suggests that semiotics can be central to new developments in communication studies that transcend the limits of anthropocentrism.

Chapter 2 – "Hermeneutics" by Johan Fornäs presents a detailed exploration of the notion of interpretation and its relation to meaning, culture and communication. To better contextualize hermeneutics, the chapter surveys essential assumptions and themes of interpretation theory together with a brief historical overview of key contributions that have marked the developmental trajectory of the field.

Chapter 3 – "Paralanguage (The Cracked Lookingglass of a Servant, or the Uses, Virtues and Value of Liminality)" – a chapter by Michael Schandorf – explores the importance of paralinguistic phenomena in human communication, especially succeeding the rise of digitally mediated communication and various forms of human–computer interactions. It considers the ways in which paralanguage can be used as a potent tool for unmasking hidden or unacknowledged theoretical assumptions about language, communication, interaction and information.

Chapter 4 – "Corpus-Methodology and Discursive Conceptualizations of Depression" by Kim Ebensgaard Jensen presents the basics of corpus-linguistic methods by providing an investigative case study of blog posts about depression. While surveying ways of identifying salient topics, discursive strategies and underlying conceptualizations within narratives, the chapter also introduces the reader to the world of depression and its linguistic encodings used by the individuals experiencing it.

Chapter 5 – "Communication in Critical Theory (Frankfurt School)" by Olivier Voirol introduces the reader to the foundational stones of critical theory as part of the late philosophical legacy of German idealism and its materialist critique. The chapter also looks at how different approaches and practices – such as psychoanalysis, cultural criticism and social research, among others – have influenced and altered the ways of thinking within critical theory.

Chapter 6 – Usha Harris' chapter entitled "Reimagining Communication in Mediated Participatory Culture: An Emerging Framework" explores the possibilities and challenges of communication in the global, digital environment together with the associated intercultural experiences. The chapter proposes a framework which actively tries to integrate diversity in all forms.

Chapter 7 – "Global Culture" by Tanner Mirrlees unpacks the complex notion of global culture by contextualizing, summarizing and critically evaluating three narrower articulations of the concept: as a whole way of life, as a universalization of a particular way of life and as the existence of cultural works beyond national borders.

Chapter 8 – "Cultural Hybridity, or Hyperreality in K-Pop Female Idols?: Toward Critical, Explanatory Approaches to Cultural Assemblage in Neoliberal Culture Industry" by Gooyong Kim presents a critical reevaluation of discourses surrounding cultural hybridity. Situated in the context of South Korean popular culture, the chapter suggests an innovative perspective grounded in Baudrillard's notion of hyperreality, while also considering the political economy of hybridization in South Korea's recent popular culture boom.

Chapter 9 – "Postcolonial Scholarship and Communication: Applications for Understanding Conceptions of the Immigrant Today" by Adina Schneeweis explores the increasingly divisive discourses and actions that parallel the process of diversification of societies. The author proposes an alternative approach informed by postcolonial and anti-colonial scholarship, which situates contemporary discrimination practices within the construction and perpetual discursive justification of otherness.

Chapter 10 – "Cyberhate, Communication and Transdisciplinarity" by Emma A. Jane and Nicole Vincent explores the notion and acts of gendered cyberhate – the harassment and abuse of women and girls within online environments. The authors suggest that pervasive features of academic scholarship may be partially responsible for these dynamics, and propose corrective measures that communications scholars can employ to help alleviate the situation.

Chapter 11 – "Political Economy of Communication: The Critical Analysis of the Media's Economic Structures" by Christophe Magis investigates media structures and their development from an economic perspective within the specific critical tradition of the political economy of communication (PEC). Drawing on the concept of commodity, it focuses on the ways in which communications extend the practices of commodification to previously non-economic realms.

Chapter 12 – Sara Monaci's "The Propaganda Machine: Social Media Bias and the Future of Democracy" presents an innovative re-conceptualization of propaganda grounded in a mix of classical approaches to the concept and ideas borrowed from critical internet studies. It emphasizes the manipulative potential of social media and information-sharing practices which can produce new forms of the propaganda machine.

Chapter 13 – "From Fans to Followers to Anti-Fans: Young Online Audiences of Microcelebrities" authored by Maria Murumaa-Mengel and Andra Siibak, extends on the notion of internet-celebrity/microcelebrity to categorize three types of young audiences: fans, followers and anti-fans. This perspective enables the exploration of young audiences' engagement with the microcelebrity-generated content and its omnipresence in contemporary youth's media routines.

Chapter 14 – Anne-Sophie Letellier and Normand Landry's "Reimagining Media Education: Technology Education as a Key Component of Critical Media Education in the Digital Era" considers how the proliferation of digital technologies initiates a need to reconsider the theoretical foundations and practical modalities of media education. The authors reevaluate the field in an attempt to systematize the development of knowledge and skills related to the usage, functioning, and governance of digital technologies.

Chapter 15 – "From Media Ecology to Media Evolution: Toward a Long-Term Theory of Media Change" by Carlos A. Scolari introduces the basics of media ecology and considers the possibilities for the emergence of a new discipline – media evolution – which considers past, current, and future transformations of the media ecosystem.

Chapter 16 – Emma Rodero's "Media Psychology" looks at the relationship between media consumption and people's behavior, perception, feelings and thoughts from the perspective of media psychology – a field which deals with the application of psychological theories and methods to the study of individuals' interactions with media and technology. It presents an interesting paradigm for conceptualizing the influence of communication processes on humans at individual or group levels. In light of this context, the chapter analyzes central concepts, theories, applications and methods of media psychology.

Michael Filimowicz
Veronika Tzankova

Other Volumes in the Series

Reimagining Communication: Experience
Reimagining Communication: Action
Reimagining Communication: Mediation

Acknowledgment

The chapter summaries presented here have in places drawn from the authors' chapter abstracts, the full versions of which can be found in Routledge's online reference for the volume.

1

REIMAGINING SEMIOTICS IN COMMUNICATION

Paul Cobley

Introduction

Any lay definition of communication is likely to put "meaning" at its very center. Such a definition, provided by humans, would no doubt take human interaction as its paradigm and the casual or formal transfer of meaning in such interactions as natural, inevitable and straightforward. The question of meaning in the scientific study of communication, by contrast, has been riven by profound difficulties. "Meaning" has been the site of an enduring dilemma regarding the possibility of it ever becoming an entity susceptible to rigorous interrogation. At the seat of this dilemma are the troublesome components that are at play in any act of meaning. If meaning is to be investigated as an organic phenomenon, circulated and disseminated by living beings, then the key to meaning is surely to be found in the dispositions of those beings as they transmit or receive meaning. Even if such beings are conceived as *traversed* by meaning, the question of whether meaning as an occurrence is left untouched in meaningful interactions is very much moot. Alternatively, it is possible that meaning is not to be considered as a phenomenon or process that traverses its participants. A corollary of this is that meaning is only to be found through the study of what it constitutes for the living beings that enact it.

Put another way, either meaning is a process or quantity that exists outside of its bearers or, on the other hand, it is to be found within – or at the very least, is inexorably bound up with – its bearers' dispositions. Any putatively scientific study of communication, one which wished to avoid speculations or assumptions about the psychology of individuals and groups, would therefore have to eliminate from the equation the conduit *for* and the terminus *of* meaning, as well as any biases accruing to them. Indeed, there has been a continuing thread through communication studies in the last century in which meaning itself has been completely, deliberately and heuristically omitted from considerations.

Probably the originator of that thread is the work of information theory and that of Claude Shannon in particular. Shannon presented a highly influential model of the communications process in which "transmission" is the key feature. An information source in the model, with a message, uses a transmitter to produce a signal, which is received, by a receiver, which delivers a concomitant message to a destination. Although Shannon (1948, 379) does concede that the matter of signification, significance or meaning is prevalent in communication, ultimately he considers it extraneous to the task of measuring information. As Lanigan (2013, 59) writes,

> The *meaning* of human interaction is the paradigm for all theories and models of communication. Yet semantics – interpreted meaning – is irrelevant for studying information as a mathematical phenomenon – signal behavior – in electrical engineering. Unfortunately, the warning by mathematician Claude Shannon (1948, 1993a, 1993b, 1993c), inventor of information theory, against drawing analogies between information and communication *processes* has been ignored for decades (Gleick, 2011, 242, 416). The meaning problem was suggested to Shannon by Margaret Mead during his first public lecture on the theory at the Macy Foundation Conference on Cybernetics held 22–23 March 1950 in New York City. In short, information theory studies the signifying physical properties of electrical signals, whereas communication theory studies the meaning of human interaction.

What Lanigan points out, here, is the tendency of meaning to impinge, broadly in an unwarranted fashion, on issues of information theory. Equally, it might be argued that information theory perspectives have impinged upon the central issues of meaning in communication theory. That is to say, there is a temptation to omit or overlook humans' (or other organisms') predispositions amidst the flow of meaning in communication.

In what follows, it will be argued that semiotics has harbored the potential to settle some of the issues featured in the dilemma of meaning, issues that have bifurcated the fields of communication studies and information theory. As an "approach to meaning," semiotics' progress has sometimes been blighted by a penchant for assuming that the participants in meaning are generally to be considered as stable entities, mere channels for meaning. Criticisms of this approach have tended to emphasize the importance of readers' or audiences' meanings which semiotics has supposedly, at different times, neglected unduly. As will be shown, such criticisms were not entirely cogent because they were focused on, at best, a mid-most target for appraising meaning: a quasi-sociological audience, stable and cognitively independent in its implementation of meaning. Meanwhile, semiotics developed a much more sophisticated perspective in which communication was not to be taken as a perdurable process but, instead, was distributed across species, ineluctably tied to cognition and capable of definition in a fashion that was reconcilable, if not of a piece with, the scientism of information theory. Meaning, as central to communication, was recast neither as a fully material fixed entity, nor as simply the sum of humans' (or other organisms') dispositions. This is not to argue that the development of a new perspective on semiotics in communication was teleological, a narrative of progress in which there is learning from past mistakes. Rather, the reimagining of semiotics has grown out of the false starts, misinterpretations and detours that have characterized semiotics' fate in the study of communication. It is instructive to consider some of those now.

Founding Studies of Meaning

One of the founding, landmark texts on the issue of meaning is closely tied to the enterprise of semiotics in the twentieth century. Following on from articles which appeared as early as 1910, C. K. Ogden and I. A. Richards published in the UK *The Meaning of Meaning* (1923), an attempt to set out a scientific investigation of meaning contributing to a "science of Symbolism" (1923, v). Interestingly, Ogden and Richards already envisage that their study implicates cognition rather than just communication since their investigations "arise out of an attempt to deal directly with difficulties raised by the influence of Language upon Thought" (v). In an eclectic mix of references, including supplementary essays by Malinowski and Crookshank, Ogden and Richards attempt a comprehensive overview, in a manner that is almost inconceivable to the contemporary academy, of words, symbols and perception in the question of meaning. In one sense, their remit is to broaden the attempt of Bréal (1900) to establish a new area of semantic studies. This involves not just a theory of the sign and reference, although Ogden and Richards do offer that. Arguably, what is interesting for the current discussion is that *The Meaning of Meaning* also attempts to address meaning as a cognitive process, as opposed to a "volitional" or communicative process (1923, 50–76). Already haunted by the figures of two of semiotics' founders, Welby and Peirce (Petrilli, 2009, 2015), in this way the volume provides an important staging post in the route to contemporary semiotics, despite the misquoting and misinterpretation of Peirce in particular (Fisch, 1986 [1978], 345).

Perhaps even more well-known than Ogden and Richards in communication studies is the foundational separation of syntactics, semantics and pragmatics in "semiotics" (Morris, 1938, 52). The division corresponds to, respectively, the relationship between signs, the relationship between signs and their objects, and the relationship between signs, their users and the general context in which signs are implemented (Cherry, 1978, 233). There have been questions over whether this subdivision is efficacious and, indeed, whether it is not undermined by itself being subject to a large series of subdivisions (see Lieb, 1971). Certainly, the division has been taken up in linguistics, where pragmatics has become a major industry, with academic journals, books, professorial appointments and degree courses, in contrast to semantics (whose star has been in decline or collapsed into pragmatics) and syntactics (which, as a named field, never really took off). Yet, the key issue for communication in this subdivision concerns not so much the division itself as where it has been put to work. Morris (1938, 3) introduced the term "semiosis" to designate the process "in which something functions as a sign." This was not a question of linguistics. Indeed, Morris was very careful to ensure that his terminology – sometimes derived from Latin, sometimes derived from Peirce – evoked semiosis beyond the human. This is not a small point since it is the junction by which one route for investigating communication has been taken and one route has been neglected.

Just two years after Morris' formulation, far from fledgling Anglophone communication theory, geographically and intellectually, Jakob von Uexküll published his "Bedeutungslehre" as part of *Treatises on Theoretical Biology and its History as Well as on the Philosophy of Organic Natural Sciences* (1940). Translated into English as "The theory of meaning" in 1982 at a moment of burgeoning interest in von Uexküll's work in semiotics, the 50+ page essay presented the question of meaning from the standpoint of non-human animals' habitats and niches. Crucial to his discussion of meaning is the observation that animals live in relation to objects. That is, there can be no "neutral object" for non-human animals; rather, such creatures are always in a relation to the objects that they encounter. Non-human animals do not theorize objects nor do they consider their mechanisms. Instead, the implications of objects are always accommodated to the specific senses harbored by the animal and the specific niche which it inhabits. From this observation arises von Uexküll's term, *Umwelt*, which is often translated from German as "environment" but in this context refers to the animal's sensorium. The notion of *Umwelt* suggests that all species live in an "objective world" that is constructed out of their own signs, the latter being the result of their own sign-making and receiving capacities. The theory of meaning that von Uexküll laid out here and in other works suggested that the key to understanding meaning was certainly to assess how it was lodged in the bearers of meaning, but also how it concerned objects that existed independently of those bearers. In addition, of course, von Uexküll showed how meaning is a cross-species phenomenon, heralding a departure from some of the anthropocentrism that characterized much of the study of communication in the twentieth century. Such anthropocentrism was particularly in question across disciplines by the time that "The theory of meaning" was published in a new translation in 2010.

As foundations for semiotics and for an understanding of meaning in communication studies, all three of these perspectives held great promise. *The Meaning of Meaning*, Morris' *Foundation of the Theory of Signs* and "The theory of meaning" all posited, in different ways, that communication was not a matter of pre-formed individual humans manipulating freely available signs. Instead, signification was presented as a cognitive process which already determined how semiosis – an array of dynamic signs rather than a single isolable sign – worked to produce meaning for the participant in that semiosis. In the latter two of these foundational texts, meaning and semiosis were presented in their accrual across species rather than just in the human. Despite the flurry of high-profile reviews of *The Meaning of Meaning* and its continued existence as a reference point for individual observations about signification (Gordon, 1997), its disparate perspectives have not really furnished scholars with the resources for an umbrella movement or specific current of thought in the academy. The syntactics/semantics/pragmatics distinction has remained a reference point; yet, as noted above, semantics has largely been collapsed into pragmatics, syntactics is not a named sub-discipline and pragmatics has been largely pursued in linguistics, with only infrequent ventures beyond (e.g., Wharton, 2009). In relevance theory, pragmatics commenced an elementary interest in cognition (Sperber and Wilson, 1986). Von Uexküll's theory of meaning implicates human and non-human animal cognition. This work was kept alive by semiotics, particularly in the writings of Thomas A. Sebeok, especially after 1979. Von Uexküll was also of interest to the followers of Deleuze and Agamben, as is evident from the editorial material that accompanies the 2010 translation of "The theory of meaning" along with *Forays into the Worlds of Animals and Humans*. What interested both sets of aficionados of the work of von Uexküll was the conception of signification beyond the realms of the human. In the mainstream of communication studies, this concern had certainly been in play since Shannon and had developed in relation to machines throughout the 1950s and after with the Macy Conferences on cybernetics (Dupuy, 2000) and in relation to animals in the growth of animal communication studies from the late 1950s onwards (Maran et al., 2011). Yet where cybernetics did embrace other species' communication in addition to machine communication, it did so in terms of systems rather than in respect of meaning processes.

The unifying perspective on these pressing matters for communication was only to arrive in the guise of contemporary semiotics. What impeded the arrival of that perspective constitutes an instructive set of circumstances in respect of communication studies in general.

Semiotics in Communication Studies: False Starts, Misinterpretations and Detours

Most accounts of semiotics in communication studies trace the former's beginnings to the work of the Swiss linguist, Ferdinand de Saussure. Interestingly, the foregrounded topic in such tracing is "meaning," a term and concept that Saussure fastidiously avoided in his *Cours de linguistique générale* (*Course in General Linguistics* 1916; translated into English in 1959 and 1983) Based on the notes of his students, for Saussure died before he could conceive the volume, Saussure's *Cours* projects "semiology," "a science *which studies the role of signs as part of social life*" (1983, 15; italics in the original). Rather than just tracking the ways in which signs have been used to refer from objects from one epoch to the next, semiology was to institute a "synchronic" interrogation of the very conditions upon which signs operate. Despite this call for a general sign science, Saussure focused on the isolated linguistic sign, a "two-sided psychological entity," not a "link between a thing and a name, but between a concept and a sound pattern" (1983, 66). The sound pattern and concept he named, respectively, the *signifiant* and the *signifié*, noting that they were bound in a manner that was not pre-ordained but, rather, *arbitrary*.

Effectively, this arbitrariness is the crucial component in the forging of a code theory from Saussure's *Cours*. Such a theory was by no means inevitable: indeed, since the discovery of Saussure's original notes for his course at the University of Geneva, his entire *oeuvre* has undergone a reassessment which indicates that semiology could have been somewhat different (see Sanders, 2004; Harris, 2006; Bouissac, 2010). Yet, his emphasis on the language system (*langue*) underlying sign use – the sum of differences that occur between linguistic signs – meant the *Cours* became embroiled in the question of meaning. For Saussure, there was a need to eschew the idea of "meaning" because of the temptation to assume a natural connection between signs and their objects. What Saussure insisted, instead, was that signs exist in a system of "values" which are generated by each sign's encompassing of an arbitrary relation between sound pattern and concept which then stands in opposition to all other signs. Actual instances of linguistic communication facilitated by *langue*, and which Saussure gave the name *parole*, are the beneficiaries of these oppositions and differences. Put another way, instances of *parole* came to be seen as generated by the underlying system of differences that was *langue*. Although the *Cours* itself did not formulate the matter in such simplistic terms and although early semiology tried to account for some of the vagaries of the *langue/parole* couplet, it was nevertheless tempting to see signification as "coded," with a system allowing certain codings to exist and not others.

One of the pioneers of semiology whose work is still considered standard in undergraduate courses in communication study was Roland Barthes. *Elements of Semiology* (1964, translated into English 1967) and *Mythologies* (1957, translated into English 1973) enacted Saussurean theory with reference to, respectively, the "fashion system" and general instances of popular culture. In the latter volume, specific "mythologies" in popular culture – the haircuts of the Roman characters in Mankiewicz's film of *Julius Caesar*, wrestling, *steack frites*, striptease, the face of Greta Garbo, the New Citroën and the brain of

Einstein – were presented as instances of *parole* emanating from a basis in a general "myth" or *langue*. In short, this kind of semiology promoted the scrutiny of surface phenomena in order to reveal deeper, hidden agendas or, alternatively, attempted to reveal the "code" beneath the manifest "message."

The influence of this early incarnation of semiology should not be underestimated. It almost made Saussurean sign theory synonymous with the analysis of everyday phenomena, transforming quotidian trifles into complex *texts* to be decoded by competent readers. The idea of the text, a complex systematic whole inculcating specific readers or audiences, passed from the initial formulations of Barthes (1977a [1971]) and Lotman (1982 [1977]) into the vocabulary of communication study and literary criticism and then into common parlance. The notion of the text might be the most enduring contribution of semiotics to communication study. Certainly, it was fundamental and, for all the talk of the singular sign in commentaries on semiotics, the conception of semiosis, the text or a collection of signs is dominant in properly informed discussions by semioticians. As Umberto Eco writes, "a single sign-vehicle conveys many intertwined contents and therefore what is commonly called a 'message' is in fact a *text* whose content is multileveled *discourse*" (1976, 57). The "message" is clearly associated with information and its concern with the process of transmission, independent of any accrual of meaning. The "text" does have affinities with the concept of the "message," therefore; yet it also features the undertones of what Barthes (1977a [1971]) calls the "work" or the opus which suffused the humanities and framed messages in terms of authorial intent, richness of allusion and, often, the possibility of full transmission or pure communication.

Nevertheless, the principal bearing of the concept of the text is toward a synchronic, disinterested perspective concerned not with the value or import of any communication but, rather, with the mechanics by which it constructs meaning. Furthermore, as the other semiotician and inaugurator of the concept of text, Juri Lotman (1982), stressed, as an entity, the text is, in its very nature, *for someone* and can become *for someone*. It invites, in its very fabric, specific modes of reading; as such, the text *presupposes* the reader. As Barthes (1981, 42), was to put it, with slightly more of a literary slant,

> If the theory of the text tends to abolish the separation of genres and arts, this is because it no longer considers works as mere "messages," or even as "statements" (that is, finished products, whose destiny would be sealed as soon as they are uttered), but as perpetual productions, enunciations, through which the subject continues to struggle; this subject is no doubt that of the author, but also that of the reader. The theory of the text brings with it, then, the promotion of a new epistemological object: the reading (an object virtually disdained by the whole of classical criticism, which was essentially interested either in the person of the author,

or in the rules of manufacture of the work, and which never had any but the most meagre conception of the reader, whose relation to the work was thought to be one of mere projection).

As is so often the case in semiotics – and is certainly so in its more glotto-centric variant, semiology – this quote presents a theory in which the reader or reading is to be considered a component of meaning.

The method of uncovering the "myths" of popular culture, plus the ideological positions of the reader that they assumed, became so influential that by 1971 Barthes had almost disowned it. In an essay looking back at *Mythologies* (1977b, [1971]; see also Cobley, 2015), he suggested that casual or folk myth criticism had become so pervasive that it was practically becoming a myth itself. This did not prevent Barthes' early semiology remaining on the syllabi of degree courses in subjects like communications, cultural studies and media studies, where first-year undergraduates would learn to criticize what now seem some of the more brazen ideological claims of popular culture. Indeed, semiological myth criticism is ideally suited to first-year undergraduate syllabi: it provides a simple method and enables the student to expose some of the fallacies that are circulated in everyday life. Yet, in the study of communication, it is supplemented by and extended in a more varied approach. So, often, semiology is part of a package which includes such methods as quantitative content analysis, theories of readership, media policy study, theories of ideology and so forth. What is often overlooked is that, irrespective of the need to be supplemented, in and of itself semiological myth criticism is burdened and impeded by its own promulgation of the idea that communications are, first, codes that are easily discerned and, second, produce specific readings but are susceptible of decoding by fully autonomous humans. As a theory of meaning, it is somewhat wanting.

By contrast, Eco's influential volume, written in English in 1976 as *A Theory of Semiotics* and based on his *Trattato di semiotica generale* (1975), offered a much more nuanced understanding of coded meaning. By way of a discussion of key milestones in twentieth-century sign theory, including Peirce, Ogden and Richards and Hjelmslev, plus an informed discussion of contemporary communication and information theory, Eco showed that "code" ultimately implied determinate meaning but also allowed for flexibility in certain areas. He begins his discussion of codes with the example of an engineer in charge of a water gate between two mountains who needs to know when the water level behind the gate is becoming dangerously high. The engineer places a buoy in the watershed; when the water rises to danger level, this activates a transmitter which emits an electrical signal through a channel which reaches a receiver downriver; the receiver then converts the signal into a readable message for a destination apparatus. Thus, under the designation "code," the engineer has four different phenomena to consider:

a) a set of *signals* ruled by combinatory laws (bearing in mind that these laws are not naturally or determinately connected to states of water – the engineer could use such laws to send signals down the channel to express passion to a lover);

b) a set of states (of the water); these could have been conveyed by almost any kind of signal provided they reach the destination in a form which becomes intelligible;

c) a set of behavioral responses at the destination (these can be independent of how a) and b) are composed);

d) a rule coupling some items from the a) system with some from b) and c) (this rule establishes that an array of specific signals refers to specific states of water or, put another way, a syntactic arrangement refers to a semantic configuration; alternatively, it may be the case that the array of signals corresponds to a specific response without the need to explicitly consider the semantic configuration).

(Eco, 1976, 36–37)

For Eco, it is only the rule in d) which can really be called a code. Nevertheless, he notes the combinatory principles that feature in a), b) and c) are often taken for codes. Examples include "the legal code," "code of practice," "behavioural code." Most importantly for the study of meaning in communication, Eco emphasizes that "code" is a "holistic" phenomenon in which a rule binds not just the sign-vehicle to the object to which it refers but also binds it to any response that might arise irrespective of the reference to the object becoming explicit. So a), b) and c) are to be taken as "*s*-codes" – systems or "structures" that subsist independently of any communicative purpose. They can be studied by information theory, but they only command attention from communication science when they exist within a communicative rule or code, d) (Eco, 1976, 38–46).

What can be seen from this is that the "behavioural response" or, in more anthropocentric communication terms, the interpretation, is built into the semiotic definition of code proper offered here. Arguably, Eco had been an adherent of this premise more explicitly and even earlier than Barthes, for he had written on the principle at length in his book *Opera aperta* (1962) which went into numerous editions and appeared in an English version in 1989. Much of Eco's later work on the relations of text, interpretation and over-interpretation (for example, 1990a) stemmed from the interests first formulated in *Opera aperta*. Already, there were several analyses of television in that volume, as well as a chapter on openness, information and communication, topics to which he was to return in *A Theory of Semiotics*. The influence of Eco's combination of semiotics, information and communication theory and the various formulations he germinated was to be profound for communication studies (Corner, 1980; Jin, 2011). The influence began, as is often the case, on a local level. Eco was invited to give a paper at the Centre for Contemporary Cultural Studies at Birmingham University, featuring a semiotic inquiry into the television message, which he wrote up for publication in the Centre's Occasional Papers series (Eco, 1972). Along with his paper asking "Does the public harm television?" (Eco, 1973, later re-published in Eco, 1994), Eco's insights had considerable impact on subsequent papers in the series, such as Dave Morley's "Reconceptualising the media audience: towards an ethnography of audiences" (1974) and, most crucially, Stuart Hall's "Encoding and decoding in the television discourse" (1973; often reprinted in truncated form – for example, Hall et al., 1981).

Encoding and Decoding: The Text and the Reader

Hall's essay was not solely an exercise in semiotics. It was a synthesis of Eco, Barthesian semiology, Gramsci's hegemony theory, a concept of state apparatuses derived from Althusser, plus themes in the sociology of Frank Parkin – and it focused, particularly, on violence on television. From Parkin came the inflection in sociological terms of the reader originally conceived by semiotics. Thus, Hall posited "dominant," "negotiated" or "oppositional" readings – that is: a reading position that accepts the instilled codes that dominate the media text; one that accepts some of the codes but rejects or is unsure of others; or one that strenuously rejects them. Indeed, the typology echoed Lotman's (1974, 302) "non-understanding, incomplete understanding, or misunderstanding" in communication through texts. Later work by Morley on the reception of television programs (1980) explicitly implemented the dominant/negotiated/oppositional trichotomy as did Hobson's (1982) qualitative study of UK soap opera viewers. The general encoding/decoding model of Hall, with its program for quasi-semiotic investigation of the message or text or encoding of media output coupled with studies of specific audiences' differential decodings, inspired work in the field for over two decades. Subsequent studies included Morley (1986, 1992), Ang (1984, 1991, 1996), Radway (1984) Seiter et al. (1989), Lull (1990), Gray (1992), Gillespie (1995), Hermes (1996), Nightingale (1996) plus a revivified part-uses and gratifications cross-cultural study of *Dallas* viewers by Liebes and Katz (1993).

What is characteristic of much of this work is that it sought to provide a fuller picture of reader or audience activity than that which was afforded by the semiotic "reading off" of audience positions from the configuration of the text. In truth, Hall's encoding/decoding model was still very much a "reading off"; but it was hoped that when ethnography of real audiences was introduced into the equation, the possibility of moving towards a more watertight appraisal of texts' bearings (especially political ones) was available to the field. These studies effectively aimed to procure greater depth in the understandings of audience responses to texts, a "thicker" description (Geertz, 1993) in ethnographic terms, which would reveal the political co-ordinates of meaning and the text/reader interaction. As will be seen, below, this strand of work embodied a critique of semiotics for being excessively fixated on the meaning of the text to the detriment of the meaning whose potential was actualized by readers. Put another way, the discussion of codes was thought to be in need of superseding by the ethnographic extension of the encoding/decoding model. As John Corner (1980) observed at the inception of such work:

> It cannot be denied that the term [code] has been used most frequently in some of the most exciting and suggestive work to be carried out in that broad area of inquiry. Here, Hall's papers over the past ten years constitute an outstanding example. What I think can be concluded is that many instances of its present use do not deliver what is promised and sometimes obscure what it is a prime intention of any cultural research to make clear – that is, how social meanings get made.

What Corner's statement reveals, quite clearly, is a largely sociological agenda for the question of both meaning and reading – an agenda that semiotics was repeatedly criticized for not pursuing. Such an agenda is not invalid, of course, and it is capable of producing important insights. Yet, from a contemporary semiotic perspective – bearing in mind, also, that the theory of meaning in media and communication studies has not moved much further forward since this period – it is itself open to criticism. The encoding/decoding approach to such communications as those offered by television programs was rapidly coming into question with the advent of post-internet media in the early twenty-first century, especially Web 2.0. These latter putatively entail more *demonstrable* activity (interactivity, for example) which suggest that the text/audience relationship can once more be measured in terms of *use* (click-throughs, favorites folders, history, for example – see Livingstone, 2004). Thus, the reader as conceptualized by Big Data, upon whom it is much more economical to generate research, might be as complex as the reader envisaged by ethnography. Yet, arguably, what the reader in the encoding/decoding model and Big Data have in common are the concealment of affective or emotional dispositions in communication. The latter might reach into the private space of media use, but it can only read off click-throughs. This seems to suggest that associated communications research is straying further and further away from the possibility of grasping what might constitute meaning.

The other, related and extended, criticism, from a contemporary semiotic perspective, is that the ethnographic development of the encoding/decoding model simply was not readerly enough. It yielded data on what people said about their reading of texts, but took this largely at face value as "specimen" data (Alasuutari, 1995) such that the focus was on what respondents said rather than what responders *meant*. However concerted the attempt at thick description might be, it was itself forced to "read off" codes from the anthropological or sociological evidence of readership. Moreover, it was forced to do that as if the readings of audiences were fully autonomous, engagements with codes by readers who were not pre-constituted by any cognitive priming, coded or otherwise. In short, the encoding/decoding model did not really get to grips with reading or with meaning and it left a legacy for communication studies which tended to preserve the notion of code or determinate meaning.

This was not the only factor in the sealing of semiotics' fate in communication studies, but it was an important one. In the period of popularity attendant on Hall's model, semiotics had become extremely fashionable in the field of communications. In the Anglophone world, semiotics was becoming even more closely associated with communication studies than it had been with literary studies. Yet, being in fashion is perilous because it entails the grim possibility of going out of fashion. In the immediate wake of encoding/decoding, there were numerous English language primers and teaching books in communication (Fiske and Hartley, 1978; Williamson, 1978; Dyer, 1982; Fiske, 1982) which rode the wave of fashion but, in retrospect, prevented the latter-day developments of semiotics from contributing to communication studies. These texts conventionalized a series of distortions of the founding texts in semiotics, among which were: the misleading rendering of *signifiant*, *signifié* and *signe* as "signifier," "signified" and "sign," following the translation of Saussure's *Cours* into English in 1959 and before the time of Harris' superior, 1983 translation of the *Cours*; the re-orientation or misconstrual of the *signifiant* as "material" (rather than psychological), with slippages from "sound pattern" to "sound" (Culler, 1975; Coward and Ellis, 1977; Hall et al., 1981 – see Cobley, 2006); and, possibly the most damaging distortion – the clumsy and desperate conflations of Peirce and Saussure. Following Roman Jakobson's formulations on the Peircean theory of the sign (1965), taken up especially in Peter Wollen's *Signs and Meaning in the Cinema* (1969), Peirce's icon/index/symbol trichotomy of signs was cast in Saussurean guise. This was exacerbated by Hawkes (1977) book in the same UK tradition that spawned Wollen's, Fiske's *Introduction to Communication Studies* (1982) – "What Saussure terms iconic and arbitrary relations between signifier and signified correspond precisely to Peirce's icons and symbols" (1982, 46) – and Dyer (1982) where "Indexical, iconic and symbolic signs," said to enact relations between "signifier and signified" (Dyer, 1982, 99), get their own section in a chapter in which there is absolutely no mention of Peirce (see Cobley, 2019).

These last points may seem premature since Peirce has not really figured in this discussion thus far. In another way, however, they are appositely placed since Peircean semiotics has offered a much different perspective on meaning from that engendered by the encoding/decoding model. This would not be apparent if one was presented with the single trichotomy of icon/index/

symbol, grafted onto the arbitrary sign as derived from Saussure, in a euphoric dream of the possibility of uncovering the code for everything.

Toward the end of his life, Eco stated quite bluntly in an interview that he and his fellow workers in cognate fields such as communications during the 1960s and 1970s had "pissed code" (Kull and Velmezova, 2016). They had been incontinent with respect to their conviction that code could solve the mysteries of meaning. Since that time, Eco, in addition to all the other things he did, became a committed scholar of Peirce, working, notably and fittingly, on the subjects of cognition and animal communication.

Meaning, Cognition and Peircean Semiotics

Even before the encoding/decoding model was formulated and while some communication theorists were pissing code, semiotics had developed a much more ambitious project which is now beginning to bear fruit in communication studies. The fashionable moment of semiology was superseded by a broader tradition of semiotics which became ever more visible after the establishment of the International Association for Semiotic Studies in 1969. With Émile Benveniste as the first President, one of the prime movers in establishing the Association was the Hungarian polymath, Thomas A. Sebeok, who himself grappled with the question of code in semiotics. Starting with post-war communication and information theory, Sebeok devoted increasing attention to non-human communication as he developed "zoosemiotics." Initially he proceeded with a concept of code drawn from an information theory-inflected post-Saussurean linguistics. By the time of his final book (2001), however, he repeatedly made reference to the five major codes: the immune code, the genetic code, the metabolic code, the neural code and, of course, the verbal code. His other references were to codes in a very weak sense, but the first four in this clutch were significant because, as well as being the codes that Eco (1976, 21) had declared to be outside the remit of semiotics, they indicated much more expansive thinking. In considering non-human communication or semiosis at large, beyond the human, codes and meaning had been put into perspective. The big codes, in life, were largely invariant; the vagaries of meaning in the codes of human culture, by contrast were, as Hector Barbossa would say in *Pirates of the Caribbean* (2002), "more what you'd call 'guidelines' than actual rules" or, as Eco would phrase them, "*s*-codes." As early as 1972, Sebeok noted that the "need for different kinds of theory at different levels of 'coding' appears to be a pressing task" (1972, 112), thus posing a pivotal question that is cognate with communication theory minus the anthropocentric trappings: "What is a sign, how does the environment and its turbulences impinge upon it, how did it come about?" (1972, 4).

Such a non-anthropocentric outlook is of a piece with Sebeok's position as one of the foremost promoters of the work of the American logician, scientist and philosopher Charles Sanders Peirce. Peirce's semiotics was fastidiously developed – despite his "Sop to Cerberus" (see Cobley, 2019) – with a view to its applicability to all realms rather than just Saussure's target for a semiology of signs "*as part of human social life.*" The most striking difference between the sign in Saussure and that in Peirce, however, seems to be a mere formal matter: it is that the latter envisages a trichotomy consisting of a Sign or "Representamen," an Object and an "Interpretant." The three-fold sign derives from a tradition of thinking much different from that of the relatively recently developed linguistics and, indeed, the history of thought as conceived since the Enlightenment. A little like a *signifiant* being tied to a *signifié*, in Peirce's terms Representamen (a sign-vehicle) can stand for an Object (something in the mind or something in the world). This kind of relationship is frequently considered to characterize the sign, a "relation" between some ground and some terminus. Yet such a relationship, *qua* sign, had already been discovered to be false by the late Latin thinkers. Peirce's advance in sign theory, an advance that is integral to his semiotics being relevant to all nature, was to add a third term.

Peirce's third component, the Interpretant, is carefully named. It is not an "interpreter" – in other sign theories, such an entity would be an agency *outside* the sign as a whole. Instead, it carries out two functions. First, it sets up the sign relation: it is the establishment of a sign configuration involving Representamen and Object. When a finger (Representamen) points at something (Object), this is only a sign configuration if some link between the pointing digit and the something that is "pointed to" is made. This making of the link is the Interpretant. If the finger pointed but was placed behind its owner's back, concealed from anyone else in that space, then there is no sign configuration however much the finger points. Put another way, no Interpretant is produced. The second feature of the Intrepretant consists in the way that any person looking at what the finger points to is bound to produce another sign (e.g., the finger points at the painting on the wall and the onlooker says: "Vermeer"). So the Interpretant is another Representamen, "an equivalent sign, or perhaps a more developed sign" (*CP* 2.228). In Peircean semiotics, the Interpretant is that which the sign produces, its "significate effect" (*CP* 5.475): that other sign is usually – but not always – located in the mind.

There are, then, two points to note with respect to these features of Peirce's sign. The first is that it leaves no room for the usual criticism of semiology or the two-sided sign, that an interpreter or reader is needed in order to be able to understand how meaning and communication work. That requirement is already satisfied by being built in to the formulation of the sign. The second is that signs are not really isolable. When the Interpretant becomes in itself a sign or Representamen, "becoming in turn a sign, and so on ad infinitum" (*CP* 2.303), the sign exists in a *network* of Interpretants (*CP* 1.339) whose bearing is determined by prevailing circumstances. This entails that the sign or semiosis, in terms of meaning, is thoroughly contextual. It does not follow, though, that semiosis is unrelentingly malleable, as some tendencies in communications, media and cultural studies might have it. In reimagining the role of semiotics in communication, these technical aspects of Peirce's sign theory are of considerable importance. The Interpretant is an act of sign processing conceived in non-anthropocentric terms; signs and meanings are continuous; sign users are *in media res*, always within semiosis and not outside, manipulating it; and semiosis is a cognitive process such that cognition does not precede acts of sign use.

The fixity and rule-bound conception of code is loosened in Peircean semiotics which emphasizes, instead, the work of the Interpretant. Sebeok's later work, for example, even treats the term "code" as a mere synonym for "Interpretant" (see, for example, 2001, 80 and 191 n.13) as part of a pluralistic conception of codes which was coupled with an as-yet unspecified determining role of the genetic master code. Peirce's triadic version of the sign, his typologies of sign functioning and the design of his sign theory to cover all domains, provided the groundwork for Sebeok (and others) to make his work amount to an outline of the way that semiosis is the criterial attribute of life (see Sebeok, 2001; cf. Petrilli and Ponzio, 2001). Semiotics, in this formulation, is not just a method for understanding some artifacts of interest to arts and the humanities. The study of signs has been re-thought in recent decades as the human means to think of signs *as* signs, whether they are part of communication in films or novels, the aggressive expressions of animals or the messages that pass between organisms as lowly as the humble cell. As Sebeok demonstrates (1997), when one starts to conceive of communication in these places then the sheer number of transmissions of messages (between components in any animal's body, for example) becomes almost ineffable. This amounts to a major re-orientation for communication. To be sure, the communication that takes place in the sociopolitical sphere is of utmost importance: the future of this planet currently depends on it. However, the model of communication put forth by contemporary semiotics insists on the understanding that human affairs are only a small part of the proper object of communication studies. Considering the role of the communicator, for example, Self (2013, 362) has suggested that Actor Network Theory, posthumanism and contemporary semiotics have problematized the human communicator, with biosemiotics and cybersemiotics in particular manifesting "structural interpretations that do not depend upon the centrality of a human communicator to the production of meaning within the system." He adds: "Thus, at the beginning of the twenty-first century the powerful communicator remains but has been repositioned into larger constellations of social, structural, and semiotic forces that lie beyond any individual or institution."

A constellation larger still, underpinning those that Self notes, is nonverbal communication. From his zoosemiotic period onwards, Sebeok continually attempted to draw the attention of glottocentric communication theorists to the larger framework in which human verbal communication is embedded. He warned of the "terminological chaos in the sciences of communication, which is manifoldly compounded when the multifarious message systems employed by millions of species of languageless creatures, as well as the communicative processes inside organisms, are additionally taken into account" (1991, 23). Since the overwhelming amount of communication in the known universe is nonverbal, as opposed to a seemingly exponential but relatively minuscule amount of verbal communication, the massive growth in the study of nonverbal communication since the 1960s (see, for example, Weitz, 1974; Knapp, 1978; Kendon, 1981; Poyatos, 1983; Hall, 1990; Beattie, 2003; the *Journal of Nonverbal Behavior*, 1976–present) has been very much warranted, even if much of it focuses on human nonverbal communication and lacks the breadth that semiotic or even proto-semiotic (see Ruesch and Kees, 1956) studies have demanded. Also, there has been broad interest in the status of the animal and its forms of communication which has developed in the last 20 years (e.g., Baker, 2000; Fudge, 2004) as well as the accelerated interest in the relation of the human to machines evinced in posthumanism (see, for example, Wolfe, 2003, 2010). None of these, though, have been as ambitious as biosemiotics (Sebeok and Umiker-Sebeok, 1992; Hoffmeyer, 2008, 2010; Kull, 2001, 2007; Barbieri, 2007), with its laying bare that the objects of biology are thoroughly characterized by communicative processes.

Semiotics Beyond the Human: Where Is "Real" Meaning?

What the fast-growing field of biosemiotics has made clear concerns precisely the key issues that have been central to the present reimagining of semiotics in communication: that semiosis cannot be conceived separately from cognition; that semiosis is continuous across all domains of life besides the human, a fact that has consequences for how communication is theorized; that the possibility of interpretation is part and parcel of semiosis, rather than outside it; and, with reference to the *Umwelt*, meaning is neither a substance exterior to an organism nor a mysterious form within an organism. Peirce's semiotics provides important grounds for biosemiotics. As Ransdell points out in a classic essay on Peirce (1997 [1977], 168), referring specifically to the Object in Peirce's sign triad, humans have limited access to the real:

> Can we somehow get outside of our own minds, our own semiosis, to compare the real object to our idea of it to see to what extent the latter is a faithful and adequate representation of the former? Of course not. Consequently, either the real object is forever unknowable – a Kantian *Ding an sich* – or else it is that which is present to us in the immediate object when the latter is satisfactory.

The argument Ransdell makes here about "the real" being what is present in the immediate object (a technical aspect of Peirce's semiotics) is, effectively, the same one that is at the core of Jakob von Uexküll's concept of *Umwelt*. As mentioned above, an *Umwelt* is the means by which organisms capture "external reality" in response to semioses. It is the "world" of species according to their specific modeling devices, sensorium or semiotic capacity to apprehend things (von Uexküll, 2001a, 2001b). Signs grow – from the organism itself and from elsewhere, other organisms, or in feedback from itself (as in echolocation). The *Umwelt* of a species, then, is composed by the circulation and receiving, insofar as it is physically allowed by an organism's sensorium, of signs (von Uexküll, 1992, 2001a, 2001b, 2010; Deely, 2009; the essays in Kull, 2001; Brentari, 2015). Thus, species effectively inhabit "different worlds" because the character of the "world" they apprehend is determined by the semiotic resources that are available to them through their sensoria. A dog can apprehend sweetness in a bowl of sugar, but it cannot measure the amount of sugar, gain a knowledge of the history of sugar production or use the sugar in different recipes; a human can do all of these and also listen to stories about sugar, but it cannot hear very high-pitched sounds like the dog can. The human inhabits an *Umwelt* characterized by nonverbal and verbal communication according to the senses it possesses. Those senses, of course, as the example of the dog's "superior" hearing demonstrates, are not unlimited but, rather, specifically geared for the exigencies of survival.

As with Peirce's semiotics, in the concept of *Umwelt* there is the realization that beyond species' capacities of semiosis there is a world – the "real world," in one sense – which cannot be reached. In any *Umwelt*, misinterpretation of signs, overlooking of signs and signs not being 100% adequate representations of reality, maintain any species, to some extent, in a state of illusion. Interestingly, this is the point that Tomaselli (2016) makes in an attempt to reconcile the encoding/decoding model with Peircean semiotics. Considering a court case in which the South African Supreme Court brought an action against the Minister of Defence by the End Conscription Campaign (ECC) to prevent the minister producing further anti-EEC disinformation, Tomaselli argues that via the influence of Eco, a version of the Interpretant was already implicit in Hall's model. During the case it was argued that Peirce's phaneron or supersign, "encodes" all and everything that is present to the mind, including the imaginary, the fictional and the supernatural. A holistic concept,

> The phaneron involves the interpretations of both producers (conceived texts, encoding) and viewers (perceived texts, Interpretants) into a total framework of meaning (social and public texts [apartheid, anti-apartheid]) which may have little to do with the "reality" that the Minister's expert witness encountered, experienced or was responding to.
>
> *(Tomaselli, 2016, 68)*

Tomaselli's implementation of the broader framework met with success in the trial and it is important as an indicator in the argument regarding the "reality" of meaning. He points out that the expert witness on behalf of the minister had used a transmission model of communication which, in an authoritarian fashion, excluded the broad kind of semiotic analysis of meaning that the ECC's team ultimately employed successfully in terms of the case's outcome.

The point to be made, here, is not that "anything goes" in interpretation or that meaning is in the eye of the beholder. Among semioticians, Eco in particular had repeatedly made it his mission to discourage readers from getting drunk on interpretative strategies (see Eco, 1990b). The questions are: can semiosis or "the sign" be considered determinate, a coded or fixed set of co-ordinates? Or is semiosis indeterminate, fully interpretable and susceptible to the will of the reader? Implicit in the work stemming from the encoding/decoding model is a leaning toward the latter. The non-semiotic transmission model used by the Minister of Defence's team in the case clearly inclined toward the former. Furthermore, Tomaselli freely admits that the semiotic approach used in the case was employed in the service of resistance to the apartheid-era authoritarian mobilization of signification. There seems to be a definite flavor of such resistance, invariably in situations with much lower stakes than those in apartheid-era South Africa, in the ethnographic studies of reading that stemmed from Hall's work.

Yet, if the Peircean sign comprises an Interpretant, making the bearing of semiosis thoroughly contextual, does it not follow that semiosis, in the Peircean view, is one more example of indeterminacy, full interpretability and the thoroughgoing autonomy of the reader? The answer lies not just in the Interpretant's facilitating of the flow of sign to sign (semiosis) but on the sign's constitution as a *relation* rather than just a configuration of parts. What has been forgotten, for the most part, in the history of sign studies, is that semiosis inheres in *the sign relation itself* rather than in its components' reaching of a terminus. The American philosopher, John Deely, drawing on the late Latins and Peirce, has been central to the recovery of this memory, both for human semiosis and for that of other species. The sign user is not a fourth term to be added to the Peircean triad in order to make it work in an expected fashion (Deely, 2006). Nor is meaning a matter of the intersubjectivity that groups of readers with sociological characteristics in common might bring to signs. Such intersubjectivity constitutes the sign user/maker or reader as merely "being-in-between" (Deely, 2017) sign and terminus. Yet, this is not Deely's principal objection to the insufficiency of intersubjectivity as a theoretical concept. For him, echoing Aristotle, "over and aboveness" or suprasubjectivity is the defining feature of relation. A typical Deely example (2017, 15) distinguishes between intersubjective and suprasubjective relations:

> We are supposed to meet for dinner; you show up and I don't (or vice-versa), and you are annoyed until you find out that I died on the way to the dinner. At my moment of death, at the moment I ceased to have a material subjectivity encounterable in space and time, the relation between us went from being intersubjective as well as suprasubjective to being only suprasubjective; yet under both sets of circumstances I (or you) as the objective terminus of the dinner engagement remained suprasubjective (if not intersubjective!) as a constant influencing the behavior of the one still living in whom the relation retained a subjective foundation as a cognitive state provenating the relation as suprasubjectively terminating at an "other."

The sign – or semiosis – as this example shows, consists not in a determinate entity, the terminus, but in a *relation* that is indeterminate in respect of its terminus except insofar as it is understood by agents within the relation. The sign is *suprasubjective* in that its force – like that of fictions and the law – endures even when one or more of the subjects is removed.

The sign – or semiosis – is still determined by the agency of its reader, as in the settling of the relation of the meeting for a dinner date; but the suprasubjectivity of that sign relation is such that the sign is not nullified when one of the projected diners' putative autonomy is shown to be illusory by the affliction of sudden death. The sign is not suprasubjective in the way that a coded entity may be taken to be. Nor is its reliance on context the same as saying that its meaning depends on what readers think of it at a particular time. Instead, its suprasubjectivity and contextuality derive from two orders of *being* recognized by the Latin thinkers. In one set of circumstances, the relation in a sign could be of the order of *ens reale* (independent of mind for its existence), in another set the relation could be of the order of *ens rationis* (dependent on mind for its existence – see Deely, 2001, 729). As with the significations of the Minister of Defence in the case recounted by Tomaselli, semiosis can go either way – but not without the force of relation.

Conclusion: Meaning Is Real

Repeatedly in communication studies, semiotics has been criticized for being text-centered and willful in its disregard for readership. Hopefully, the latter will have been dispelled by the foregoing discussion which has shown that, certainly in its Peircean version, but time and again in other variants, semiotics has operated with an in-built consideration of readership. Yet, the question of semiotics' text-centeredness – its apparent insistence that there is something called meaning – remains. Moreover, the response to that question must be an affirmative one. Yes, semiotics does pursue a meaning that cannot be collapsed into the various readings that are available to the totality of sign users.

However, it does not pursue this for the reasons usually assumed. Semiotics is neither irrevocably text-centered nor an authoritarian master narrative. It is, instead, an anti-psychologism.

As Stjernfelt (2014, 4) makes clear, Peirce's work in particular is an ambitious anti-psychologism. Anti-psychologism, he writes (2014, 13), is basic for semiotics; it refuses to take signs as reducible to psychological phenomena. That is because psychologism tends toward relativism. Stjernfelt offers a simple, but compelling, fictional example:

> If mathematical entities were really of a purely psychological nature, then truths about them should be attained by means of psychological investigations. The upshot of psychologism might thus be that a proper way of deciding the truth of the claim that 2+2 = 4 would be to make an empirical investigation of a large number of individual, psychological assessments of that claim. So, if we amass data of, say, 100.000 individual records of calculating 2+2, we might find that a small but significant amount of persons take the result to be 3 – which would give us an average measure of around 3.999 as the result. This might now be celebrated as the most exact and scientific investigation yet of the troubling issue of 2+2 – far more precise than the traditional, metaphysical claims of the result being 4, which must now be left behind as merely the coarse and approximate result of centuries of dogmatic mathematicians indulging in armchair philosophy and folk theories, not caring to investigate psychological reality empirically.

In the same way, semiotics seeks not to reduce signs to individual mental representations. Both the sign-vehicle, its content and act of signification are considered by semiotics in their bearing as types whose tokens can be discerned in processes of cognition and communication. "If signs were only particular, fleeting and ever-shifting epiphenomena of brains and minds," Stjernfelt writes (2014, 47),

> this would not only give up signs as such as stable objects of scientific study – but it would, in turn, destroy even psychology itself along with all other sciences, because sciences, as already Aristotle realized, always intend general structures, even when they describe particular objects.

As part of the generality that Stjernfelt notes, the future of communication studies surely must lie in the embrace of a non-anthropocentric perspective on semiosis. This is not to say that the study of human communication needs to be scaled down or that local studies in the field of communication need to be replaced by more general ones. Obviously, local studies are crucial. Yet their worth, particularly in projecting for the future, will be vitiated if they proceed without cognizance of human communication's cognacy with that of the other species that inhabit the planet and who may be affected by humans' behavior.

It will be vitiated, too, by disregard for the role that will be played by machines in the global ecology of communication.

Sign users, as has been argued, are *in media res*, always within semiosis and not outside, manipulating it. In related fashion, signs and meanings are continuous and not isolable in the way that often occurs in communication studies and elsewhere. Work on networking has considered the manner in which the self is at once plugged into a potentially global system of communications yet, in the instrumentality of much communication, is rendered isolated or restricted in developing global collectivity (see, for example, Castells, 2005, 2009; van Dijk, 2012). However, such work did not predict the specific patterns of semiosis that have been witnessed in the social media age. If it is going to be possible to predict when the social media age will end or how it can develop in a fashion that will contribute to the commonweal, then this is one example where extending the understanding of the continuity of semiosis must surely be a crucial task.

An impediment to such work, as has been seen, is the positing of selves in semiosis as either cognitively pre-constituted or so pre-constituted in addition to sociologically determined in their relation to meaning. Peircean semiotics and cognitive semiotics are considerably sized and established constituencies in global semiotics and both are premised on sign action occurring not just in the social processes of communication, but in the cognitive processes inherent in members of species (see Bundgaard and Stjernfelt, 2010). That is to say, the nature of an *Umwelt* is a good predictor of the actions of the inhabitants of that *Umwelt*. In the case of the human *Umwelt*, there is an additional imperative in that the human has the capacity, particularly in the sphere of communication, to enhance that *Umwelt* for the future. By the same token, of course, the human also has the capacity for such enhancement which may negatively impinge on the *Umwelten* of other species.

Central to all of these observations is the phenomenon of meaning. What any feature of the world means to its inhabitants is shaped by personal factors, group factors, sociological factors, historical factors and so forth. Yet it is also shaped by species factors and by the nature of relation. Relation, in its very being, imputes stability to meaning. For humans, though, it is of particular importance because we can shift so rapidly from mind-dependence (where fictions can rule the lives of humans) to mind-independence (conceiving reality beyond the fictions rendered by an *Umwelt*). The action that might be taken in these simultaneously broad and stark dimensions would amount to a profound reimagining of communication.

References

Alasuutari, P. (1995). *Researching Culture: Qualitative Method and Cultural Studies*. London: Sage.

Ang, I. (1984). *Watching Dallas: Soap Opera and the Melodramatic Imagination*. London: Routledge.

Ang, I. (1991). *Desperately Seeking the Audience*. London: Routledge.

Ang, I. (1996). *Living Room Wars: Rethinking Media Audiences for a Postmodern World*. London: Routledge.

Baker, S. (2000). *The Postmodern Animal*. London: Reaktion.

Barbieri, M. (ed.) (2007). *Introduction to Biosemiotics: The New Biological Synthesis*. Dordrecht: Springer.

Barthes, R. (1967 [1964]). *Elements of Semiology* (trans. A. Lavers & C. Smith). London: Cape.

Barthes, R. (1973 [1957]). *Mythologies* (trans. A. Lavers). London: Paladin.

Barthes, R. (1977a [1971]). *'From "Work" to "Text"'* in *Image – Music – Text* (ed. and trans. S. Heath). London: Fontana.

Barthes, R. (1977b). *'Change the Object Itself' Image – Music – Text* (ed. and trans. Stephen Heath). London: Fontana.

Barthes, R. (1981). 'Theory of the text', in R. Young (ed.) *Untying the Text: A Poststructuralist Reader*. London: Routledge, pp. 31–47.

Beattie, G. (2003). *Visible Thought: The New Psychology of Body Language*. London: Routledge.

Bouissac, P. (2010). *Saussure: A Guide for the Perplexed*. London: Continuum.

Bréal, M. (1900). *Semantics: Studies in the Science of Meaning* (trans. Mrs. H. Cust). London: William Heinemann.

Brentari, C. (2015). *Jakob von Uexküll: The Discovery of the Umwelt between Biosemiotics and Theoretical Biology*. Dordrecht: Springer.

Bundgaard, P. & Stjernfelt, F. (eds.) (2010). *Semiotics*, 4 vols. London: Routledge.

Castells, M. (ed.) (2005). *The Network Society: A Cross-cultural Perspective*. London: Edward Elgar.

Castells, M. (2009). *The Rise of the Network Society*, 2nd edn. New York: Wiley-Blackwell.

Cherry, C. (1978). *On Human Communication: A Review, a Survey, and a Criticism*, 3rd edn. Cambridge and London: The MIT Press.

Cobley, P. (2006). 'Saussure: Ferdinand Mongin de: theory of the sign', in K. Brown (ed.) *Encyclopaedia of Language and Linguistics*, 2nd edn. Oxford: Elsevier, pp. 757–768.

Cobley, P. (2015). 'The deaths of semiology and mythoclasm: Barthes and media studies', *Signs and Media* (10): 1–25.

Cobley, P. (2019). 'Peirce in contemporary semiotics', in T. T. Jappy (ed.) *The Bloomsbury Companion to Peircean Semiotics*. London: Bloomsbury, pp. 31–72.

Corner, J. (1980). 'Codes and cultural analysis', *Media, Culture and Society* 2: 73–86.

Coward, R. & Ellis, J. (1977). *Language and Materialism: Developments in Semiology and the Theory of the Subject*. London: Routledge and Kegan Paul.

Culler, J. (1975). *Structuralist Poetics: Structuralism, Linguistics and the Study of Literature*. London: Routledge and Kegan Paul.

de Saussure, F. (1959 [1916]). *Course in General Linguistics* (trans. W. Baskin). New York: McGraw-Hill.

de Saussure, F. (1983 [1916]). *Course in General Linguistics* (trans. R. Harris). London: Duckworth.

Deely, J. (2001). *Four Ages of Understanding: The First Postmodern Survey of Philosophy from Ancient Times to the Turn of the Twentieth Century*. Toronto: University of Toronto Press.

Deely, J. (2006). '"To find our way in these dark woods" versus coming up Short', *Recherches sémiotiques/Semiotic Inquiry* 26 (2/3): 57–126. (The Deely essay was commissioned 11 April 2007, submitted in final form October 2007, actual publication was in January 2009. The discrepancy between publication date and actual journal date is a result of the fact that *RSSI* had fallen behind in its issues and is in a 'catch-up' mode with the issue in question).

Deely, J. (2009). 'Semiosis and Jakob von Uexküll's concept of umwelt', in P. Cobley (ed.) *Realism for the 21st Century: A John Deely Reader*. Scranton: Scranton University Press, pp. 239–258.

Deely, J. (2017). 'Ethics and the semiosis-semiotics distinction', in Special issue of *Zeitschrift für Semiotik* (ed. M. Tønnessen, J. Beevor & Y. Hendlin, 37 (3–4): 13–30).

Dupuy, J.-P. (2000). *The Mechanization of the Mind: On the Origins of Cognitive Science* (trans. M. B. DeBevoise). Princeton/Oxford: Princeton University Press.

Dyer, G. (1982). *Advertising as Communication*. London: Methuen.

Eco, U. (1962). *Opera aperta*. Milan: Bompiani.

Eco, U. (1972). 'Towards a semiotic inquiry into the television message', *Working Papers in Cultural Studies*, no. 6, Birmingham: Centre for Contemporary Cultural Studies.

Eco, U. (1973). '"Does the public harm television?", Cyclostyled paper for the Italia Prize Seminar, Venice, reprinted as "Do the public have bad effects on television?"', in *Apocalypse Postponed*. Bloomington: Indiana University Press, 1994, pp. 87–102.

Eco, U. (1975). *Trattato di semiotica generale*. Milan: Bompiani.

Eco, U. (1976). *A Theory of Semiotics*. Bloomington and London: Indiana University Press.

Eco, U. (1989). *The Open Work* (trans. A. Cancogni). Cambridge, MA: Harvard University Press.

Eco, U. (1990a). 'Unlimited semiosis and drift: pragmaticism vs. "pragmatism"', in *The Limits of Interpretation*. Bloomington and London: Indiana University Press.

Eco, U. (1990b). *The Limits of Interpretation*. Bloomington: Indiana University Press.

Eco, U. (1994). *Apocalypse Postponed*. Bloomington: Indiana University Press.

Fisch, M. (1986 [1978]). 'Peirce's general theory of signs', in K. L. Ketner & Christian J. W. Kloesel (eds.) *Peirce, Semeiotic and Pragmatism: Essays by Max H. Fisch*. Bloomington: Indiana University Press, pp. 326–351.

Fiske, J. (1982). *An Introduction to Communication Studies*. London: Methuen.

Fiske, J. & Hartley, J. (1978). *Reading Television*. London: Methuen.

Fudge, E. (2004) *Renaissance Beasts: Of Animals, Humans, and Other Wonderful Creatures*. Urbana: University of Illinois Press.

Geertz, C. (1993). 'Thick description: toward an interpretive theory of culture', in *The Interpretation of Cultures: Selected Essays*. London: HarperCollins, pp. 1–15.

Gillespie, M. (1995). *Television, Ethnicity and Cultural Change*. London: Routledge.

Gordon, W. Terrence (1997). 'C. K. Ogden and I. A. Richards' *The Meaning of Meaning* from early reception to projected revisions', in P. Bhatt (ed.) *Significations: Essays in Honour of Henry Schogt*. Toronto: Canadian Scholars Press, pp. 341–350.

Gray, A. (1992). *Video Playtime: The Gendering of a Leisure Technology*. London: Routledge.

Hall, J. A. (1990). *Nonverbal Sex Differences: Communication Accuracy and Expressive Style*. Baltimore and London: Johns Hopkins University Press.

Hall, S. (1973). 'Encoding and decoding in the television discourse', *Occasional Paper No. 7*, Birmingham: Centre for Contemporary Cultural Studies, pp. 1–12.

Hall, S. et al. (eds.) (1981). *Culture, Media, Language*. London: Hutchinson.

Harris, R. (2006) 'Was Saussure an integrationist?', in L. de Saussure (ed.) *Nouveaux regards sur Saussure: mélange offerts á René Amacker*. Geneva, Librairie Droz, pp. 209–217.

Hartley, J. (1982). *Understanding News*. London: Methuen.

Hawkes, T. (1977). *Structuralism and Semiotics*. London: Methuen.

Hermes, J. (1996). *Reading Womens Magazines: An Analysis of Everyday Media Use*. Oxford: Polity.

Hobson, D. (1982). Crossroads: *The Drama of a Soap Opera*. London: Methuen.

Hoffmeyer, J. (2008). *Biosemiotics. An Examination into the Signs of Life and the Life of Signs*. Scranton: Scranton University Press.

Hoffmeyer, J. (2010). 'Semiotics of nature', in P. Cobley (ed.) *The Routledge Companion to Semiotics*. London: Routledge, pp. 29–42.

Jakobson, R. (1965). 'Quest for the essence of language', *Address to the American Academy of Arts and Sciences*, February 10, published in *Diogenes* 51 (1966): 21–37.

Jin, H. (2011). 'British cultural studies, active audiences and the status of cultural theory: an interview with David Morley', *Theory, Culture and Society* 28(4): 124–144.

Kendon, A. (1981). *Nonverbal Communication: Interaction and Gesture*. Berlin: Mouton.

Knapp, M. L. (1978). *Nonverbal Communication in Human Interaction*. London: Thomson.

Kull, K. (ed.) (2001). *Jakob von Uexküll: A Paradigm for Biology and Semiotics* special issue. *Semiotica* 134 (1/4).

Kull, K. (2007). 'A brief history of biosemiotics', in M. Barbieri (ed.) *Biosemiotics: Information, Codes and Signs in Living Systems*. New York: Nova, pp. 1–25.

Kull, K. & Velmezova, E. (2016). 'Umberto Eco on biosemiotics', paper presented at the 16th Gatherings in Biosemiotics, Charles University, Prague, 7 July.

Lanigan, R. (2013). 'Information theories', in P. Cobley & P. Schulz (eds.) *Theories and Models of Communication*. Berlin: de Gruyter, pp. 59–84.

Lieb, H. H. (1971). 'On subdividing semiotics', in Y. Bar-Hillel (ed.) *Pragmatics of Natural Languages*. Dordrecht: Reidel, pp. 94–119.

Liebes, T. & Katz, E. (1993). *The Export of Meaning: Cross-cultural Readings of Dallas*. 2nd edn. Oxford: Polity.

Livingstone, S. (2004). 'The challenge of changing audiences. Or, what is the audience researcher to do in the age of the internet?', *European Journal of Communication* 19 (1): 75–86.

Lotman, J. (1974). 'The sign mechanism of culture', *Semiotica* 12 (4): 301–305.

Lotman, J. (1982 [1977]). 'The text and the structure of its audience', *New Literary History* 14 (1): 81–87.

Lull, J. (1990). *Inside Family Viewing: Ethnographic Research on Television's Audience*. London: Routledge.

Maran, T. et al. (2011). 'Introduction', in Maran et al. (eds.) *Readings in Zoosemiotics*. Berlin: de Gruyter Mouton, pp. 1–22.

Morley, D. (1974). *Reconceptualising the Media Audience: Towards an Ethnography of Audiences*. Birmingham: Centre for Contemporary Cultural Studies, pp. 1–13.

Morley, D. (1980). *The Nationwide Audience*. London: B. F. I.

Morley, D. (1986). *Family Television*. London: Commedia.

Morley, D. (1992). *Television, Audiences and Cultural Studies*. London: Routledge.

Morris, C. W. (1938). *Foundations of the Theory of Signs*. Chicago: University of Chicago Press.

Nightingale, V. (1996). *Studying Audiences: The Shock of the Real*. London: Routledge.

Ogden, C. K. & Richard, I. A. (1923). *The Meaning of Meaning: A Study of the Influence of Language upon Thought and of the Science of Symbolism* (with supplementary essays by B. Malinowski & F. G. Crookshank). London: Kegan Paul, Trench, Trubner and Co.

Peirce, C. (1931–1958). *The Collected Papers of Charles Sanders Peirce*. Vols. I–VI [Charles Hartshorne & Paul Weiss (eds.)], Vols. VII–VIII [Arthur W. Burks (ed.)] Cambridge, MA: Harvard University Press.

Petrilli, S. (2009). *Signifying and Understanding: Reading the Works of Victoria Welby and the Signific Movement*. Berlin: de Gruyter Mouton.

Petrilli, S. (2015). 'Sign, meaning, and understanding in Victoria Welby and Charles S. Peirce', *Signs and Society* 3 (1): 71–102.

Petrilli, S. & Ponzio, A. (2001). *Thomas A. Sebeok and the Signs of Life*. Cambridge: Icon.

Poyatos, F. (1983). *New Perspectives in Nonverbal | Communication*. Oxford: Pergamon.

Radway, J. A. (1984). *Reading the Romance: Women, Patriarchy and Popular Culture*. Chapel Hill and London: University of North Carolina Press.

Ransdell, J. (1997 [1977]). 'Some leading ideas of Peirce's semiotic' www.iupui.edu/ ~arisbe/menu/library/aboutcsp/ransdell/LEADING.HTM (last accessed 8 November 2018. This is a lightly revised version of the paper originally published in *Semiotica* 19 (1977): 157–178).

Ruesch, J. & Kees, W. (1956). *Nonverbal Communication: Notes on the Verbal Perception of Human Relations*. Berkeley: University of California Press.

Sanders, C. (ed.) (2004) *The Cambridge Companion to Saussure*. Cambridge: Cambridge University Press.

Sebeok, T. A. (1972). *Perspectives in Zoosemiotics*. The Hague: Mouton.

Sebeok, T. A. (1979). 'Neglected figures in the history of semiotic inquiry: Jakob von Uexküll', in *The Sign and Its Masters*. Austin: University of Texas Press.

Sebeok, T. A. (1991). 'Communication', in *A Sign is Just a Sign* Bloomington: Indiana University Press, pp. 22–35.

Sebeok, T. A. (1997). 'The evolution of semiosis', in R. Posner et al. (eds.), *Semiotics: A Handbook on the Sign-Theoretic Foundations of Nature and Culture*. Berlin: de Gruyter, pp. 436–446.

Sebeok, T. A. (2001). *Global Semiotics*. Bloomington: Indiana University Press.

Sebeok, T. A. & Umiker-Sebeok, J. (eds.) (1992). *Biosemiotics: The Semiotic Web 1991*. Berlin: Mouton de Gruyter.

Seiter, E. et al. (1989) *Remote Control: Television, Audiences and Cultural Power*. London: Routledge.

Self, C. (2013). 'Who?', in P. Cobley & P. Schulz (eds.) *Theories and Models of Communication*. Berlin: de Gruyter, pp. 351–368.

Shannon, C. E. (1948). A mathematical theory of communication. *The Bell System Technical Journal* 27 (July/October), 379–425 and 623–656.

Sperber, D. & Wilson, D. (1986). *Relevance: Communication and Cognition*. Oxford: Blackwell.

Stjernfelt, F. (2014). *Natural Propositions: The Actuality of Peirce's Doctrine of Dicisigns*. Boston, MA: Docent Press.

Tomaselli, K. (2016). 'Encoding/decoding, the transmission model and a Court of Law', *International Journal of Cultural Studies* 19 (1): 59–70.

van Dijk, J. (2012). *The Network Society. The Social Aspects of New Media*. 3rd edn. London and Thousand Oaks: Sage.

von Uexküll, J. (1940). 'Bedeutungslehre', Leipzig: Verlag von J. A. Barth.

von Uexküll, J. (1982). 'The theory of meaning', *Semiotica* 42 (1): 25–82 (Translation).

von Uexküll, J. (1992). 'A stroll through the worlds of animals and men: a picture book of invisible worlds', *Semiotica* 89 (4): 319–391.

von Uexküll, J. (2001a). 'An introduction to umwelt', *Semiotica* 134 (1/4): 107–110.

von Uexküll, J. (2001b). 'The new concept of umwelt: a link between science and the humanities', *Semiotica* 134 (1/4): 111–123.

von Uexküll, J. (2010). *Forays into the Worlds of Animals and Humans with a Theory of Meaning* (trans. J. D. O'Neill, Introduction by D. Sagan, Afterword by G. Winthrop-Young). Minneapolis: University of Minnesota Press.

Weitz, S. (ed.) (1974). *Nonverbal Communication: Readings with Commentary*. Oxford: Oxford University Press.

Wharton, T. (2009). *Pragmatics and Non-verbal Communication*. Cambridge: Cambridge University Press.

Williamson, J. (1978). *Decoding Advertisements*. London: Marion Boyars.

Wolfe, C. (ed.) (2003). *Zoontologies: The Question of the Animal*. Minneapolis: University of Minnesota Press.

Wolfe, C. (2010). *What is Posthumanism?* Minneapolis: University of Minnesota Press.

Wollen, P. (1969). *Signs and Meaning in the Cinema*. London: BFI.

2

HERMENEUTICS

Johan Fornäs

Hermeneutics is a philosophical reflection on interpretation: how people make meaning, for instance, when reading books or looking at pictures. Though it deals with the foundations of human culture, it still does not always have the central place it deserves within communication theory. One of its most productive branches – Paul Ricoeur's critical hermeneutics – has been undeservedly neglected, not least in relation to recent modes of mediated communication. There is hence in contemporary media studies a need for greater recognition of the hermeneutical foundations of communication.

I first define interpretation, explain why it is central to any reimagining of communication, and clarify its relations to neighboring concepts like meaning, culture and communication. I then discuss key assumptions and themes of interpretation theory, including how to locate and reconstruct meaning. Third, a brief historical overview of different hermeneutic contributions traces the transition from Romanticist to critical hermeneutics. The hermeneutic circle is described as a model of how meaning is made but also studied. Finally, some words on the limits of interpretation, in response to antihermeneutic challenges from posthumanists and new materialists.[1]

Interpretation, Meaning, Communication and Culture

The hermeneutic perspective is first of all about *interpretation*. The word "interpret" derives from Latin and refers to explaining, uncovering or indeed just making the meaning of something – to elucidate, decipher or understand something. Interpretation is everywhere in human life. Interpreting is necessary. It is at the very foundation of human life and culture (Heidegger, 1927/1962).

In a narrow sense, interpretation can be a rather specific activity. One example is doing *translations* between languages, for instance from Swedish to English, to sign language or braille, or from written to spoken language. If I make notes in Swedish and translate them directly into English speech, I simultaneously do a double translation from visual to oral and from Swedish to English. Another special case is when a stage or film director or actor "translates" a written manuscript into a theater show or a movie. Here, the text is *performed* in such a way that it conveys how the performer has understood the meaning of the play. Similarly, a conductor or musician interprets a musical composition by transforming notes in the score to aural sounds, again in such a way that the listener is meant to get an idea of how the performer understands the work in question. The process can always be inverted: in the reverse direction, a scholar may record a performance and then *transcribe* its sounds into musical notes, similar to when someone transcribes a recorded interview into a written transcription.

Working as a professional interpreter usually refers to one of these specialized activities. However, interpretation is not limited to such separate cases or professionals. It is in fact activated whenever people communicate with words, music, images, scents or whatever phenomenon that can be given meaning. Interpretation is linked to but not identical with perception. Interpretation is when humans use their senses to understand the sensations as bearing some kind of meaning. Interpretation in this general sense is not just a translation from one language to another, though there is such an element involved when people give meaning to something and, for instance, move between the word "sun" and the physical object in the sky. This general operation of interpretation is at the basis of those special versions effected by translators and performers, since some kind of understanding of a work is the basis for it being translated or performed.

What then is *meaning*? It is when people experience things as *pointing* toward something else, outside of themselves, something that is absent but made virtually present in people's minds. Meaning is made when interacting people make repeated associations between something materially present and something absent that is referred to and understood as meaningful. If a bunch of people in their daily practice start to use the combination of visual signs comprising the written word "sun" to signify the most strongly shining object in the sky, then this word has received that meaning. By interpreting something, this something is used and understood as being not just material things anymore, but as constituting texts, as webs of signs and symbols that in turn are combined into larger works and discourses, and that carry meaning for those who claim to understand them. Instead of just being arbitrary black dots on a white piece of paper, the letters s, u and n have been transformed into a word in the English language, where it means the same thing as for instance a drawing of a circle surrounded by beam lines pointing out in all directions. Even when it is night or cloudy, and the "real" sun is absent from our current experience, the verbal or pictorial sign "points" toward this absent phenomenon and makes it virtually present in our minds.

Such links between a material object that is present and functions as a *text* and something else that is physically absent but called forth symbolically by how the text is interpreted are the result of collective efforts over time. Humans develop interpretive communities that cultivate traditions of interpretation, making it easier to decipher a particular text by placing it in a specific genre. Meanings are the result of social conventions developing historically. Understanding textual meanings demands a complicated process of interpretation, even though in much of ordinary everyday life it occurs almost instantly and without any conscious thought. When somebody cries "help" not much thought is necessary to know that somebody is in danger: this interpretation is never acknowledged to result from a combination of language skills that explain what "help" means and social norms that tells what to do in that situation.

Making meaning is a basic human operation that makes possible a wide range of further operations. Through interpretation, people can imagine the future or remember the past, explore the possible or the impossible, construct imaginary worlds of utopian or dystopian fantasy, construct identities and make plans. From a hermeneutic perspective, interpretation is a practice that defines human *culture*. Culture derives from a Latin word for cultivation, related to cult and colony but also including some kind of care. Throughout history, a number of different definitions of this concept have evolved, and now co-exist in an often-confusing way (Williams, 1976, 87–93; Fornäs, 2017, 11–84). One is the undifferentiated "ontological" concept of culture as *cultivation*, distinguishing human societies from nature. This was later developed into the anthropological concept of culture as *lifeforms*, allowing for a plurality of co-existing subcultures. This in turn was rather different from the aesthetic concept of culture, comprising all the *arts* and the various cultural institutions and professions. In the twentieth century, a fourth main concept of culture was established, the hermeneutic one, defined precisely as *meaning-making* or signifying practice (Williams, 1981, 12f, 206f). This is widespread in many areas today, and it serves well to also highlight the earlier ones. For instance, different ways of understanding the social world distinguish one (anthropological) culture from another, and (aesthetic) art forms can be regarded as specialized laboratories for exploring the reach and limits of meaning-making.

Culture as a meaning-making practice is thus based on interpretation. It involves two equally important and interwoven aspects. One is what might be seen as a vertical dimension of *imagination*: the act of relating a sign to its meaning, building a bridge between the present and the absent. The other main pillar is the horizontal dimension of *communication*: the necessary interaction between different human subjects and between different texts, whether they are actually co-present or just implicitly relating to each other. Meaning is never made by a single, autonomous subject relating to a singular, independent text, but always rely on relations between plural subjects and texts. Being "machineries of meaning," media of communication serve as the specific technology of culture (Hannerz, 1992, 3, 26f; Fornäs, 1995, 135–141).

Making meaning enables people to represent their world: to present its hidden features and make them symbolically present. *Representations* represent something else: by presenting themselves they represent or symbolically re-create something that is physically absent, making it virtually present in its actual absence. A politician can thus represent her voters, or a symbol can represent the socially agreed meaning it is used to stand or "stand in" for (Williams, 1976, 266ff; Hall, 1997; Fornäs, 2017, 54f). Elected political assemblies speak for an absent group that gets an indirect voice through this representation. More generally, texts, musics or images depict something else: standing for an object or an idea that they seem to make present in people's minds. In cultural studies, Stuart Hall and others use this notion of representation to investigate how communication media represent reality in ways that intrinsically link the political to the aesthetic.

A common way to subdivide communication is to distinguish unidirectional mass media from interactive social media, often regarding the former as hierarchic and potentially authoritarian and the latter as more egalitarian and liberating. John Durham Peters (1999, 33–36) has similarly distinguished two main modes of communication, though with the opposite evaluation. On the one hand, Socrates's mode of *dialog*, where two co-present partners meet and interact; on the other hand, Jesus Christ who in his Sermon on the Mount represented the mode of *dissemination*, where a message is left open for anyone to interpret at a distance from the speaker. These examples made it possible to see that dialog (and by implication interpersonal social media) could well contain forms of manipulation when one speaker does not give up until the other has given the expected response, whereas there may sometimes be a certain freedom of interpretation in the dissemination affected by mass media, allowing for oppositional readings. This argument is enlightening, but I would from a hermeneutic perspective like to suggest two modifications.

First, instead of just turning the ladder of values upside down, both sides should be regarded as having ambivalent potentials. Dialog can be a creative and free exchange between equals or a manipulative effort to force the other to agree with me. Dissemination can on the other hand be either a public and open gift for others to use as they like, or a hierarchical relation between a powerful sender and submissive consumers. It all depends on the specific situation, context or setting of the communicative event, and on the kind of interface that each genre of media use constructs. Instead of idealizing one of these modes of communication and demonizing the other, whether sticking to the dominant hierarchy or just turning it upside down, it is better to cultivate an ambivalent sensitivity for the affordances (potentials and limitations) of both of them.

Second, it is important to see that bridging these two main poles, there is a third mode of communication, which is actually the hidden foundation and core moment of both the others – the central hub linking them with each other, but often itself remaining in oblivion. It deserves to be defended against any attempt to disregard its key role. It is always involved in either of the other two modes, but can also exist as a separate activity. This is the practice of meaning-making *interpretation*, matching the key hermeneutic definition of culture, combining consciousness and communication, phenomenology and technology, experience and structure. The dynamic intersection of imagination and mediation that defines culture can certainly forge links between people, with dialog and dissemination as two core modes, but there is also a third possibility. There is always mediation, but not necessarily between two subjects. Contrary to a common intuition, the indispensable actor in processes of communication is not the writing "sender" who physically produces the message or text, but the reading "receiver" who uses it to construct meanings. Meaning is not primarily made by the producer of the text, but by its interpreting user or reader. The author is just the first such reader, whose interpretation of the text is however not the only possible, nor necessarily the most valid one. The author produces the text, but the reader produces its meanings.

Communication and culture are often understood as based on a transfer of contents in chains from senders to receivers. Such a *transmission* view of communication gives textual production a key position, as it tends to equate the transfer of texts with a parallel transfer of meanings. This may certainly in many cases be fully relevant. However, it fails as a basic model, since there are important cases where it does not hold. Contrasting to this, a cultural or ritual view associates communication with sharing, communion and community (Carey, 1989/1992, 18). It is interpretation that is the primary act of communication. From a modern hermeneutic point of view, interpretation, meaning, culture and communication are thus closely interwoven.

A Communicative Philosophy of Communication

The term *hermeneutics* may perhaps relate to the antique Greek god Hermes, who served as messenger between gods and men. Hermeneutics is at any case a philosophy of translation, connection and communication: a theory of interpretation that reflects upon how all kinds of texts mediate between people, across time and space, and between actual, virtual and imaginary realities. A text is a web of symbols or signs, created by signifying practice and inviting interpretation: to explore the universe of meaning surrounding a text within a context defined by an interpretive community of some kind. To interpret is to follow the traces of a text, and to explore its range of meanings. To interpret is neither to make any arbitrary individual associations, nor to pin down one single, given meaning of a text, but rather to investigate and inhabit the worlds of meaning that the text hints at within an intersubjective context. It is to give oneself in to "the intended meaning of the text" which is not "the presumed intention of the author" but rather of the text itself, as it "seeks to place us in its meaning": this intention of the text is "the direction that it opens up for thought," indicating that meaning-making "must be understood in a fundamentally dynamic way," as an ongoing becoming or coming-to-meaning (Ricoeur, 1971/2008a, 117).

Where is meaning? Where do meanings emerge? Different theories suggest different locations. It is tempting to look for its origin in the *author* of a text, for instance, a speaker on the phone, a film director or a website manager. In dialogs as well as with mass media, we tend to ask what the producer of each statement, novel, image or musical piece might have intended to convey to us. This invites a kind of psychological inquiry into the speaker's or writer's *intentions* with a text. For example, novelists are often asked what they intended to say with their works, as if their own opinions were the truest possible answer to asking for the core meaning of a narrative. Older hermeneutics, back in the eighteenth and nineteenth centuries, took this for granted and aimed at placing themselves in the author's shoes and reconstructing his lifeworld, regarding the text as primarily a clue to the interior mind of its producer. However, though this approach is very common, not least in oral conversation, it is misguided. Often, authors are not quite aware of what their texts might mean to others, and they express feelings they were never conscious of having themselves. Not least with the invention of writing, texts can be dissociated from authors and used in new contexts where their meanings seem to diverge considerably from whatever the author might have had in mind. These new meanings cannot simply be regarded as untrue. For instance, even though the swastika once in South Asia might have signified divine life energy of some kind, since the Nazis adopted it, it really has changed meaning and is today globally associated (also) with fascist ideologies.

A second answer is then that the meaning derives from the *text* itself. In the mid-twentieth century several branches of formalist literary theory emphasized the autonomy of the text. Much of semiotics (with Fernand de Saussure, Roland Barthes and Umberto Eco and others) stressed the autonomy of signs, locating meaning in their intertextual form structures, in the relation between textual elements. When analyzing any media text, it must be more important to check how it is actually structured than to ask for what producers intended. However, this may lead to a problematic formalism that overstates the autonomy of texts and misses the influence of different kinds of contexts. If the text was completely self-sufficient, it would probably just have one univocal meaning, but that is rarely the case – instead, many reception studies prove that different people can interpret the same TV series or news report in a different way.

Then, how about these *contexts*? It is perhaps a certain spatial, temporal or social context that in practice determines what a text has to say. Texts bound to some religious context, used in institutionalized rituals, all confirm a certain belief, and they can change their meaning if transferred elsewhere, for instance, if a verse from the Bible or the Koran is read as a secular love poem. In cultural studies, such suggestions have sometimes been made. Still, it seems hard to sustain that it does not matter at all what text it is for what meanings people derive from them.

A fourth possibility would of course be to point at the receiver, the listener, viewer or *reader* in the media audience. Some audience research has gone in this direction, not least at the height of so-called postmodernism, when the freedom of individuals to interpret media messages was emphasized. It is true that in shifting context different text users construct different meanings. However, caution is advisable, since a reader can after all not freely make any arbitrary meaning: the text may have a wide set of interpretive potentials, but it does set certain limits.

From a modern hermeneutic viewpoint, it is the contextually framed encounter between texts and subjects that determine textual meanings. Ricoeur repeatedly stressed that meaning should not be sought *behind the text*, in the intentions of its author. Instead, meaning always emerges *in front of the text*, between text and reader or user. It is always the encounter of text and reader that produces meaning. The first reader may sometimes be the author, but then as recipient rather than as producer.

A human intention to communicate is required, but it needs not necessarily reside in a text-maker, only in the meaning-making interpreter. "Indeed, it is the reader – or rather the act of reading – that, in the final analysis, is the unique operator of the unceasing passage" that goes from the world of the author to the world of the reader through the mediation of the world of the text (Ricoeur, 1981, 28). It is usually thought that the act of making a text is the primary one, but actually it is the act of interpreting texts that is the foundation of culture. In a sense it is not the production but the reception which comes first and is the primary cultural operation. The author of a text may be its first reader, but it is also possible to interpret natural things as texts even when there is no human text-maker at all. Or rather, it is the interpreter who makes the text as text, i.e., interprets a web of objects as having meaning and thus recognizes something material as a text that points to something else, outside of itself: its meanings. Instead of any variant of the transmission model of communication, the basic model of culture should recognize this primary interface of text and reader, in which meanings emerge between them, and where the author is just a particular kind of reader.

When a text has no human sender, the subject constructs an imaginary or virtual author: "God," "Mother Nature" or perhaps one's own consciousness, when interpreting dreams or creative fantasies. One example of this mechanism is when John Durham Peters (2015, 254–260) reads for instance clouds as mediating texts. People can ascribe meaning to such natural phenomena, though there is (from a non-religious perspective) nobody who made the clouds with an intent to have them interpreted; no actual "sender" with a communicative intention exists. In this case, there is just a decoding receiver and no sending subject. I agree with Peters that this is also a form of communication, but it is neither dialog nor dissemination. It too combines imagination and mediation – only not between people, but between subjects and texts in contexts.

Interpretation is a form of communication too. It makes common by sharing, as meaning-making is never done by a single, autonomous individual subject. It is always based on intersubjective sharing of texts and meanings, allowing for a fusion of horizons between text and subject, i.e., an interchange between text and subject, in which meaning grows. The key model of culture as mediation should therefore not be the chain of transmission of a message from sender to receiver, since that is just one special case of meaning-making mediation, dependent on a particular division of labor between producers and consumers. Interpretation deserves being acknowledged as a third mode of communication, beyond dialog and dissemination but at the same time a shared and central element of them both, as it is the foundation of culture, defined as meaning-making practice. A better basic model than any chain of communication is therefore the cultural triangle, where plural subjects interact with plural texts in plural contexts to make meanings, identities and social worlds (Fornäs, 2017, 180–192). This mediating process unfolds by the creative intersectional dynamics that makes meaning out of materiality.

Interpretation can exist by itself, but it also enables dialog as well as dissemination. In interpersonal dialog, a collective text is produced by mutual exchange between two or more communicators who continuously interpret and respond to each other's statements. In dissemination, an institutional actor monopolizes textual production as an authorial function and directs it to a more or less anonymous audience who can still speak back indirectly, by either closing their eyes and ears, refusing to receive the message, or interpreting against the grain, refusing the authorial intention by an oppositional reading according to the "hermeneutics of suspicion" which Ricoeur (1965/1970, 32–36) found particularly developed in Marx, Nietzsche and Freud.

Hermeneutics is not in itself a designated method, but it can be used to inspire methodological development. It provides a philosophical reflection on what people do when they make meanings. As such it sheds light on meaning-making in ordinary everyday life practices, for instance, in media use, but also how interpretation works in research, including in media and communication studies. Rita Felski (2015, 33) concludes that "the neglect of the hermeneutic tradition in Anglo-American literary theory is little short of scandalous," as hermeneutics "simply is the theory of interpretation and leaves room for many ways of deciphering and decoding texts." In particular Paul Ricoeur's critical version of hermeneutics opens up to such combinations with a wide range of other, more specific tools for mapping, analyzing, explaining and understanding texts. Hermeneutics reflects on what media phenomena might mean, how they are interpreted by different actors and which potential meanings they convey. It is a philosophy of communication in its cultural sense of meaning-making, focusing on how meaning is made in processes of communication. By being able to integrate other perspectives as moments of explanation, it also itself has a communicative character, rather than closing itself off from other theories.

Everyone always interprets – this is inescapable in any human society. We ascribe meaning to what we see, hear or otherwise experience, and cannot but incidentally cancel this urge to understand, even though we sometimes try to escape. The natural sciences are engaged in interpretation as well, trying to make sense of laboratory experiments, astronomic observations or statistics. Today's advanced physics demands a highly complicated interpretive competence. However, the human sciences are bound to a *double hermeneutic*, as what they study is human society and culture which is already itself making meanings in ways that atoms or stars are not. In human sciences like the humanities and social sciences, this meaning-making is not just a knowledge tool but a focused topic of research. In contrast to molecules or stones, human beings, texts and communities are subjects and actors in meaning-making processes. What human sciences study is already interpreted by those whom they study. Physical theories need not relate to how nature interprets or describes itself, but cultural researchers are from the start engaged in a reflexive conflict of interpretations where they are not the only ones who try to understand what is happening.

In Ricoeur's simultaneously critical and communicative version of hermeneutics, meaning-making must take detours through various forms of more or less systematic mappings of a text. These can engage lots of different methods and theories, for instance, semiotics, structural analysis, discourse analysis or media archaeology. Some of these have been rather hostile toward older hermeneutic traditions, but to Ricoeur, they can be integrated as tools toward deeper and richer understanding: as supports rather than as alternatives to interpretation.

This makes it a bit difficult to distinguish hermeneutics from other theories of communication. Take *semiotics*, for example. It is the theory of signs, but since a sign is a token that has meaning, semiotics may well serve to assist hermeneutic exploration of how and what webs of signs might mean. There are differences in perspective, but no strict dividing lines. While semiotic analysis often focuses on the micro-level of interrelated individual signs, hermeneutics studies larger textual units like whole works and even genres. Ricoeur also describes semiotics as considering only the "internal laws" of a text, while hermeneutics on the contrary strives to "reconstruct the set of operations by means of which a work arises from the opaque depths of living, acting, and suffering, to be given by an author to readers who receive it and thereby change their own actions" (Ricoeur, 1981, 17). Semiotic theory sees both the production and the reception of texts as irrelevant, focusing exclusively on the interior of a text, while hermeneutics "seeks to reconstruct the whole arc of operations by which practical experience is turned into works, authors, and readers" (Ricoeur, 1981, 18). According to Ricoeur, there is neither an inside nor an outside to the work but rather a concrete process that links intratextual structures to the contexts of production and reception. "We must stop seeing the text as its own interior and life as exterior to it. Instead we must accompany that structuring operation that begins in life, is invested in the text, then returns to life" (Ricoeur, 1981, 28). In this view, semiotics like other modes of structural analysis tends to be more statically focused on the internal relations of webs of signs, while hermeneutics strives to be more sensitive to the dynamic processes of interpretation. However, since these dynamics centrally include the mediation by texts, semiotics may well serve as a technique to analyze these texts.

To sum this up so far, everybody – including all scholars – interpret, whether aware of it or not. Hermeneutics is a philosophical reflection on how this works in practice, rather than a distinct method for interpreting. Semiotics, discourse analysis, media archaeology or big data mapping may have other intentions, but they offer operations that can actually deepen and widen our understanding of the world. Even antihermeneutic new materialists who strive to escape meaning-making actually illuminate the reach and limits of precisely such dynamic signifying processes.

From Romanticist to Critical Hermeneutics

Hermeneutics was born out of a scholarly praxis where religious and philosophical texts from the Bible or the Greek classics had to be translated and explained to later generations of readers in other regions. *Philology*, the "love of words," was an interpretive and editing practice that studied literary texts and other written records handed over through tradition or in libraries, establishing whether they were authentic and what might have been their original form, and not least determining their meaning. Such classical texts had to be translated between languages but also made intelligible for readers generations apart from when they had been written. Techniques developed for how to find ways of explaining textual meanings across substantial temporal as well as geographical distances. Out of such specialized practice grew insights into the operations of translation and interpretation that could be generalized from their original applications to literature and other written language into all conceivable meaning-making practices and media forms. There is still a logocentric bias in much hermeneutic discourse, as written words remain a central form of mediation, but the resulting theory of interpretation can with certain adjustments be applicable also to other symbolic modes, such as images or music. The modern standard history of hermeneutics revolves around a series of male German scholars, deeply influenced by *German Romanticism.*

In the early nineteenth century, *Friedrich Schleiermacher* (1768–1834) was one of the first and most important scholars who widened the scope and strived to present a more general approach to interpretation. He defined this general hermeneutic as the art to correctly *understand* the talk and especially the writing of someone else. He thought there was a single, correct literal meaning to a text, and that this meaning was determined by the author and his *intentions.* There were therefore two tasks for the hermeneutic inquiry: (1) the grammatical interpretation of the written words within the relevant language system, which, for instance, poetry could also expand and creatively renew; and (2) the psychological interpretation, looking for what the author might have wanted to say with his words, understood as emanating from the author's life: including his belonging to a national and cultural identity as well as to a specific time epoch. The good reader must put himself into the life situation of the author, and thus into his mind, so as to understand the text even better than the author himself. Schleiermacher's theory and method of interpretation thus stressed understanding and community, where the reader went into a clinch with the text by projecting himself into the author and his time. Meaning-making involved a process comprising three core terms: life experience, expression and understanding, and this interpretive process had its basis already in everyday interaction, though its highest form was the interpretation of artworks (Schleiermacher, 1838/1998). Compared to medieval traditionalism, this represented a modern secularization and individualization: it was no longer God but an individual author that was the origin and source of textual meaning.

To Schleiermacher, understanding a text demanded that the reader moved in circles between the parts and the whole. Even to understand a single sentence you had to go back and forth between the individual words and the whole sentence: analyzing how sounds or letters add to form the word, but at the same time also being aware of the total expression in which the word is just a part. Meanings are formed from both directions, and the interpreter therefore needs to move in both directions at the same time. This idea of the *"hermeneutic circle"* involves a hierarchy of levels, from single words and their sounding or written constituents to whole sentences, works and even historical and social contexts. Such a hermeneutic circle would be an evil circle preventing any understanding at all if understanding and meaning are regarded as static: as if you either understood a text or you did not. If understanding a text demands that you understand single word units in order to understand the whole, while the meaning of each word is in turn determined by its larger context in the whole work, then the true meaning would remain forever unattainable. However, hermeneutics is not about ready-made meaning but about meaning-making as a dynamic process. The hermeneutic circle is not a vicious circle but rather a good and productive one, since interpretation is not a given fact but a process of meaning-making practice. By moving between parts and wholes, and between texts and contexts, the understanding of meanings will grow and mature. I will soon return to how Ricoeur further developed these ideas.

In the next key step, *Wilhelm Dilthey* (1833–1911) presented and built on Schleiermacher's work, actually coining the term the hermeneutic circle. He defended the humanities against the overwhelmingly successful natural sciences, by arguing that the forms of knowledge produced by the former had something in common that was radically different from those of the latter. According to Dilthey, the natural sciences *explain* (*Erklären*) the physical world, while the human sciences *understand* (*Verstehen*) the human world, based on Schleiermacher's triplet of experience, expression and understanding (Dilthey, 1910/2002). Dilthey stressed that meaning was always contextual, so that to interpret any text one has to know its historical and social background. However, since the interpreter also lives and is embedded in a specific context, he has to try and transpose himself into the author's position in another historical time, re-experiencing what he might have felt when he wrote the text. Interpretation is thus a method of feeling with others in order to reconstruct the author's intended meanings. The fundamental gap between the explanations offered by natural sciences and the understanding of the human sciences expressed a Romanticist approach that made strict divisions between body and mind, nature and culture. Meaning remained bound to psychological intentions of individual subjects, even though they had to be mediated by physical texts in order to enable communication with others, as well as self-reflection.

In the twentieth century, the phenomenologist *Martin Heidegger* (1889–1976) broke with the intentionalism and psychologism shared by his forerunners. To him, interpretation and understanding are at the foundation of human existence, and rather than a special method for reading certain kinds of texts, hermeneutics is a philosophy reflecting on what that our constant dependence on textual mediation means for human existence in general. He developed the idea of the hermeneutic circle by, for instance, showing that nobody approaches a new text innocently: even in the first step toward interpretation, the interpreter must always have certain pre-understandings or prejudices, that allow people to interpret phenomena in a preliminary way, before then through closer interaction with the text deepen and nuance that understanding (Heidegger, 1927/1962). It is the meaning of the text, rather than any author's intention, that is the basis and aim of interpretation. One facet of the circle is that to understand an artwork, one must understand art in general, and vice versa. It is always the work rather than its producer that we should try to understand.

Hans-Georg Gadamer (1900–2002) shared this de-psychologizing focus on the text rather than its author, as well as the necessity of prejudices, or prejudgements, derived from language, concepts, ideas and experiences. To him, the hermeneutic circle embraces a series of new understandings of a whole reality that are born through explorations of the details of human existence. Gadamer (1960/2004) adopted Dilthey's sharp division between the instrumental methods of natural sciences and the humanities' half intuitive search for truth by understanding. In contrast to distancing scientific techniques, interpretation to Gadamer entails a fusion of horizons where the reader's lifeworld is stepwise enriched by overlapping the world of the text. The horizon of the past is the tradition to which the reader belongs, but it is always in motion as the new is interpreted within the frames of the old but simultaneously the old is re-interpreted in light of the new, in an endlessly flowing process. In media studies, the concept of remediation can be regarded as such an example, as new media both remind of the old ones and strive to overcome their limiting deficits, so that people first use old concepts and habits to deal with the new and then reinterpret the old media as forerunners of the new ones (Bolter & Grusin, 1999).

To Gadamer, the meaning of a text is never once and for all given, but always something that can and must be further developed over time. The use of a text by new subjects in new contexts gradually changes its meaning. The history of a text's production (including its author's experiences and intentions) does not determine its meaning: it rather evolves through the history of its impact or reception (*Wirkungsgeschichte*). All texts have more or less substantial gaps inviting further elaborations and interpretations: ongoing reading fill out such voids and modify the resulting meaning, making texts into works, dynamic products of human creative activity. Gadamer argued that understanding always includes an element of *application*, whereby the interpreted text is put into practical use when dealing with problems in the contemporary context of the interpreter. Interpreting a law text is for instance always linked to applying the law to a specific conflict situation, in order to make a legal judgment. There is a similar moment of application also in literary reading or indeed any interpretation of media texts, as all media users apply their interpretations to their own current situation. These ideas open up for a moment of practice and action that could potentially be fruitfully related to other media-relevant schools of thought, such as pragmatism, speech-act theory, performance theory and reception theory in the so-called reader-response tradition, from I. A. Richards to Wolfgang Iser, Hans-Robert Jauss, Stanley Fish and Norman Holland. Janice Radway (1984) is one of too few who have applied this tradition in studying media use.

This is where too many histories of hermeneutics stop. By letting Dilthey and Gadamer represent hermeneutics at large, this tradition becomes vulnerable to a number of serious critiques, not least concerning its applicability to current media and communication phenomena. This is a great pity, since a most decisive turn was made by the French philosopher *Paul Ricoeur* (1913–2005), whose multifaceted and openly dialogic work sheds light on strong and weak points in his predecessors. Ricoeur in fact dealt with several such contestations and developed hermeneutics in a way that avoided the troubles identified by a series of antihermeneutic critics, thus saving hermeneutics by responding to reasonable objections and making productive use of critical perspectives to strengthen interpretation theory. Ricoeur has suggested three main such revisions: (1) to bridge the gap between explanation and understanding, (2) to stop searching for textual meaning in the author's intentions and (3) to allow for a dynamic plurality of meanings rather than a unique "correct" meaning for each text.

(1) Ricoeur himself contrasts earlier Romanticist hermeneutics with his own *critical hermeneutics*. From his perspective, understanding a text is not necessarily agreeing with it. Interpretation strives to understand what a text says, but this does not preclude even the strongest critique against it. Ricoeur (1965/1970, 32–36) distinguished two branches: the ordinary hermeneutics of listening understanding which seeks to faithfully reconstruct textual meanings, and the hermeneutics of suspicion, with Karl Marx, Friedrich Nietzsche and Sigmund Freud as significant representatives. To this male canon could well be added feminist critics like Mary Wollstonecraft. They all read texts against the grain, in opposition to the authors' intentions. Such critical reading is not alien to hermeneutics – on the contrary, critique may well sharpen interpretation, just as much as understanding makes criticism more effective too. In a debate with Jürgen Habermas, Ricoeur (1971/2008c, 263–299) thus argued for a dialectic between critique and understanding, rather than a divisive polarity.

Ricoeur saw structural analysis not as a substitute for interpretation, but as a necessary way of understanding texts better, going "beyond antithetical opposition" and instead regarding them as complementary (1971/2008a, 115):

> to explain is to bring out the structure, that is, the internal relations of dependence that constitute the statics of the text; to interpret is to follow the path of thought opened up by the text, to place oneself en route toward the *orient* of the text.
>
> *(Ricoeur, 1971/2008a, 117)*

Interpretation is in process: a "dialectic of explanation and understanding" (Ricoeur, 1976, 71, 74), in which explanation through structural analysis serves as necessary "mediation between two stages of understanding" (Ricoeur, 1976, 75); "between a naïve interpretation and a critical one, between a surface interpretation and a depth interpretation," so that explanation and understanding are located "at two different stages of a unique hermeneutical arc" (Ricoeur, 1976, 87).

Instead of rejecting structuralism or the ideology critique of Frankfurt School critical theory, Ricoeur certainly identified some of their limitations but all the same argued for making productive use of their methods and perspectives in order to sharpen rather than abandon the processes of interpretation. In the 1960s and 1970s, various forms of structural analysis were the most obvious examples of such critical and distancing tools of explanation that could improve rather than nullify understanding. One may well regard semiotics, narratology and discourse analysis as equally helpful methods of explanation when they are put in the service of understanding. I have elsewhere at length also argued that a similar response can be made to the antihermeneutic challenges posed by various posthumanist and new materialist authors. One need not choose between materiality and meaning but should rather be aware of the dynamic process where the former gives rise to the latter (Fornäs, 2017, 204–215).

Against the "antithetical opposition" that the earlier, Romanticist hermeneutics of Dilthey (and partly also Gadamer, though he was closer to Ricoeur) constructed between understanding and explanation, Ricoeur saw them as necessary constituents of a shared hermeneutic arc from surface understanding through explanation to deeper understanding. At Dilthey's time, it might have felt urgent to defend a humanities territory against the hegemony of the victorious natural sciences and their positivist colonization of all other disciplines. Gadamer continued to separate the "truth" of understanding from the "methods" of science. Ricoeur (1973/1981, 60f; see also Fornäs, 2017, 210) refused this separation and argued that understanding and explanation worked together on both sides of knowledge production. He criticized Gadamer's key work on "truth AND method" for actually favoring (interpretive) truth OR (scientific) method, while Ricoeur advocated developing truth (understanding) THROUGH method (explanation).

(2) If the abolition of this divide was the first of Ricoeur's breaks with Romanticism, he shared a second step with Gadamer. Face-to-face dialog often strives to understand what the other wishes to say. In the arts, it is also quite common to believe that the secret meanings of a work are to be found in the mind of its author. Schleiermacher bravely declared that the interpreter aims to "understand a writer better than he understood himself," but not by criticizing him, only by illuminating unconscious aspects of its production (Gadamer, 1960/2004, 191), and still the search for meaning remained bound to an effort to understand its author. Like Heidegger, Gadamer instead acknowledged the relative autonomy of the text and looked for meaning there rather than in any motivation of the text producer. He made it legitimate to decipher involuntary meanings that neither the author nor the text itself

intended. Fully accepting critical interpretations and explanatory distanciation, Ricoeur (1971/2008a, 108) fulfilled that move and advocated a "depsychologization of interpretation," arguing that "to read a book is to consider its author as already dead and the book as posthumous" (1971/2008a, 103). "In other words, we have to guess the meaning of the text because the author's intention is beyond our reach. Here perhaps my opposition to Romanticist hermeneutics is most forceful" (Ricoeur, 1976, 75). Reading a text is different from entering a mutual dialog, and the distanciation in time and space that writing and reading allow for requires an interpretive effort that goes beyond asking for the sender's intentions. "The text is the very place where the author appears. But does the author appear otherwise than as first reader?" (Ricoeur, 1977/2008, 105). To interpret is not to place oneself in the mind of an author but "within the sense indicated by the relation of interpretation supported by the text" (Ricoeur, 1977/2008, 119).

> Nothing has more harmed the theory of understanding than the identification, central in Dilthey, between understanding and understanding others, as though it were always first a matter of apprehending a foreign psychological life behind a text. What is to be understood in a narrative is not first of all the one who is speaking behind the text, but what is being talked about, the *thing of the text*, namely, the kind of world the work unfolds, as it were, before the text.
>
> *(Ricoeur, 1977/2008, 127)*

Inspired by Ricoeur, Rita Felski (2015, 173) argues that hermeneutics, as "a philosophy of relation," "casts texts and readers as cocreators of meaning." In Ricoeur's (1976, 87) own words: "The sense of a text is not behind the text, but in front of it. It is not something hidden, but something disclosed." Meaning is not primarily made behind the text, by its author. Instead, it emerges in the contextualized encounter between the text and its reader: between media texts and those who use them: book readers, television viewers, music listeners or webgame participants. The model of a linear chain from sender to receiver suits commercial or authoritarian media production rather well, where production and reception are institutionally separated. For understanding meaning-making in general, however, a better model is the cultural triangle where human subjects, texts and contexts are three cornerstones between which meanings are made and linked to either texts, people (forming identities) or situations (creating meaningful social worlds).

(3) In his transition from Romanticist to critical hermeneutics, Ricoeur made a third revision, this time again with some support in Gadamer's view on reception history adding new meaning to a text. If meaning emerges between the text and its reader, then new readers in new contexts can always add new layers of meaning. There is no fixed meaning content hidden in the

text but meaning-making is rather an ongoing dialogic and often conflictual process. "This opportunity for multiple readings is the dialectical counterpart of the semantic autonomy of the text. [...] Hermeneutics begins where dialogue ends" (Ricoeur, 1976, 32). Each interpretation can always be questioned and challenged by new and better ones, in a never-ending *conflict of interpretations*, in which there is never a last word to be said: "Or, if there is any, we call that violence" (Ricoeur, 1971/2008b, 162).

Hermeneutic Spirals

The use of media texts thus results in a series of competing interpretations, each of which involves a circular or pendulum movement in three steps from surface understanding through explanation to depth understanding. Over time, the whole process results in a hermeneutic spiral. In the first step, a provisional understanding suggests itself, in a creative guess on the background of one's pre-understanding and initial impression of a text. This is the moment where you let yourself be touched by the text and let it transform you, change your perspective in some way. Then, structural analysis, semiotics, discourse analysis or any other method to analyze texts may be used to validate and modify the initial guesses. This finally leads to a fuller and richer understanding supported by critical investigation of the text in question.

The hermeneutic circle originally related to how the meaning of the whole and of its parts shed light on each other, how "the presupposition of a certain kind of whole is implied in the recognition of the parts. And reciprocally, it is in construing the details that we construe the whole" (Ricoeur, 1976, 77). This circle need not be a vicious, but rather a productive one. Through "the conflict between competing interpretations," each new interpretation must prove itself to "not only be probable, but more probable than another interpretation": "there is always more than one way of constructing a text" but "it is not true that all interpretations are equal. The text presents a limited field of possible constructions" (Ricoeur, 1976, 79). Ricoeur certainly also contrasts understanding and explanation, but instead of separating them into two domains (science/ humanities), he shows how they are always intertwined.

> If the objective meaning is something other than the subjective intention of the author, it may be construed in various ways. [...] To construe the meaning as the verbal meaning of the text is to make a guess. But [...] if there are no rules for making good guesses, there are methods for validating those guesses we do make.
>
> *(Ricoeur, 1976, 76)*

Strictly speaking, explanation alone is methodical. Understanding is instead the nonmethodical moment that, in the sciences of interpretation combines with the methodical moment of explanation. This moment precedes, accompanies, concludes, and thus envelops explanation. Explanation, in turn, develops understanding analytically. This dialectical tie between explanation and understanding results in a very complex and paradoxical relation between the human sciences and the natural sciences. Neither duality nor monism, I should say.

(Ricoeur, 1977/2008, 138)

The hermeneutic spiral adds several other dialectical movements to that between whole and part: between understanding and explanation, listening and suspicion, appropriation and distanciation, guessing and validating, synthesis and analysis. In everyday life as well as in media studies, there is never a direct way to understand meanings, but always a necessity to take detours to follow the mediations that are so essential to human communication. Hermeneutics is a philosophy of communication that fully acknowledges its necessary detours and mediations.

To sum up, all research involves moments of interpretation through which meanings are (re)constructed in relation to texts. An amount of imagination is required to make the first guesses at what a text might mean, but then also an amount of careful investigation to validate those guesses and find support in the textual material itself. Some kind of structural analysis can determine the formal relations between textual elements (syntax, grammar) as well as their material constitution (phonetics, texture), while a contextual approach can find out how the text is used (pragmatics, application). In this way, interpretation is to explore the actual and potential cloud or universe of meanings surrounding a text in its context of use. It develops in an historical process, going backward to the archaeology of textual production, including authorial intentions, life-worlds and structures of feeling which are important even though they do not firmly determine meanings, and forwards to the teleology of reception, in which new meaning potentials can always be added through consecutive conflicts of interpretation.

Dynamic Borderlands

It is important not to forget about the limits of interpretation. Meaning is everywhere, in the sense that people cannot stop making meaning as long as there is human subjectivity, society and culture. Yet this does not mean that everything is always already meaningful. Making meaning is indeed an ongoing process, which implies that there is a fluid borderland between meaning and not-yet-meaning. Rather than regarding meaning as a fixed entity to map, the focus should be on meaning-making as a dynamic endeavor that takes place wherever human subjects meet texts in shifting social contexts. This also means that meaning is by itself not everything there is. Against a kind of culturalist hubris, there is a need to more humbly accept that making meaning is a creative process that is never fully achieved. All is not meaningful, but all can be pulled into signifying practices and thus become cultural phenomena to be interpreted. Hermeneutics is a theory of interpretation that follows and deepens such work, in a dialectical movement between materiality and meaning.

Positioned on the boundary of cultural discourse, the human body, "while *in* language, is never fully *of* language," says Judith Butler (1993, 67). Materiality is not "always already language"; yet "language and materiality are not opposed, for language both is and refers to that which is material, and what is material never fully escapes from the process by which it is signified"; "neither can materiality be summarily collapsed into an identity with language," since "the process of signification is always material" (Butler, 1993, 68).

In the borderlands of the universe of meaning, the spirals of interpretation thus never stop advancing. Hermeneutics remains a foundation for all humanities and social sciences, and for all cultural research and media studies, since meaning-making is central to all the phenomena they study (in contrast to the natural sciences), but also because all research and knowledge production is based on that kind of interpretive approach that hermeneutics helps us become more aware of.

How then to deal with the contemporary questioning of meaning and interpretation? A "hard" antihermeneutic position within posthumanist new materialism and media archaeology wants to get rid of meaning and interpretation in favor of some kind of direct access to materiality. In shifting terms, this was once suggested by Susan Sontag (1966/2009), Michel Foucault (1969/2002), Karen Barad (2003), Nigel Thrift (2008) and Hans Ulrich Gumbrecht (1994), among others. However, most such critics soon end up in a "soft" version, according to which meaning-making only needs to be supplemented by a heightened awareness of the multiple modes of materiality involved in this process. Such criticism should be taken seriously, as hermeneutics – like any other scholarly tradition – is never finished but always needs to be further developed and sharpened (Fornäs, 2017, 219f).

Besides a stress on the dynamic border-crossings of signifying practices, I also think there is a need for two other developments that have so far only begun. One is to get away from the logocentric bias toward literature and other written verbal texts, and to embrace in a more nuanced manner the different ways in which media users make meanings from images or sounds. There may be room for visual hermeneutics or music hermeneutics, just as there is visual and music semiotics, or at least the theory of interpretation needs to consider carefully whether meaning-making work in some respect otherwise in different symbolic modes.

Second, there may also be a need to overcome the cognitivist bias, by developing an "affective hermeneutics" that takes emotions and feelings as seriously as ideas and thoughts (Ahmed, 2004, 1–13; Felski, 2015, 176–182; Thrift, 2008, 175–184). One could perhaps argue that an affect emerging from someone encountering a piece of music is also a kind of interpretation, pointing to a state of mind that was absent but is called forth by the sounding tones.

Hence, to conclude, interpretation or meaning-making can be viewed as the most fundamental mode of communication, which unites dialog and dissemination, meaning and materiality, words and other symbols, cognitions and emotions. A critical hermeneutic understanding of culture and communication goes beyond widespread linear ideas and is better understood as a triangular intersection between subjects, texts and contexts. That is why hermeneutic reflection on meaning-making, interpretation and culture deserves a central place in contemporary media and communication studies.

Note

1 Thanks to Tereza Pavlickova for insightful feedback.

References

Ahmed, S. (2004). *The Cultural Politics of Emotion*. Edinburgh, UK: Edinburgh University Press.

Barad, K. (2003). Posthumanist Performativity: Toward an Understanding of How Matter Comes to Matter. *Signs: Journal of Women in Culture and Society*, 28(3), 801–831.

Bolter, J. D. & Grusin, R. (1999). *Remediation: Understanding New Media*. Cambridge, MA/London, UK: MIT Press.

Butler, J. (1993). *Bodies That Matter: On the Discursive Limits of 'Sex'*. New York/London, UK: Routledge.

Carey, J. W. (1989/1992). *Communication as Culture: Essays on Media and Society*. New York/London, UK: Routledge.

Dilthey, W. (1910/2002). *The Formation of the Historical World in the Human Sciences*, Selected works, vol. 3. Princeton, NJ: Princeton University Press.

Felski, R. (2015). *The Limits of Critique*. Chicago, IL/London, UK: The University of Chicago Press.

Fornäs, J. (1995). *Cultural Theory and Late Modernity*. London, UK: Sage.

Fornäs, J. (2017). *Defending Culture: Conceptual Foundations and Contemporary Debate*. Basingstoke, UK/New York: Palgrave Macmillan.

Foucault, M. (1969/2002). *The Archaeology of Knowledge*. London, UK: Routledge.

Gadamer, H.-G. (1960/2004). *Truth and Method* (2nd ed.). London, UK/New York: Continuum.

Gumbrecht, H. U. (1994). A Farewell to Interpretation. In H. U. Gumbrecht and K. L. Pfeiffer (Eds.) *Materialities of Communication* (pp. 389–404). Stanford, CA: Stanford University Press.

Hall, S. (1997). The Work of Representation. In S. Hall (Ed.) *Representation: Cultural Representations and Signifying Practices* (pp. 13–64). London/Milton Keynes, UK: Sage/The Open University.

Hannerz, U. (1992). *Cultural Complexity: Studies in the Social Organization of Meaning.* New York: Columbia University Press.

Heidegger, M. (1927/1962). *Being and Time.* Oxford, UK: Blackwell.

Peters, J. D. (1999). *Speaking into the Air: A History of the Idea of Communication.* Chicago, IL: The University of Chicago Press.

Peters, J. D. (2015). *The Marvelous Clouds: Toward a Philosophy of Elemental Media.* Chicago, IL: University of Chicago Press.

Radway, J. (1984). *Reading the Romance: Women, Patriarchy, and Popular Literature.* Chapel Hill, NC/London, UK: University of North Carolina Press.

Ricoeur, P. (1965/1970). *Freud and Philosophy: An Essay on Interpretation.* New Haven, CT/London, UK: Yale University Press.

Ricoeur, P. (1971/2008a). What is a Text? Explanation and Understanding. In P. Ricoeur (Ed.) *From Text to Action: Essays in Hermeneutics, II* (pp. 101–120). London, UK/New York: Continuum.

Ricoeur, P. (1971/2008b). The Model of the Text: Meaningful Action Considered as a Text. In P. Ricoeur (Ed.) *From Text to Action: Essays in Hermeneutics, II* (pp. 140–163). London, UK/New York: Continuum.

Ricoeur, P. (1971/2008c). Hermeneutics and the Critique of Ideology. In P. Ricoeur (Ed.) *From Text to Action: Essays in Hermeneutics, II* (pp. 263–299). London, UK/New York: Continuum.

Ricoeur, P. (1973/1981). The Task of Hermeneutics. In P. Ricoeur (Ed.) *Hermeneutics and the Human Sciences: Essays on Language, Action and Interpretation* (pp. 43–62). Cambridge, UK: Cambridge University Press.

Ricoeur, P. (1976). *Interpretation Theory: Discourse and the Surplus of Meaning.* Fort Worth, TX: Texas Christian University Press.

Ricoeur, P. (1977/2008). Explanation and Understanding. In P. Ricoeur (Ed.) *From Text to Action: Essays in Hermeneutics, II* (pp. 121–139). London, UK/New York: Continuum.

Ricoeur, P. (1981). Mimesis and Representation. *Annals of Scholarship,* II(3), 15–32.

Schleiermacher, F. D. E. (1838/1998). *Hermeneutics and Criticism and Other Writings.* Cambridge, UK: Cambridge University Press.

Sontag, S. (1966/2009). Against Interpretation. In P. Ricoeur (Ed.) *Against Interpretation and Other Essays* (pp. 3–14). London, UK: Penguin Books.

Thrift, N. (2008). *Non-Representational Theory: Space, Politics, Affect.* London, UK: Routledge.

Williams, R. (1976/1988). *Keywords: A Vocabulary of Culture and Society.* London, UK: Fontana Press.

Williams, R. (1981). *Culture.* London, UK: Fontana Press.

3

PARALANGUAGE (THE CRACKED LOOKINGGLASS OF A SERVANT, OR THE USES, VIRTUES AND VALUE OF LIMINALITY)

Michael Schandorf

Paralanguage: A Nonlinear History

The term "paralanguage" is typically credited to Trager (1958), who offered it as a way to conceptualize all of those aspects of (paradigmatically spoken) communication that are not obviously grammatical but nevertheless have an obvious impact on the meaning and significance of an utterance, such as vocal volume, pitch and rhythm (prosody), which could also include "voice qualities" such as regional, ethnic or socioeconomic accent, as well as gender differences in vocal inflection. Historically, "paralanguage" has generally been subsumed by "nonverbal communication" because such vocal elements of language and communication are deeply intertwined with, for example, facial expression, gesture (and kinesics more generally), and the use of or adaptation to physical space (proxemics). However, the twenty-first-century advent of conversational text and image messaging, digital communication technologies capable of accommodating interaction in both voice and video, and the burgeoning development of artificial intelligence agents capable of responding to voice commands and spoken interaction (e.g., Apple's Siri, Google's Assistant, and Amazon's Alexa), not to mention the gestural character of touchscreen interfaces, have all made research on paralinguistic phenomena newly relevant – even when the term itself is not used and much of the research is hidden away in corporate laboratories and proprietary technologies. While there are good – or at least theoretically justifiable – reasons for the relative absence of the term "paralanguage," there are also justifiable arguments for its revival, or at least a reconsideration of its value.

Speech, Language and the Transmission of Information

The idea of paralanguage began as a response to a problem. Beginning in the 1940s, as the cybernetic movement applied information theory to problems and processes of "communication and control," it was noticed that quite a lot of information was conveyed in channels beyond the strictly verbal. This was a problem for communication researchers because the new scientific study of communication (in conscious, explicit distinction to the humanistic study of language in texts) was based, fundamentally, on the intentional transmission of information in the form of "signs," typically conceptualized as vocalized *words* (or subvocalized words in the case of reading and writing): "communication" was grounded in *speech*.

While human communication clearly involves more than language (and more than speech), the structural (i.e., grammatical) properties of language provide a robust framework for the investigation of human communication as the transmission and exchange of meaning in information (as signals, signs or representations). Combined with the insistence that the possession of language is a defining characteristic of human being itself, human language – despite being by far the most complex form of communication known – has conventionally been taken to be the basic paradigmatic model for communication broadly. The term "paralanguage" was offered as a means of incorporating communicative actions, or *speech acts/events* that were not strictly linguistic, within the core paradigm of "language."

Obviously, embodied aspects of communication such as tone of voice, laughing or crying, etc., not to mention facial expression, physical activity and technological or object-related modifications (including, for example, makeup, clothing or a bullhorn), can have a significant effect on the production and reception of the verbal message. But the historical assumption has been that the "real" message – the "real" information – is contained in the words, which, though they may be qualified by connotation, must ultimately be decoded to their appropriate denotations and from their grammatical relations.

But this means that, given the basic assumptions of communication as the exchange of information in the form of messages and in order for the message to be optimally effective, whenever a sender sends a message to a receiver, they must send along with that message some kind of information that delineates the boundaries of the message as well as information about what kind of information the message provides. Gregory Bateson (1972), who was instrumental in bringing the cybernetic movement into the social sciences, called this information about information *metacommunication*. "Eventually," and, in fact, relatively quickly all things considered, it "came to be seen that what message is regarded as 'meta' to another is a purely relative matter. [...] As one of Bateson's colleagues, Jay Haley, expressed it years later: 'There was no "message" but only metamessages qualifying one another'" (Kendon, 1990, 25–27).

The recognition of the complex interplay between denotation and connotation in the metamessages and infinitely variable contexts of communication – as opposed to a strict understanding of *language* as a grammatical system – presented serious challenges to the development of the young empirical science of communication in the mid-twentieth century. For example:

> [V]ocal and bodily expressions are never repeatable in exactly the same proportions and therefore defy experimentation. Like sentences in human language, they are infinite in variety. The application of scientific methods and experimentation is destructive to the spontaneity and communicative value in emotional expression. It changes the data.
>
> *(Ritchie, 1975, 81)*

This is a problem common, at least to some extent, to all forms of controlled experimentation, but it is of particular concern in the investigation of "paralanguage" and "metacommunication" because the situational subtleties and performative complexities of "multimodal" interaction are precisely what these ideas point to.

Paralanguage and/or Nonverbal Communication?

One response to the problem was to ignore it, more or less (as conventional linguistics did by retreating into idealized models of "well-formed sentences" and away from the messiness of language "performance"). Another was the development of the study of "nonverbal" communication (beginning with Ruesch & Kees, 1956). Ironically, one of the most important early contributors to this field (through his development of "kinesics"), Ray Birdwhistell "used to say that the study of nonverbal communication would be like the study of non-cardiac physiology" (Kendon & Sigman, 1996, 231; Birdwhistell argued for a multimodal or multidimensional approach to communication rather than a verbal/nonverbal dichotomy). Nevertheless, nonverbal communication quickly developed into a robust, multidisciplinary field of investigation that included dedicated university courses, conferences and academic programs. (Other foundations of the field include Crystal's [1969] work on prosody, Ekman and Friesen's [1969] on "nonverbal leakage," and Friesen and Ekman's [1978] *facial action coding system*; for historical overviews of nonverbal communication research, see Knapp, 2006; Bressem, 2013; Manusov, 2016). By the 1980s interest in nonverbal communication had even expanded into a thriving popular fascination with "body language," which tended to follow the research by framing nonverbal communication, implicitly or explicitly, in terms of interpersonal power and influence, reflecting the cybernetic emphasis on "communication and control" (e.g., Henley, 1977).

The marked term "*non*verbal," however, is an indication of just how bound this effort was to the traditional model's privileging of language, which is the implication of Birdwhistell's quip: the problem of nonverbal communication could – and *should* – be addressed in a manner analogous to that of verbal communication. Given that words are signs that carry information, the obvious solution to the problem of the nonverbal or the paralinguistic was simply to classify, quantify and organize the "words" (or "signs," or "representations") and the "grammar" of nonverbal or paralinguistic communication (e.g., Poyatos, 1993).

The term "nonverbal" has dominated this effort, rather than "paralanguage," which was generally relegated to specifically *vocal signs* other than words (although even explicitly verbal expressions have sometimes been classified in terms of "paralinguistic information," as in the case of a "Freudian slip"; cf. Brown et al., 1985). Trager's (1958) original conception of "paralanguage" was offered as referring to the "voice qualities" and "vocalizations" that occur in language (where language is defined as an "act of speech"). "Paralinguistics," then, comprised the "study of all those vocal phenomena which are separate from language, but in which language is embedded" (McQuown, 1971; Leeds-Hurwitz, 2005).

But drawing the line between vocalizations (as signs as information) and the embodiments of those vocalizations in the organs of speech, in facial expression, and in bodily performance turns out to be far from straightforward. Gradually though sporadically, when used at all, the term "paralinguistic" has generally come to be used to indicate particular senses of the nonverbal understood in direct relation to specifically linguistic phenomena, sometimes including non-vocal behavior in face-to-face interaction, such as facial expression and gesture, and occasionally in relation to quasi-verbal phenomena in digitally mediated communication (e.g., Carey, 1980; Schandorf, 2013; Pavalanathan & Eisenstein, 2016).

Paralanguage and "Languaging": The Problem of Definition (and Operationalization)

For their part, most linguists have understood "paralanguage" to be, by definition, extrinsic to their field, if not vaguely oxymoronic given their inherent equation of language with communication, or at least their assumption that the very idea of the latter is dependent upon the former in human communication. Sociolinguists and linguists interested in phonemic variation and morphology, whose territories include phenomena that might be labeled "paralinguistic" (e.g., interjections), typically account for such things in terms of pragmatics, and thus as grammatical phenomena. Wharton (2009), for example, dismisses the term "paralanguage" precisely because it is sometimes used to refer to non-linguistic vocal communication while at other times it is used to refer to all aspects of communication "that are not a part of language *per se*" (5–7).

The problem of definition is persistent. On the one hand, intonation or prosody (being *vocal*) can be understood as obvious examples of "paralanguage," and facial expression and upper body gestures, particularly of the hands, are understood to have a clear and direct relation to the use (or "performance") of language without being technically linguistic (i.e., grammatically necessary to the transmission of meaning). On the other hand, a basic tenet of the contemporary investigation of co-speech gestures (following, for example, the work of David McNeill, 1992, 2005) is that such gestures are *part of language*, not a separate form of "body language" and not "para-" to language as a system of communicative action. This problem of definition, then, is just as much a question of how *language* is to be defined (and what "languaging" is taken to encompass) as it is a problem of what counts as *paralanguage*.

Even for communication researchers, the limitations and problematic assumptions built into the idea of paralanguage were identified fairly early (see Abercrombie, 1968), and the term was never used as extensively as "nonverbal." The term "verbal" was understood to encompass more than "language" while being less methodologically constraining than "linguistic." In other words, the verbal/nonverbal distinction allowed communication researchers to differentiate themselves from the narrow purview of the linguists, and was far more accommodating for this purpose than language/paralanguage.

Nevertheless, the marked term "*non*verbal" explicitly privileges language just as much as the term "*para*language." Even Mary Ritchie Key, who produced one of the few key texts on paralanguage, while insisting on the interdependence of the verbal and the nonverbal, reminds her readers,

> not to think of any of these elements of facial expression as isolated entities, but as parts of the accompaniments and supplements to speech events. They may act as sentence markers, enforcers, or contradictory indicators, but all contributing to the message.
>
> *(Ritchie Key, 1975, 90)*

"Message" in this case simply *means* the verbal or at very least verbalizable (and typically, the *intended*) message – the message carried by signs (or "representations," or "cues," on the model of words) which transmit information.

"Paralanguage," in Theory

A few intrepid souls have resisted the devaluation of the term "paralanguage," importantly including Fernando Poyatos (1993), who has attempted a broad integration by conceiving communication as a tripartite system of *language–paralanguage–kinesics*. This goes beyond the argument for the interdependence of the verbal and the nonverbal by claiming that the "speech event" or act of communication is always and inevitably embodied. Poyatos has invaluably illustrated the many ways in which even written language is inherently bound to bodily enactment – both in its production and in its reception and interpretation (for an introduction, see Poyatos, 2013). In many ways Poyatos's work serves as an important forerunner to more recent accounts of the embodiment of communication (for a summative collection of this work see Poyatos, 2002a, 2002b, 2002c). However, like many others, it is also importantly limited by underlying assumptions derived from the information-theoretic and cybernetic foundations of communication studies.

The idea of paralanguage, in fact, is particularly valuable for its ability to bring these assumptions to the surface. At the most basic and general level, "In order to communicate, animals must create perceptible signals. For human communication, the predominant way in which signals are produced is by moving parts of our bodies," which includes the vocal tract (Wilcox, 2013, 785) – speech can and has been understood as "vocal gesture" (e.g., Park, 1927; Paget, 1930; Kendon, 2004; Engelland, 2014; *inter alia*). Because of the range of phenomena that can be labeled "paralinguistic" (thanks to the privileging of a strictly grammatical concept of language), and the fact that much of this physical behavior of "signal production" is often not under explicit, conscious control, paralanguage has been characterized as information unconsciously "leaked" (as in Ekman & Friesen [1969] or, in Goffman's terms, "given off" [1956] or "exuded" [1981]) by a speaker). But:

> [W]hat is this nonverbal paralinguistic information? What is being leaked nonverbally? It is information that the careful observer sees the agent [i.e., the speaker] must have in order to act as he does. That is, it is information that guides the agent's behavior. Yet he himself appears not to be aware of having this information. It is information about the purposive character of the act, information that the observer needs in order to understand adequately, such as information about its motives and aims. Paralinguistic information is very different from accidental nonverbal information (due to physical limitations of the speaker's musculature, etc.). In a sense, paralinguistic information allows us potentially to know more than the speaker about his own intentions.
>
> *(Brown et al., 1985, 151)*

These authors argue, therefore, that the problem with the idea of paralanguage is with the idea of the "unconscious," which generates conceptual paradoxes. But we might just as easily argue that the problem is with the idea of *information*.

In the empirical investigation of communication, broadly, "communication" *just is* the transmission (and sometimes the exchange) of information. The idea of "communication" is thus inextricably bound to the idea of "information," and from this position the conceptualization of "language" is intimately bound to both. But these constitutive interrelations are typically taken wholly for granted, despite the widely varying definitions and uses of these terms as applied in different situations and for different purposes. This makes them a blind spot that is illuminated in interesting ways by the idea of "paralanguage." The use of the term "paralanguage," for example, demands both a definition of the term "language" and the differentiation of "language" from "communication." Attempts to systematically investigate paralanguage and nonverbal communication quite quickly present a number of challenges to these seemingly simple differentiations.

These conceptual tensions are bound to the very beginnings of communication as a discipline. In Norbert Wiener's canonical presentation of cybernetics ("the study of communication and control"), communication is defined as the transmission of information, and any system for the production ("encoding") and reception ("decoding") of signals carrying information can be thought of as a "language" (given its systemic, "coded" character), whether "in the animal [including the human animal] or the machine":

> Language, in fact, is in one sense another name for communication itself, as well as a word used to describe the codes through which communication takes place. [...] Birds communicate with one another, monkeys communicate with one another, insects communicate with one another, and in all this communication some use is made of signals or symbols which can be understood only by being privy to the system of codes involved.
>
> *(Wiener, 1954, 74)*

For Wiener, relying on Claude Shannon's definition of information as "negative entropy" (Shannon & Weaver, 1949), communication ultimately relies on the accurate reception ("decoding") of transmitted information from the signals that carry it (followed by the interpretation of the meaning, "carried by" that information, in relation to the appropriate shared "code").

But quite a lot is assumed in that word "accurate" – including, for example, *intention* (which is integral to what is understood as "conscious" behavior). One of the basic properties of nonverbal and paralinguistic communication is their liminal position in relation to conscious and unconscious, and/or intentional and unintentional, action. Apart from information "leaked" or "given off," we most certainly can and do modulate our voice or adjust our facial expression, for example, for more effective communication. (Much of the study and practice of rhetoric in the Renaissance was devoted to what can be called the effective and accurate paralinguistic encoding of the message in gesture, cf. Wollock, 2013). But how much of such expression is or can be under conscious control in particular situations, and how might an accurate (and situationally sensitive) determination be made?

The traditional and received model of communication (based on a semiotic model of language wedded to cybernetic theory) remains mired in a "conduit" metaphor (Reddy, 1979) which presents the process of communication as a simple, linear, Russian doll-like functional hierarchy: "meaning" is wrapped in "information" contained in a package of "message" which is transmitted within some "channel" and/or "mode" (such as language) which is shaped by some "genre" or "modality" determined by a governing "code" (or similarly nested codes) within the confines and requirements of a specific situation. The simple, hierarchical linearity of this reduction explains much of its persistence.

However, packaging the process of communication in this neat little box of boxes makes very big assumptions – or ignores – the underlying problems of, for example, the intentionality that motivates communication in the first place. The black box of "meaning" at the center is directly and substantively bound to the black box of "intention" that shapes and is shaped by the governing or functional situation. But that connection is *broken*, not explained, by this linear model of communication. The non/verbal and para/language dichotomies, from the perspective of the standard model, are ultimately fairly simple and certainly reductive questions of channel or mode. But in practice, carving those lines is not at all simple, and any drawing of lines between conscious and unconscious or intentional and unintentional communicative action – just as much as a distinction between verbal and nonverbal or language and paralanguage – not only assumes a particular definition of "language" distinct from "communication," but makes specific assumptions about the definition and character of "information" and its relation to "meaning."

Cracked Categories → Reflected Assumptions in Suspect Service

A general resistance to confronting and articulating such assumptions is one reason why even the term "nonverbal communication" isn't as popular as it used to be (though see, e.g., Matsumoto et al., 2016). The deep interrelation of the verbal and nonverbal is now typically assumed (often and increasingly within a more multimodal conception of what communication involves), while intentionality – as much as language itself – is now generally understood to be both far less categorically determinable and far more of a social and interactional phenomenon than a property of a rational, individual agent generating information as messages. For the same reasons, the term "paralanguage" is rare: accounts of language as social cognition and action (rather than strictly a system, grammatical or otherwise, for the encoding and decoding of information) incorporate phenomena, behavior, or "cues" that might have been labeled "paralinguistic," within the broader *act* of "languaging." (Müller et al., 2013a, 2013b, for example, cover, in two volumes comprising 172 chapters, studies of gesture specifically, embodied physical interaction broadly, functional and pragmatic approaches to interaction, cognitive and neurophysiological perspectives on embodiment and communication, a variety of disciplinary and methodological perspectives including cognitive anthropology and ethnography, and cultural and cross-cultural communication and interaction, as well as the grammar of and gestures in sign languages, and *more* – all without the categorical use of the term "paralanguage" and only limited use of the term "nonverbal.")

But while the term "paralanguage" is not that useful as an analytical category, it *is* quite useful as a lens capable of magnifying the boundaries of our theoretical assumptions and their ensuing constraints. Again, "paralanguage" demands a definition of "language" and the differentiation of "language" from "communication," which in turn relies on assumptions about the character and functions of "information." Use of the term "paralanguage" draws attention to our basic assumptions about what it is that we are doing when we investigate language, communication, information and even "meaning." And this is valuable right now for at least two reasons.

The first is that our assumptions about language have historically been strongly and determinatively anthropomorphic in ways that have broad (and in the Anthropocene, critical, crucial and vital) ethical implications. Despite Norbert Wiener's broad gloss of communication as "language" (which was, in practice, typically taken to be metaphorical), "language" has conventionally been understood to be – and in some quarters is still often adamantly proclaimed and viciously defended as – an exclusive ontological property of human being, a defining characteristic that determines our difference from "lesser" forms of life. Nevertheless, in recent decades the line between human language and animal "languages" has been steadily blurring. The evidence of grammatical properties in nonhuman communication systems grows by the day (e.g., Ten Cate, 2014; Taylor et al., 2017). The work of Catherine Hobaiter and colleagues (e.g., 2014, 2017), for example, has uncovered vast and sophisticated repertoires of meaningful gesture and vocalization in wild chimpanzees. Dolphins are known to refer to one another with "signature whistles," i.e., *names* (King & Janik, 2013; Janik & Sayigh, 2013). The way that we define "language" – what we allow to be authorized and authenticated as "linguistic" – has real and important implications and ramifications for how we understand and value communication and interaction broadly, and thus even for the very idea of "relation" upon which communication and interaction are built.

This is directly related to the second reason that "paralanguage" can be a valuable theoretical tool: the idea of "paralanguage" reinforces the inherently and inevitably embodied character of communication, which reinforces the (social, rhetorical and ethical) agency of communicators or interactants (as "the power to affect and be affected," following Bennett, 2010). What is specifically excluded from discussion of communication without the "nonverbal," just as much as discussion of language without the "paralinguistic," is *voice*, both as the physical organ of expression and as a marker and index of personhood. (The word "person" derives from the ancient Greek theater's *personare* – the mask that the voice *sounds through*.) Frederick Ruf (1997) argues that *voice* marks the nexus of embodiment, sociality and con-text – the constitutive components (or dimensions) of communication as action, which are reduced out of the standard model of communication. The recognition of voice requires not just hearing but *listening*, and thus requires moral and ethical acknowledgment (Idhe, 2007). Studies of communication that ignore the "nonverbal," and studies of language that ignore the "paralinguistic," leave those communicating – the "subjects" of (i.e., those subjected to – as much as those excluded from) study and discussion – both literally and metaphorically, *voiceless*. They (we) are not simply mis(re)presented, they are denied the self-determination of their own presence when the embodiments of paralanguage are devalued or dismissed.

These ethical imperatives are of particular concern in, among other things, studies of mediated communication and digital media and technology, which have arguably come to dominate the contemporary study of communication. In contrast to interactional trends in the study of language and communication, studies of digital media – thanks as much to the roots of such study in information theory, cybernetics and semiotics as to the comparatively more reduced and structured forms digital data – remain mired in the reductive assumption that all communication *just is* information. Thus "information" is both derived from and privileges situational function understood through categories of "communication and control."

But for this reason, "data" tends to dehumanize, to reduce out ethical concerns in favor of functional goals. The reduction of experience and expression to data and information eviscerates *voice* (all too often, by design), locating and interpellating people within determinative systems. On the one hand, the power of information systems to reify determinative categories of information has become a defining feature of twenty-first-century life. On the other hand, digitally networked communication and information systems present a fecund field of combat where different categorical (i.e., ideological) systems compete for influence, dominance and control of societies, communities and the bodies that comprise them.

Making this argument for attention to the "paralinguistic" or "nonverbal" in (academic) language, as I'm doing here, neither contradicts nor weakens the claim: language provides conceptual structure – but it inevitably makes available many interpenetrating and protean conceptual structures at once, shaped and evoked by different definitions in response to the needs (e.g., material, physical, psychological, social, affective, functional) of communicators or interactants. The idea of "paralanguage," by revealing and magnifying the liminal boundaries of our foundational theoretical concepts, speaks for what is silenced in the reduction of life to data, of people to information, and of (social, political and moral) agency to determinative, functional channels (e.g., of consumption or of ideological performance).

Unfortunately, the ways we understand and operationalize our definitions of language, communication and information too often reinforce or contribute to the control or suppression of voices – even when the intention is to reveal them. The basic methodological mode of "coding and counting" too often embodies a retreat from the ethical implications of information, even when it is the "paralinguistic" that is being coded: "sentiment analysis" is mobilized to reduce and channel living voices to functional outcomes; the study of spoken interaction and "persuasive technology" are bent to build better voice assistants and to drive marketing platforms and "engagement," with improvements of such systems being directly related to the effective mobilization of paralinguistic elements and "cues" (as data points). This informationalization of *voice* constitutes the weaponization of paralanguage – where communication becomes control. The "user" is compensated with access to other "users," or the ability to control a device or navigate a system by voice command, and this can make a person feel "powerful." But that power is extremely and increasingly narrow and often serves to limit agency and personhood in very strict ways; it is functional in a way that we might name "para-ethical."

Pointing to Possibilities beyond the Cracked Mirror

More than half a century after the first appearance of the term, "paralanguage" can begin to serve as a response to the problem of communication and all it entails, not by insisting on the distinction between language and "paralanguage," but by reminding us that language is inevitably embodied in and providing meaning *only* by that which is indicated in the prefix "para-." Without the embodied "para-" there can be no language, whether systemic or specifically enacted/performed: without the "meta-" there is no communication; it is the con-text that provides the text with meaning – the background that articulates the focus (Grossberg, 1992). This is a far different set of assumptions than those grounding typical analytical methods of "coding and counting," which rely on predetermined or functionally generated categories re-presented in and as data and information.

The problems of communication are cast in an entirely different light when "information" is understood not as transmitted and exchanged, but as generated in interaction within (and as) the "para-" and "meta-" *con-text*: this is information as verb – *inform*ing the space of relations (Schandorf, 2019). For this reason and others, the study of digital media and digitally mediated interaction provides a valuable – and largely undeveloped – opportunity to explore the liminal boundaries and the crucial limitations of our foundational concepts – and of our methods in the study of communication. Because digitally mediated communication exists obviously, explicitly and functionally as digital information (which, in its early years, was explicitly conceptualized as *dis*embodied), it has been seen as the fulfillment of the cybernetic prophecy of mind-becoming-computer – but that prophecy was largely *self*-fulfilling, reliant on its own categorical determinations. The temptation to reduce communication to information, because of its digital form (and because of the relative ease of doing so), is a temptation that can be profitably resisted, but it requires strenuous rethinking of what the digital does.

One opportunity is presented by the "paralinguistic" phenomena of digital media that have consistently fascinated both researchers and the public alike, including most recently emoji (e.g., Stark & Crawford, 2015), animated gifs (e.g., Tolins & Samermit, 2016), and other imagistic or non-textual forms, platforms and interfaces of digitally mediated interaction (though Poyatos [1993] is among those who have long argued that even punctuation is inherently *gestural*). Similarly, sociolinguistic studies of digital interaction have given valuable attention to phenomena such as code-switching, orthographic variation and attitudinal or modal expression (including irony and hyperbole) in online language (Schandorf [2019] provides a historical review of research on digital media interaction). All of these pragmatic and "paralinguistic" forms of interaction point to the inevitably and inescapably embodied character of digital interaction because so much of its "nonverbal" communication is performed in strictly or (until very recently) predominantly textual (e.g., typed) forms. The importance of the "paralinguistic" in digital interaction is arguably even more vital than those Poyatos has demonstrated in written literature because so much of our daily, mundane interaction now takes place in digital spaces. However, such phenomena or performatives are often derided or devalued as "merely phatic" because of their tenuous link to the explicitly "coded" textual message – the "real" information that constitutes the "message."

This is a misrepresentation, and a mistake. But it is a misrepresentation that has been honestly inherited from the founders of communication research. We can remember, however, that the assumptions and goals of those founders need not be our own. Their con-text was very different. The foundations of communication research were a product of Cold War imperatives and motivations. Today we remain trapped by the ideological constraints of "communication and control" – even in our common accounts and mundane practices of social media.

One reason that it has been so easy to weaponize digitally mediated interpersonal interaction for the purposes of propaganda and (culture war) social and political disruption is that the entire edifice of communication research, and thus the most widely understood account of communication as well as the functional and operational assumptions built into our communication systems, are deeply grounded in analytical methods designed to reveal and propagate social and political influence and control (even in interpersonal interaction) – which entails the de-voicing of communicators even in the elucidation of their (our) "speech." We will not get out of this mess until we can face the inherent contradictions and eviscerations built into our foundational concepts of what communication is.

"Paralanguage" as an analytical category – as a "mode" or "channel" of information transmission – will not help much. But as a theoretical tool that demands conceptual coherence, the idea of paralanguage can help to return *voice* to the investigation and practice of communication. In this way, paralanguage offers a response to a much more vital set of problems than the one that originally prompted its coining: the problem of "information," which is fundamentally not simply a functional problem, but an *ethical* problem – a problem of *meaning*.

Conclusion: Turning Cracked Lines into Liminal Spaces – Paralanguage as a Tool for Reimagining

The idea of paralanguage began as a response to the problem of vital yet non-linguistic (non-grammatical) elements of spoken communication. The idea of paralanguage thus contributed to the development of the field of nonverbal communication, which has addressed this problem by, essentially, stripping much of the vitality of communication away in a reduction to "codes" and the encoding and decoding of information as signals and signs. While the problem remains as vital as ever, the strain against the de-voicing of communicators is readily apparent in (among other things) the vast range of "paralinguistic" and nonverbal phenomena that have been part of digitally mediated interaction from the beginning, such as emoticons (and other expressive uses of punctuation and orthography), emoji, animated gifs and other forms of non-linguistic utterance and meaning. People have always and continue to resist constraints on expression and interaction – on communicative agency – in the sharing of meaning, repeatedly demonstrating that there is nothing "para" about the "paralinguistic" in interaction. By pointing toward the embodied character of all communication and interaction, the idea of paralanguage can serve as a valuable theoretical tool for revealing our theoretical assumptions, biases and limitations in the study of communication, and perhaps help us to move past those linear limitations built of "communication and control."

References

Abercrombie, D. (1968). Paralanguage. *British Journal of Disorders of Communication*, *3*(1), 55–59.

Bateson, G. (1972). *Steps to an ecology of mind*. New York, NY: Ballatine Books.

Bennett, J. (2010). *Vibrant matter: A political ecology of things*. Durham, NC: Duke University Press.

Bressem, J. (2013). 20th century: Empirical research of body, language, and communication. In C. Müller, et al. (Eds.), *Body–language–communication: An international handbook on multimodality in human interaction* (Vol. 1, pp. 393–416). Berlin, Germany: Mouton De Gruyter.

Brown, B., Warner, C. & Williams, R. (1985). Vocal language without unconscious processes. In A. Siegman & S. Feldstein, *Multichannel integrations of nonverbal behavior* (pp. 149–193). Hillsdale, NJ: Lawrence Erlbaum.

Carey, J. (1980). Paralanguage in computer mediated communication. In N. Sondheimer (Ed.), *The 18th annual meeting of the association for computational linguistics and parasession on topics in interactive discourse: Proceedings of the conference* (pp. 67–69). Philadelphia, PA: University of Pennsylvania.

Crystal, D. (1969). *Prosodic systems and intonation in English*. Cambridge, UK: Cambridge University Press.

Ekman, P. & Friesen, W. (1969). Nonverbal leakage and clues to deception. *Psychiatry, 32*(1), 88–106.

Engelland, C. (2014). *Ostension: Word learning & the embodied mind*. Cambridge, MA: MIT Press.

Friesen, W. & Ekman, P. (1978). *Facial action coding system: A technique for the measurement of facial movement*. Palo Alto, CA: Consulting Psychologists Press.

Goffman, E. (1956). *The presentation of self in everyday life*. Edinburgh, UK: University of Edinburgh Social Sciences Research Centre.

Goffman, E. (1981). *Forms of talk*. Philadelphia, PA: University of Philadelphia Press.

Grossberg, L. (1992). *We gotta get out of this place: Popular conservatism and postmodern culture*. New York, NY: Routledge.

Henley, N. (1977). *Body politics: Power, sex, and nonverbal communication*. Englewood Cliffs, NJ: Prentice-Hall.

Hobaiter, C. & Byrne, R. W. (2014). The meaning of chimpanzee gestures. *Current Biology, 24*, 1596–1600.

Hobaiter, C., Byrne, R. W. & Zuberbüller, K. (2017). Wild chimpanzees' use of single and combined vocal and gestural signals. *Behavioural Ecology and Sociobiology, 71*, 96. doi:10.1007/s00265-017-2325-1

Ihde, D. (2007). *Listening and voice: Phenomenologies of sound* (2nd ed.). Albany, NY: State University of New York Press.

Janik, V. M. & Sayigh, L. S. (2013). Communication in bottlenose dolphins: 50 years of signature whistle research. *Journal of Comparative Physiology A, 199*, 479–489.

Kendon, A. (1990). *Conducting interaction: Patterns of behavior in focused encounters*. Cambridge, UK: Cambridge University Press.

Kendon, A. (2004). *Gesture: Visible action as utterance*. Cambridge, UK: Cambridge University Press.

Kendon, A. & Sigman, S. (1996). Ray L. Birdwhistell (1918-1994). *Semiotica, 112*, 231–261.

King, S. & Janik, V. (2013). Bottlenose dolphins can use learned labels to address one another. *Proceedings of the National Academy of Sciences, 110*(32), 13216–13221.

Knapp, M. (2006). An historical overview of nonverbal research. In V. Manusov & M. Patterson (Eds.), *The Sage handbook of nonverbal communication* (pp. 3–19). Thousand Oaks, CA: Sage.

Leeds-Hurwitz, W. (2005). The natural history approach: A Bateson legacy. *Cybernetics and Human Knowing, 12*(1–2), 137–146.

Manusov, V. (2016). A history of research on nonverbal communication: Our divergent pasts and their contemporary legacies. In D. Matsumoto, H. Hwang & M. Frank (Eds.), *The APA handbook of nonverbal communication* (pp. 3–15). Washington, DC: American Psychological Association.

Matsumoto, D., Hwang, H. & Frank, M. (2016). *The APA handbook of nonverbal communication*. Washington, DC: American Psychological Association.

McNeill, D. (1992). *Hand and mind: What gestures reveal about thought*. Chicago, IL: University of Chicago Press.

McNeill, D. (2005). *Gesture and thought*. Chicago, IL: University of Chicago Press.

McQuown, N. A. (1971). Natural history method: A frontier method. In A. R. Mahrer & L. Pearson (Eds.), *Creative developments in psychotherapy* (pp. 430–438). Cleveland, OH: Case Western Reserve University Press.

Müller, C., Cienki, A., Fricke, E., Ladewig, S. H., McNeill, D. & Bressem, J. (2013a). *Body–language–communication: An international handbook on multimodality in human interaction* (Vol. 2). Berlin, Germany: Mouton De Gruyter.

Müller, C., Cienki, A., Fricke, E., Ladewig, S. H., McNeill, D. & Teßendorf, S. (2013b). *Body–language–communication: An international handbook on multimodality in human interaction* (Vol. 1). Berlin, Germany: Mouton De Gruyter.

Paget, R. (1930). *Human speech: Some observations, experiments, and conclusions as to the nature, origin, purpose and possible improvement of human speech*. New York, NY: Harcourt Brace.

Park, R. (1927). Human nature & collective behavior. *American Journal of Sociology*, *32*(5), 733–741.

Pavalanathan, U. & Eisenstein, J. (2016). More ☺, less:) The competition for paralinguistic function in microblog writing. *First Monday*, *21*, 11.

Poyatos, F. (1993). *Paralanguage: A linguistic and interdisciplinary approach to interactive speech and sound*. Amsterdam/Philadelphia, PA: John Benjamins.

Poyatos, F. (2002a). *Nonverbal communication across disciplines. Volume I: Culture, sensory interaction, speech, conversation*. Amsterdam, the Netherlands: John Benjamins.

Poyatos, F. (2002b). *Nonverbal communication across disciplines. Volume II: Paralanguage, kinesics, silence, personal and environmental interaction*. Amsterdam, the Netherlands: John Benjamins.

Poyatos, F. (2002c). *Nonverbal communication across disciplines. Volume III: Narrative, literature, theater, cinema, translation*. Amsterdam, the Netherlands: John Benjamins.

Poyatos, F. (2013). Body gestures, manners, and postures in literature. In C. Müller, et al. (Eds.), *Body–language–communication: An international handbook on multimodality in human interaction, vol. 1* (pp. 287–300). Berlin, Germany: Mouton De Gruyter.

Reddy, M. (1979). The conduit metaphor. In A. Ortnony, *Metaphor and thought* (pp. 284–324). Cambridge, UK: Cambridge University Press.

Ritchie Key, M. (1975). *Paralanguage and kinesics: (Nonverbal communication)*. Metuchen, NJ: Scarecrow Press.

Ruesch, J. & Kees, W. (1956). *Nonverbal communication: Notes on the visual perception of human relations*. Berkeley, CA: University of California Press.

Ruf, F. (1997). *Entangled voices: Genre and the religious construction of the self*. New York, NY: Oxford University Press.

Schandorf, M. (2013). Mediated gesture: Paralinguistic communication and phatic text. *Convergence*, *19*, 319–344.

Schandorf, M. (2019). *Communication as gesture: Media(tion), meaning, and movement.* Bingley, UK: Emerald.

Shannon, C. & Weaver, W. (1949). *A mathematical theory of communication.* Urbana, IL: University of Illinois Press.

Stark, L. & Crawford, C. (2015). The conservatism of emoji: Work, affect, and communication. *Social Media + Society.* July–December. doi: 10.1177/2056305115604853.

Taylor, C. E., Brumley, J. T., Hedley, R. W. & Cody, M. L. (2017). Sensitivity of California Thrashers (*Toxostoma redivivum*) to song syntax. *Bioacoustics, 26*(3), 259–270.

Ten Cate, C. (2014). On the phonetic and syntactic processing abilities of birds: From songs to speech and artificial grammars. *Current Opinion in Neurobiology, 28,* 157–164.

Tolins, J. & Samermit, P. (2016). GIFs as embodied enactments in text-mediated conversation. *Research on Language & Social Interaction, 49*(2), 75–91.

Trager, G. L. (1958). Paralanguage: A first approximation. In D. Hymes, *Language in culture and society: A reader in linguistics and anthropology* (pp. 274–289). New York, NY: Harper & Row.

Wharton, T. (2009). *Pragmatics and Non-verbal Communication.* Cambridge, UK: Cambridge University Press.

Wiener, N. (1954). *The human use of human beings: Cybernetics and society.* Boston, MA: Houghton Mifflin.

Wilcox, S. (2013). Articulation as gesture: Gesture and the nature of language. In C. Müller, et al. (Eds.), *Body–language–communication: An international handbook on multimodality in human interaction* (Vol. 1, pp. 785–792). Berlin, Germany: Mouton De Gruyter.

Wollock, J. (2013). Renaissance philosophy: Gesture as universal. In C. Müller, et al. (Eds.), *Body–language–communication: An international handbook on multimodality in human interaction* (Vol. 1, pp. 364–377). Berlin, Germany: Mouton De Gruyter.

4

CORPUS-METHODOLOGY AND DISCURSIVE CONCEPTUALIZATIONS OF DEPRESSION

Kim Ebensgaard Jensen

Introduction

The prospect of analyzing large data masses of texts on depression by depression patients is simultaneously very attractive and daunting, for it allows us to address how people conceptualize the illness, but how can one manually and qualitatively close-read and analyze such large amounts of data? This concern is understandable, but misguided. First, the data do not have to be analyzed exclusively qualitatively. If looking for patterns in how the illness is conceptualized, we must perform qualitative *and* quantitative analyses, as the latter allows us to determine if a conceptualization is patterned. Corpus-linguistic methodology is one methodological framework that allows for this type of analysis. Second, in addition to close-reading, there is distant-reading (textual analysis of large masses of textual data) which has recently gained popularity within the humanities. Distant-reading corresponds to zooming out to get the "big picture" and has proven useful in detecting patterns within and beyond the individual text. Unlike the humanities at large, corpus linguistics has since its inception made use of distant-reading before this mode of reading even had a name. Tognini-Bonelli (2010, 19), while not using the actual terms, captures the difference between the two modes in her discussion of text-reading versus corpus-reading.

> A text is read horizontally for content as a whole, a unique event, an individual act of will, an instance of *parole*, and one coherent communicative event. In contrast, a corpus is read fragmentedly and vertically for formal patterning and repeated events to provide insights into *langue* as a sample of social practice and not a coherent communicative event.

Third, humanists now have several digital tools at their disposal which can help perform complex quantitative, qualitative, and transformational analyses in a matter of minutes and even seconds.

Corpus linguists have generated countless valuable insights, and corpus linguistics has thrived and evolved within empirical linguistics, but it is only now that researchers in other disciplines have started to appreciate its potential. This chapter presents an illustrative corpus-based study (which is admittedly more simplistic than what you will see in most research papers within corpus linguistics) of a collection of blog posts on depression by depression patients so as to show its applicability in the analysis of discourse and its interdisciplinary potential.

What Is Corpus-Methodology?

Corpus linguistics is perhaps best described as a data-driven methodological framework for linguistic inquiry that relies on digital data and tools (Kirk, 1996, 250–251). Corpora are databases of naturally occurring language, typically described as "large and principled collection[s] of natural texts" (Biber et al., 1998, 4). Corpora are compiled in accordance with criteria determined by the research question and must be representationally balanced; as Anderson and Corbett (2009, 4) write, corpora are "motivated, created with a linguistic purpose in mind." A corpus is *per se* finite and limited to what it contains, and one cannot draw conclusions about what is not documented in the corpus.

Corpus linguistics has evolved in tandem with computer technology, but the use of corpora actually predates the computer age. In what is jokingly called "corpus linguistics BC" ("*before computers*") (Svartvik, 2007), archives of citation slips were used as data for lexicographical research. The "BC"-era ended in the 1960s as computer mainframes were introduced at universities, enabling researchers to easily perform analyses.

Corpus linguistics has an extensive body of publications, most of which are empirically oriented. For instance, Biber et al. (1998), McEnery and Hardie (2012), McEnery and Wilson (2001), Kennedy (1998) and Sinclair (1991) all discuss and exemplify important empirical aspects of corpus linguistics, as do the contributions in O'Keefe and McCarthy (2010). Many publications present and discuss analytical techniques (e.g., Rayson & Garside, 2000; Stefanowitsch & Gries, 2003). Other important publications apply corpus-analysis within specific fields of linguistic inquiry such as literary stylistics (Mahlberg, 2013), discourse studies (Partington, 2004) and sociolinguistics (Baker, 2010). Lastly, there is a host of studies that have yielded important results and insights into, for instance, cultural differences between British and American English (Leech & Fallon, 1992), grammaticalization of semi-modals (Krug, 2000), discourse and homosexuality (Baker, 2005), discursive representation of refugees in the UK press (Gabrielatos & Baker, 2008) and register variation (Biber & Conrad, 2009).

Corpus-Linguistic Analysis

Computational tools and corpus linguistics are inseparable (Kirk, 1996, 252), and digitization has long been among the defining features of corpora (Baker et al., 2006, 48; Gries, 2009, 7–8). Methodologically, the corpus analyst studies patterns of language use within the corpus, using systematic retrieval and analysis of usage-events (occurrences of the linguistic phenomenon in question) within the corpus. Consequently, corpus linguistics is "the study of language based on examples of 'real life' language use" (McEnery & Wilson, 2001, 1) and is often characterized as being particularly objective (Berez & Gries, 2009, 158–159). Kirk (1996, 253–254) argues that corpus-methodology has a number of scientific advantages in the form of falsifiability, completeness, simplicity, strength and objectivity.

That said, corpus-analysis is not entirely void of interpretation, as the data need to be interpreted and contextualized. Corpus-methodology employs observation and quantification of usage-patterns across usage-events, often by addressing complex association patterns, defined by Biber et al. (1998, 5) as "the systematic ways in which linguistic features are used in association with other linguistic and non-linguistic features." While quantification of usage-patterns requires statistical analysis, the identification of phenomena to be quantified calls upon classificatory qualitative analysis (Berez & Gries, 2009, 158).

Corpus linguists have an impressive arsenal of analytical techniques at their disposal. Some were developed within corpus linguistics, and some were adopted from other fields such as text-mining and information science. The fundamental technique, however, is concordancing which consists of identifying all usage-events of a particular phenomenon and listing every instance with some surrounding linguistic context. Table 4.1 shows an excerpt of a concordance.

This excerpt stems from a concordance with *depression* as the search term (the full concordance contains 239 occurrences of the word). Concordances enable identification of association patterns in the immediate linguistic context. For instance, we can see that *depression* is coordinated with *anxiety* several times. If this is indeed a usage-category of the word *depression*, the next step would be to quantify it. There are 23 instances of the category in the corpus; that is 9.6% of all instances of the word *depression*. Then the analyst would interpret the findings: does the fact that 9.6% of all occurrences of *depression* appear in coordination with *anxiety* suggest that depression and anxiety often go hand in hand? To answer this question, it might be necessary to study trends in the diagnosis of depression and anxiety; that is, the analyst must consider insights from a research field beyond linguistics to gain an understanding of the use of the word *depression*.

The concordancer is a digital tool designed to generate concordances within seconds.[1] Most concordancers can also perform collocational analysis (the analysis of words and expressions that are used in the company of one another), word

TABLE 4.1 Excerpt of a concordance

...		
was tough and I had a deep period of	**depression**	– again baking was something I did to e
a service Mind runs following the merger with	**depression**	Alliance last year. We're so grateful
. Later on, when I became more open about my	**depression**	and about my mental illness, and after
the positive form in the road, away from the	**depression**	and all of the rubbish of that, and tow
as it is associated with lower rates of	**depression**	and anxiety and it is easy for me to
I have BPD (Borderline Personality Disorder),	**depression**	and anxiety. As well as my mental healt
course. I've done the My Generation course,	**depression**	and anxiety course. They give you techn
3,000s fundraising trek. I've experienced	**depression**	and anxiety from my early teens, but my
Surviving Freshers week (and beyond) with	**depression**	and anxiety Having suffered from a
...		

cluster analysis (the analysis of common strings of words that a particular word appears in), n-gram analysis (the analysis of word strings as such), dispersion analysis (the analysis of the distribution of a linguistic form across the corpus) and keyness analysis (the identification of words that are significantly frequent in one corpus and infrequent in another). Some analyses, however, must be performed manually while others require specialized software or computer scripts.[2]

Corpus Linguistics and Theory

Corpus linguistics is theory-neutral:

> Corpus linguistics is not like psycholinguistics, sociolinguistics, or geolinguistics, in which the theories and methods of neighboring subject areas are applied to and correlated with linguistics data, the variation in which can usefully be explained by invoking the criteria of the respective subject areas. Nor is it like phonology or lexicology, with a *-logos* 'written word' or 'knowledge' element about the sounds or vocabulary of the language. Nor is it like semantics or pragmatics, in which there is a similar focus on the *-ics* 'written word' or 'knowledge' element about the formal or contextual meaning of utterances. Corpus linguistics does not align itself with any of these other *-linguistics*, *-ologies*, or *-ics*. As

> a methodology for research, corpus linguistics is in a class of its own. In particular, it foregrounds data and methodology, analysis, and interpretation. If the *-istics* analogy is valid at all, then corpus linguistics has a methodological bias referring to the use of corpora and computers as tools for analysis and to the use of the results as a basis for interpretation in the study of any such aspect of language.
>
> *(Kirk, 1996, 250–251)*

Corpus-methodology has been applied extensively within and across several linguistic disciplines (e.g., Leech & Fallon, 1992; Krishnamurthy, 2000; Krug, 2000; Ooi, 2000; Partington, 2004; Baker, 2005; Baker, 2006, 2010; Gries & Stefanowitsch, 2006; Wulff, 2008; Fina, 2011; Andersen & Bech, 2013; Mahlberg, 2013). It has also proven useful in disciplines beyond linguistics. For instance, McEnery and Baker (2016) apply corpus-methodology in the study of texts in social history, while Solan and Gales (2016) apply it in legal studies.

In connection with the theory–analysis relationship, three overarching approaches can be identified within corpus linguistics (Tognini-Bonelli, 2001; McEnery & Hardie, 2012, 5–6; Cheng, 2012, 187–188):

- Corpus-based: deductive and top-down, as the starting point is a theory or hypothesis which is tested against corpus data.
- Corpus-driven: inductive and bottom up, as the starting point is the observation of patterns in a corpus and, out of this observation, grows a hypothesis or theory.
- Corpus-illustrated/corpus-informed: "The type of research that we would like to place under the heading corpus-illustrated basically considers usage-events a data set for the selection of examples, in the sense that usage materials complement or supplement introspective data for theoretical hypotheses".

(Tummers et al., 2005, 234–235)

Despite its theory-neutrality, corpus-methodology is easily positioned with regard to the functionalism–formalism divide within linguistics. Formal linguistics operates with a sharp distinction between competence (language as a system of structures within the individual) and performance (language use) with competence unidirectionally determining performance. Moreover, formal linguistics focuses on competence exclusively, isolating it from usage contexts. In this perspective, then, studying performance is useless in learning about the language system. In contrast, functional linguistics is interested in the functions of language as a social, communicative, cognitive or even survival-ensuring tool, and, here, studying how language is actually used is important, because it is through observing usage that we gain insights into the functionality of language. Corpus-methodology is clearly incompatible with formalism while easily aligned with functionalism, and it is particularly compatible with usage-based linguistics (Kemmer & Barlow, 2000), which is a functionalist theoretical orientation which holds that competence emerges from performance via cognitive processes of conventionalization and generalization (Hopper, 1998, 156). Thus, it makes sense to investigate patterns of actual language use as a means of studying the language system itself. Moreover, language is not decontextualized; on the contrary, if a particular pattern of language use is associated with a particular context, contextual knowledge – even if it is non-linguistic – becomes part of the language system as association patterns (Biber et al., 1998).

Life With Depression

This section presents a corpus-based study of blog posts on life with depression. Three analyses will be presented: 1) a keyness analysis to identify the overall aboutness of the corpus, 2) a simple analysis of lexicogrammatical features of the verb *help* and 3) a corpus-based discussion of two metaphorical conceptualizations of depression.

Data and Method

The corpus consists of posts from Mind.org.uk's blog section *Your Stories*. We will call this corpus MinDS (*Min*d depression *s*tories). Posts on Mind.org.uk have CMS-tags that allow readers to select posts while filtering out others. The criterion for inclusion of a text in MinDS is that the post be tagged with the "depression" tag, which indicates posts that are mainly about depression. MinDS contains 61 texts (August 22, 2012, to August 20, 2018) with a word token count *sans* metadata of 49,435 and a word type count *sans* metadata of 4,459. The posts primarily offer personal narratives by depression patients, but also include posts by next-of-kin and healthcare professionals as well as one transcript of an interview. Table 4.2 provides an overview of MinDS' structure.

MinDS is a specialized mini-corpus. A mini-corpus is a corpus smaller than 1 million words, and a specialized corpus is one that represents a particular aspect of language use, such as a specific domain, genre, medium or authorship.

TABLE 4.2 MinDS

Text type	Texts	Words*	Portion of corpus**
Post by depression patients	54	42,033	85.03%
Post by others	6	4,506	9.11%
Podcast transcript	1	2,896	5.58%
Totals	61	49,435	100%

* Tokens; *sans* metadata
** Based on number of word tokens

This study falls under the rubric of internet linguistics (Crystal, 2005), which Crystal (2005, 1) defines as "the synchronic analysis of language in all areas of internet activity," also suggesting that there is a diachronic dimension to Internet linguistics. With the exception of the one interview, the data in MinDS are written texts, Crystal (2001, 3) characterizes blogging as follows:

> From a linguistic point of view, what we see in blogs is written language in its most 'naked' form – without the interference of proofreaders, copy-editors, sub-editors, and all the others who take our written expression and standardize it, often to the point of blandness. It is the beginning of a new evolution of the written language, and a new motivation for child and adult literacy.
>
> *(Crystal, 2001, 3)*

Moreover, most posts in MinDS are subjective texts. A subjective text focuses on the writing subject and features explicit linguistic markers attitude, experience and the like (Closs Traugott & Dasher, 2002; Roldán Riejos, 2004). Despite this subjectivity, bloggers aim at engaging the audience (Myers, 2010, 77–94), so there is also an element of what Closs Traugott and Dasher (2002) call intersubjectivity. Intersubjective texts are characterized by more interpersonal communication and contain interpersonal social-referential markers (Roldán Riejos, 2004, 41).

Keyness and Aboutness

With a short text, aboutness, or subject matter, is identifiable through close-reading, but, with large data sets, digital reading methods are preferable. One popular type of digital aboutness identification is keyness analysis: the corpus in question (also called the target corpus) is compared to another more general corpus (then called the reference corpus). Statistically, keyness analysis consists of identifying items that are significantly frequent in the target corpus and infrequent in the reference corpus. For each item, a keyness score is calculated: the higher the keyness the more key the item is to the target corpus. Identifying items with high keyness can give us an idea of the aboutness of the target corpus:

> Some keywords (such as proper nouns) reveal information about the content of a corpus of text; others (such as closed class items) can tell us about particular stylistic choices, while others can be indicative of cultural keywords … Keywords can also help to act as signposts for discourse, ideology or argumentation.
>
> *(Baker & Ellece, 2011, 67)*

Our starting point is a keyness analysis of MinDS, with BrE06 (a 1,000,000-word corpus of general British English) as the reference corpus. Table 4.3 provides a list

TABLE 4.3 Top 100 keywords in MinD with BE06 as reference corpus

Rank	Frequency	Keyness	Keyword	Rank	Frequency	Keyness	Keyword
1	2,705	5,215.2	i	51	49	93.82	running
2	900	1,885.75	my	52	375	92.34	but
3	239	1,214	depression	53	109	88.15	day
4	521	940.75	me	54	54	87.91	lot
5	221	868.33	mental	55	19	86.77	baking
6	142	481.19	myself	56	59	78.7	able
7	163	455.47	feel	57	58	78.43	someone
8	198	431.63	health	58	159	77.39	do
9	127	313.24	mind	59	49	75.53	bad
10	425	300.32	t	60	44	75.35	try
11	73	280.84	anxiety	61	102	74.69	don
12	875	275.66	it	62	53	74.08	couldn
13	123	264.28	felt	63	88	71.84	going
14	69	260.79	illness	64	340	70.46	have
15	139	243.74	really	65	21	69.94	diagnosed
16	85	238.44	feeling	66	198	69.34	when
17	126	223.21	help	67	83	69.29	something
18	64	209.76	helped	68	177	68.69	time
19	110	191.82	things	69	31	68.38	experiences
20	122	175.88	ve	70	19	66.85	depressed
21	152	166.41	m	71	110	65	because
22	45	166.33	feelings	72	78	64.72	got
23	761	165.29	was	73	43	64.37	difficult
24	199	162.43	people	74	15	62.93	mentally
25	842	162.05	that	75	44	61.53	anyone
26	97	149.62	am	76	21	60.53	struggling
27	465	148.21	you	77	123	59.38	how
28	71	137.42	problems	78	51	59.26	others
29	254	137.25	so	79	25	58.39	journey
30	141	135.12	life	80	193	57.31	what
31	33	133.82	recovery	81	47	56.31	wanted
32	28	126.18	counselling	82	42	54.25	everyone
33	24	124.44	stigma	83	34	52.59	happy
34	28	123.81	medication	84	47	51.94	university
35	57	119.92	talk	85	26	50.26	battle
36	234	115.98	can	86	16	49.41	learnt
37	101	115.56	didn	87	50	49.02	thing
38	236	113.95	about	88	8	49	cbt
39	24	111.8	blog	89	40	48.74	alone
40	194	109.93	like	90	16	48.39	gp
41	62	107.94	started	91	14	48.28	ashamed
42	31	106.92	therapy	92	11	47.99	hardest
43	79	106.16	better	93	41	47.8	getting

(Continued)

TABLE 4.3 (Cont).

Rank	Frequency	Keyness	Keyword	Rank	Frequency	Keyness	Keyword
44	1,682	105.5	to	94	53	47.54	days
45	21	103.66	loneliness	95	41	47.02	experience
46	120	99.33	know	96	21	46.01	ill
47	38	98.63	thoughts	97	15	46	anxious
48	135	98.58	get	98	36	45.34	understand
49	60	98.07	friends	99	9	44.89	bipolar
50	48	97.93	talking	100	49	44.11	wasn

of the top 100 keywords in MinDS; it was generated using the standard settings in Anthony (2018):

- Overall statistic: log-likelihood (4-term)
- Threshold: $p < 0.05$ (+ Bonferroni)
- Effect size measure: Dice coefficient
- Effect size threshold: All values

We find several high-scoring words from the domains of HEALTHCARE and MENTAL HEALTH, including *GP, counselling, therapy, health, illness, anxiety, mental, recovery, bipolar, depression* and *depressed*. Other domains on this list are COGNITION (e.g., *know, thought, understand*), EMOTION (e.g., *felt, feel, feeling*), SOCIAL LIFE (e.g., *talk, talking, friends, loneliness, alone*), CHALLENGES (e.g., *problems, hardest*), SHAME (e.g., *stigma, ashamed*), FIGHTING/WAR (e.g., *struggling, battle*) and TRAVELING (e.g., *journey*). This gives us a bird's eye view of the topics that the posts deal with and what the bloggers presumably find salient and important to write about. These include not just healthcare-related aspects but also social, emotional and cognitive aspects of depression.

Reflecting the subjective nature of the texts, first-person pronominal forms, such as *I, me, my* and *myself* are quite prominent on the list. We also encounter the second person pronoun *you* on the list, indicating textual intersubjectivity. Here, we witness one of the dangers of keyness analysis: it does not take into account textual range (the number of texts in a corpus that an expression is distributed over). The 465 instances of *you* have a range of 51, and, including all second-person forms, there are 604 instances distributed over 53 texts. Second person forms are not equally distributed in MinDS as seen in Figure 4.1.

Many second person forms appear in five texts and are not evenly distributed across MinDS (in fact, these five texts account for no less than 39.57%): MPD ("Mind podcast – depression"), ALD ("A letter to depression"), SFWBDA ("Surviving freshers week (and beyond) with depression and anxiety"), 10THCD ("10 things that have helped me cope with depression") and MLD ("My letter to

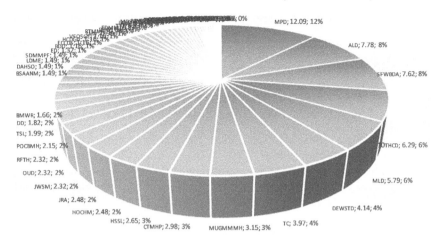

FIGURE 4.1 Proportional distribution of second-person forms across texts in MinDS

depression"). Three of these deviate from the norm established in MinDS: MDP is a transcript of an interview, and ALD and MLD are both fashioned as letters addressed at depression. SFWBDA and 10THCD, while conforming to the norm better, both offer first-person accounts with advice aimed at the reader.

The Lexicogrammar of *Help*

Help has a keyness score of 223.21. The word can be used as a verb, a noun or an adjective; all three uses figure in MinDS with the verbal use constituting 78.10% of all instances of *help*, the nominal use constituting 21.43%, and the adjectival use constituting 0.48%.

A word cluster analysis identifies multiword connections in which a word appears. In this particular case, a word cluster analysis was performed in which three-word clusters were identified with *help* being the leftmost word.

Several instances of *help* occur in the causal *help X V* (1) and *help X to V* (2) constructions.[3] In addition to these structures, *help* also figures in the related causal constructions *help to V* (3) and *help V* (4):

(1) I felt vulnerable to start with; yet the urge to give back and to support others through similar experiences to mine, was enough to help me over-come that.
(2) I don't intend on going into the finer details in this blog about what hap-pened to me in words, but I want to talk about how painting has helped me to come to terms with my mental illness.
(3) I've always written poetry as a way of helping to explain things to myself.
(4) In turn this helps combat some of the feelings of isolation that can be so overwhelming when you're having a bad day.

TABLE 4.4 Top 30 three-word clusters with *help* at the left edge

Rank	Frequency	Word cluster	Rank	Frequency	Word cluster
1	12	helped me to	16	1	help and speaking out
2	4	helped me through	17	1	help – a little
3	3	help and support	18	1	help – i wanted
4	3	helps me to	19	1	help and respect
5	2	help me to	20	1	help and risk
6	2	help myself get	21	1	help and whether
7	2	help them to	22	1	help at times
8	2	help with my	23	1	help because after
9	2	help, so i	24	1	help bring in
10	2	helped me cope	25	1	help but asking
11	2	helped me manage	26	1	help but spot
12	2	helped me realise	27	1	help but wince
13	2	helped me recover	28	1	help but you
14	2	helped me with	29	1	help control my
15	2	helped us to	30	1	help ease some

In all four constructions, there is an underlying causal-semantic relation of ENABLEMENT (Johnson, 1987, 47). ENABLEMENT is a causal relation between two situations such that one, the ENABLING FORCE, makes the other, the ENABLED SITUATION, possible. This usage-category accounts for 55.5% of all instances of the verbal use of *help*. An analysis of ENABLING FORCE expressors, such as *the urge to give back and to support others through similar experiences to mine* in (1), where they are classified into quantifiable semantic categories reveals types of experience that are conceptualized as having such an enabling effect.

The ACTIVITY category (activities such as painting, cooking, writing, baking, jogging and rugby) has the highest frequency. The second most frequent category is that of TREATMENT (psychiatric, psychological and medical treatment). This category is followed by WRITER which covers instances of self-reference in which the bloggers present themselves as ENABLING FORCES. The next category is COMPANY (friends, family, colleagues and partners), followed by HEALTHCARE AGENT (various types of healthcare professional). A similar analysis of the ENABLED SITUATION reveals that the semantic classes of RECOVERY, COPING, COGNITION, COMMUNICATION and MANAGEMENT are the most frequent ones:

Help X to V and *help X V* encode the ENABLED AGENT (the AGENT in the ENABLED SITUATION) in the X-position, allowing us to measure the referential nature of the X-position. A simple quantitative analysis shows that the ENABLED AGENT primarily encodes self-reference.[4]

This admittedly simplistic analysis of verbal uses of *help* suggests that the bloggers tend to position themselves as ENABLED AGENTS who need some kind

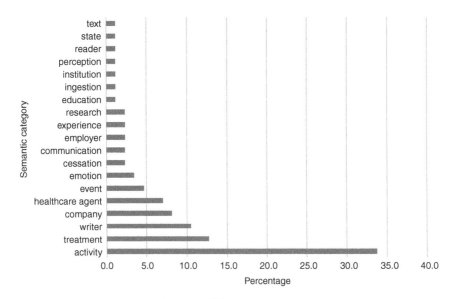

FIGURE 4.2 Semantic categorizations of the ENABLING FORCE

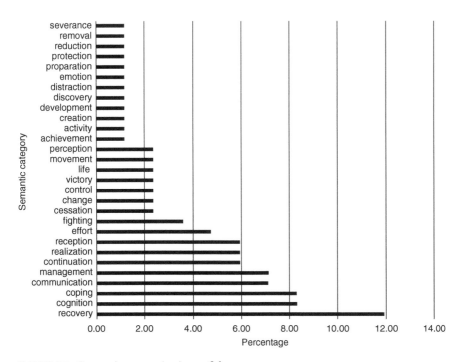

FIGURE 4.3 Semantic categorizations of the ENABLED SITUATION

of assistance or impetus to engage in situations like the recovery process, coping with and managing depression, communicating with other people and so on and so forth.

Two Metaphors

We will now explore two recurring metaphorical conceptualizations in MinDS, represented in Table 4.3 by the keywords *journey* and *battle*. Following conceptual metaphor theory (Lakoff & Johnson, 1980), we define metaphors as the conceptualization of one domain (the target domain) in terms of another domain (the source domain). Metaphors are often reflected linguistically in the use of expressions associated with the source domain in talking or writing about the target domain. The structure of the source domain is projected, or mapped, onto the target domain.

Journey Metaphor

Journey occurs 25 times in 15 texts; 23 instances (92%) encode a DEPRESSION-AS-A-JOURNEY metaphor. Consider the following examples:

(5) A weight was immediately lifted and there began another journey of seeing my psychiatrist regularly, cognitive behavioural therapy and trying to find the right medication.

(6) I was diagnosed with anxiety and depression back in 2013, but I feel that I had been suffering with mental illness for many years before that. At the beginning of my journey I felt isolated, and although I had my family around me who understood and supported me, I didn't know anyone who was going through the same thing as me.

(7) I have found since being diagnosed with depression, that sometimes the smallest (almost throwaway) comment from a friend can have a huge impact and be massively helpful – sometimes one sentence can make something click in your head and help you through your journey.

While the details differ, all three examples share an underlying metaphorical structure, which is based on a mapping from a JOURNEY source domain onto a DEPRESSION target domain, in which the DEPRESSION PATIENT is conceptualized as a TRAVELER. The JOURNEY domain arguably draws on the PATH image schema (Johnson, 1987)[5] in which a PATH stretches from a STARTING POINT to a DESTINATION. In the JOURNEY domain, then, the TRAVELER TRAVELS along this PATH. The following example of metaphorical use of *journey* specifies the STARTING POINT and DESTINATION though *from X to Y*:

(8) The journey from mental ill health to television appearance and relatively good mental health has been a long one ...

In this example, the DESTINATION lies beyond DEPRESSION in the form of good mental health combined with a television appearance, while the STARTING POINT is DEPRESSION itself. Below are further examples of this conceptualization:

(9) There are many great places you can start your journey to recovery. Take that first step today.

(10) A journey to recovery through art.

(11) My journey to recovery wasn't easy. I was faced with stigma and discrimination by so many people.

Example (9) suggests some variation regarding the mapping of the STARTING POINT. There may also be variation in terms of what maps onto DEPRESSION. In the following instance, the PATH itself is mapped onto DEPRESSION:

(12) The internet, and social media specifically, has been heavily influential during my eighteen month journey through a major depressive episode and PTSD.

Here, *depression* serves as the prepositional complement of *through*, indicating a conceptualization of DEPRESSION being the LANDSCAPE that the PATH runs through.

So far, we have explored instances where the lexeme *journey* occurs; scrutinizing other words in the same domain, we find further instantiations of this conceptualization:

(13) ... but we just need to keep taking the positive fork in the road, away from the depression and all of the rubbish of that, and towards something positive because there's always something positive.

Here there is a focus on the PATH itself as the blogger conceptualizes the RECOVERY PROCESS as a ROAD leading away from DEPRESSION toward RECOVERY.

The bloggers tend to conceptualize DEPRESSION as a JOURNEY such that RECOVERY is the DESTINATION of the JOURNEY while DEPRESSION (or some aspect thereof) is either the STARTING POINT or the LANDSCAPE TRAVERSED. Schive (2017) proposes that this metaphor is a rejection of the static conceptualization of mental illness adopted in mainstream discourses on mental illness. Conceptualized this way, depression becomes a dynamic and momentary stage rather than a permanent defining feature.

Fighting Metaphor

Conceptualizing RECOVERY as a BATTLE is not uncommon. For instance, Semino et al. (2016) observe this metaphor in cancer discourse. The metaphor is also used in MinDS in conceptualizing the experience of depression.

Battle occurs 33 times in the corpus in 13 texts and is exclusively used to metaphorically encode the experience of depression:

(14) It's a battle I'm still trying to win; a debilitating, exhausting battle. A battle to carry on fighting, to rediscover purpose, to regain happiness and enjoyment. A battle to carry on living. It's also a battle I never thought I would face.

Battle can be used as a noun or as a verb. In MinDS, the former is dominant with a percentage of 60.6%. In 40% of instances of the noun use, the ENEMY or ANTAGONIST is overtly encoded (typically via a post-modifying preposition phrase as in *battle with X* or *battle against X*), while it is encoded in 92.2% of instances of the verb use (typically via a direct object or a preposition phrase headed by *with*). There seems to be a tendency to conceptualize either the DEPRESSION (15) or the writer's own MIND as the ENEMY (16)

(15) I want to share my experience of how a team sport like rugby has helped me through some of my personal battles with mental health and stay focused at often difficult times in my life.
(16) For as long as I can remember, I have been fighting a battle with my mind.

The two may occur together:

(17) … I am a little fed up of the battle and the strength needed to battle this illness, it is so, so tiring to battle your negative thoughts every single day.

The former is encoded in *the strength needed to battle this illness* while the latter is encoded in *to battle your negative thoughts every single day*. Table 4.5 accounts

TABLE 4.5 Distribution of enemy mappings

Mapping	Verbal use	Nominal use
depression-as-enemy	66.6%	62.5%
mind-as-enemy	25.1%	37.5%
Other	8.3%	0.0%

for the distribution of mappings of the ENEMY component from the BATTLE domain onto components in the DEPRESSION domain.

DEPRESSION-AS-ENEMY is dominant, which may be a reflection of the more general conceptualization of recovery from an illness as a battle against the illness. MIND-AS-ENEMY is still somewhat frequent and may be an instance of what Schive (2017) defines as a metaphorical conceptualization of DEPRESSION as TAKING CONTROL of the SUBJECT'S MIND and turning the SUBJECT'S MIND against the SUBJECT.

Applicability of Corpus-Methodology

The analyses presented here are admittedly simplistic, and fully-fledged corpus studies normally involve more complex techniques. Nonetheless, our three analyses have yielded some patterns that, if nothing else, are worth investigating in more detail. Importantly, those patterns and the insights that the investigation has generated are based on observations of *real* language use, *real* communication.

Given the amount of textual data, many of these patterns would probably not have been identified through close-reading, as processing all 61 texts would be not only time-consuming but also very challenging to the analyst simply because of the human error factor. The computer, while of course dependent on the instructions of the analyst, is not prone to making such human errors. As Kirk (1996, 153) points out,

> [u]sers of the computer, being supplied with all the tokens of a particular phenomenon, are expected to deal satisfactorily with all of them, exhaustively, without exception or remainder, thus complying with an explicit principle of "total accountability" of the data in question.

In our study, each instance of every single word in the corpus is included in the keyness analysis, and, in our frequency analyses of the semantic categories in the *help*-constructions, all instances of each category were taken into consideration. The possibility for total accountability is scientifically attractive (and crucially important) in connection with subsequent interpretations and theorizing.

Final Remarks

Corpus-methodology is inseparable from computational analysis, as digital data and techniques are at the heart of the framework. Thus, corpus-methodology should prove useful and attractive within the humanities, including communication studies.

The case study, while simplistic, has illustrated the applicability of corpus-methodology in textual analysis, as we explored conceptualizations of depression within the MinDS corpus. A keyness analysis of the aboutness of MinDS revealed a range of salient topics in the corpus and also brought to the fore subjective and intersubjective features in the form of second and third personal pronominal forms. The keyness analysis also leads to a lexicogrammatical analysis of *help*: the word often occurs in constructions that signal ENABLEMENT, suggesting a positioning of the bloggers as ENABLED AGENTS who need ASSISTANCE or an IMPETUS in engaging in certain situations. We also explored two metaphorical conceptualizations of the experience of recovering from depression: a JOURNEY-based metaphor and a FIGHTING-based metaphor.

Such analysis, even if it were more exhaustive and systematic than what is resented here, would not lead to the ultimate and final understanding of how depression patients conceptualize life with the illness. However, it should illustrate how corpus-methodology can be of valuable assistance to researchers with an interest in the reflection of experiences of mental illness the language of people with first-hand experience of the illness.

Notes

1 See McEnery & Hardie (2012, pp. 37–48) for an overview of the evolution of concordancer technology.
2 For more on analytical techniques in corpus linguistics, see McEnery and Hardie (2012), Cheng (2012), Anderson and Corbett (2009), Jones and Waller (2015), Lindquist (2009) and Gries (2009).
3 Constructions are conventionalized pairings of form and function that are entrenched routines within a speech community (Croft, 2005, p. 274).
4 In *help X to V*, first person reference in the X-position has 80%, while second person reference has 4% and 3rd person reference has 16 %. In *help X V*, first person reference has 55.7%, second person reference has 14.3%, and third person reference has 19.0%.
5 An image schema is a basic, generalizing cognitive representation of simple and fundamental bodily experience; ENABLEMENT, which was discussed as a basic force-dynamic relation, is also a type of image schema.

References

Andersen, G. & Bech, K. (Eds.). (2013). *English corpus linguistics: Variation in time, space and genre*. Amsterdam: Rodopi.

Anderson, W. & Corbett, J. (2009). *Exploring English with online corpora*. Houndsmills: Palgrave MacMillan.

Anthony, L. (2018). *AntConc 3.5.2*. Waseda: Waseda University.

Baker, P. (2005). *Public discourses of gay men*. London: Routledge.

Baker, P. (2006). *Using corpora in discourse analysis*. London: Continuum.

Baker, P. (2010). *Sociolinguistics and corpus linguistics*. Edinburgh: Edinburgh University Press.

Baker, P. & Ellece, S. (2011). *Key terms in discourse analysis*. London: Continuum.

Baker, P., Hardie, A. & McEnery, T. (2006). *A glossary of corpus linguistics*. Edinburgh: Edinburgh University Press.

Biber, D. & Conrad, S. (2009). *Register, genre, and style*. Cambridge: Cambridge University Press.

Berez, A.L. & Gries, S. Th. (2009). In defense of corpus-based methods: A behavioral profile analysis of polysemous *get* in English. In S. Moran, D. S. Tanner & M. Scanlon (Eds.), *Proceedings of the 24th Northwest Linguistics Conference* (pp. 157–166). Seattle, WA: Department of Linguistics.

Biber, D., Conrad, S. & Reppen, R. (1998). *Corpus linguistics: Investigating language structure and use*. Cambridge: Cambridge University Press.

Cheng, W. (2012). *Exploring corpus linguistics: Language in action*. London: Routledge.

Closs Traugott, E. & Dasher, R. B. (2002). *Regularity in semantic change*. Cambridge: Cambridge University Press.

Croft, W. (2005). Logical and typological arguments for Radical Construction Grammar. In J.-O. Östman (Ed.), *Construction grammar: Cognitive grounding and theoretical extensions* (pp. 273–314). Amsterdam: John Benjamins.

Crystal, D. (2001). *Language and the Internet*. Cambridge: Cambridge University Press.

Crystal, D. (2005, February 18). The scope of Internet linguistics. Paper presented at the American Association for the Advancement of Science Meeting.

Fina, M. E. (2011). What a *TripAdvisor* corpus can tell us about culture. *Cultus, 4*, 59–80.

Gabrielatos, C. & Baker, P. (2008). Fleeing, sneaking, flooding: A corpus analysis of the discursive construction of refugees and asylum seekers in the UK Press 1996–2005. *Journal of English Linguistics, 36*(1), 5–38.

Gries, S. (2009). *Quantitative corpus linguistics with R: A practical introduction*. London: Routledge.

Gries, S. & Stefanowitsch, A. (Eds.). (2006). *Corpora in cognitive linguistics: Corpus-based approaches to syntax and lexis*. Berlin: Mouton de Gruyter.

Hopper, P. (1998). Emergent grammar. In M. Tomasello (Ed.), *The new psychology of language: Cognitive and functional approaches to language structure* (pp. 155–175). Mahwah, NJ: Lawrence Erlbaum.

Johnson, M (1987). *The body in the mind*. Chicago, IL: Chicago University Press.

Jones, C. & Waller, D. (2015). *Corpus linguistics for grammar*. London: Routledge.

Kemmer, S. & Barlow, M. (2000). Introduction: A usage-based conception of language. In M. Barlow & S. Kemmer (Eds.), *Usage-based models of language* (pp. vii–xxviii). Stanford, CA: Stanford University Press.

Kennedy, G. (1998). *An Introduction to Corpus Linguistics*. London: Longman.

Kirk, J. M. (1996). Review of K. Aijmer & B. Altenberg (Eds.) (1991). *English corpus linguistics: studies in honour of Jan Svartvik*. London: Longman (*Journal of English Linguistics*, 24, 250–258).

Krishnamurthy, R. (2000). Collocation: from *silly ass* to lexical sets. In C. Heffer, H. Sauntson & G. Fox (Eds.), *Words in context: A tribute to John Sinclair on his retirement* (pp. 31–47). Birmingham: Department of English, University of Birmingham.

Krug, M. (2000). *Emerging English modals: A corpus-based study of grammaticalization*. Berlin: Mouton de Gruyter.

Lakoff, G. & Johnson, M. (1980). *Metaphors we live by*. Chicago, IL: Chicago University Press.

Leech, G. & Fallon, R. (1992). Computer corpora – What do they tell us about culture? *ICAME Journal, 16*, 29–50.

Lindquist, H. (2009). *Corpus linguistics and the description of English*. Edinburgh: Edinburgh University Press.

Mahlberg, M. (2013). *Corpus stylistics and Dickens' fiction*. London: Routledge.

McEnery, T. & Baker, H. (2016). *Corpus linguistics and 17th century prostitution: Computational linguistics and history*. London: Continuum.

McEnery, T. & Hardie, A. (2012). *Corpus linguistics*. Cambridge: Cambridge University Press.

McEnery, T. & Wilson, A. (2001). *Corpus linguistics* (2nd ed.). Edinburg: Edinburgh University Press.

Myers, G. (2010). *The discourse of blogs and wikis*. London: Continuum.

O'Keefe, A. & McCarthy, M. (Eds.). (2010). *The Routledge handbook of corpus linguistics*. London: Routledge.

Ooi, V. B. (2000). Asian or Western realities? Collocations in Singaporean-Malaysian English. In J. M. Kirk (Ed.), *Corpora galore: Analyses and techniques in describing English* (pp. 73–89). Amsterdam: Rodopi.

Partington, A. (2004). Corpora and discourse, a most congruous beast. In A. Partington, J. Morley & L. Haarman (Eds.), *Corpora and discourse* (pp. 11–20). Bern: Peter Lang.

Rayson, P. & Garside, R. (2000). Comparing corpora using frequency profiling. In *Proceedings of Workshop on Comparing Corpora of ACL 2000* (pp. 1–6). Hong Kong.

Roldán Riejos, A. M. (2004). Strategic features of ESP from a socio-cognitive perspective. *Ibérica*, 7, 33–51.

Schive, N. (2017). *The discourse of depression: A study of the discursive otherization of depression patients and self-identification by depression patients*. MA-thesis, Department of English, Germanic and Romance Studies, University of Copenhagen.

Semino, E., Demjén, Z. & Demmen, J. (2016). An integrated approach to metaphors and framing in cognition, discourse, and practice, with an application to metaphors for cancer. *Applied Linguistics*, *39*(5), 625–645.

Sinclair, J. (1991). *Corpus concordance collocation*. Oxford: Oxford University Press.

Solan, L. C. & Gales, T. (2016). Finding ordinary meaning in law: The judge, the dictionary or the corpus? *International Journal of Legal Studies*, *1*(2), 253–276.

Stefanowitsch, A. & Gries, S. (2003). Collostructions: Investigating the intraction between words and constructions. *International Journal of Corpus Linguistics*, *8*(2), 209–243.

Svartvik, J. (2007). Corpus linguistics 25+ years on. In R. Facchinetti (Ed.), *Corpus linguistics 25 years on* (pp. 11–25). Amsterdam: Rodopi.

Tognini-Bonelli, E. (2001). *Corpus linguistics at work*. Amsterdam: John Benjamins.

Tognini-Bonelli, E. (2010). Theoretical overview of corpus linguistics. In A. O'Keefe & M. McCarthy (Eds.), *The Routledge handbook of corpus linguistics* (pp. 14–27). London: Routledge.

Tummers, J., Heylen, K. & Geeraerts, D. (2005). Usage-based approaches in cognitive linguistics: A technical state of the art. *Corpus Linguistics and Linguistic Theory*, *1*(2), 225–261.

Wulff, S. (2008). *Rethinking idiomaticity: A usage-based approach*. London: Continuum.

5

COMMUNICATION IN CRITICAL THEORY (FRANKFURT SCHOOL)

Olivier Voirol

The « Generic » Model of Communication

Key authors of German idealism such as von Humboldt, Schiller and Hegel, developed the idea of a relation between the subject and the world, and between the subjects themselves, as the fundament of the subject-constitution. Through the active mediation of language, the subject constitutes itself, on the one hand, in his relationship with the world and, on the other hand, it simultaneously constitutes the world, the latter defined by its objective form and its contingent properties (Humboldt, 1960–1981). The acting subject constitutes itself in this manner, but so does the world, which is the « object » of the aforementioned relations. Such a conception considers neither the primacy of the world over the subject, nor the primacy of the subject over the world, but the *primacy of the mediation*, which constitutes both parts actively at the same time. As a result, the relationship becomes central in this philosophical thinking: the mediation process makes possible the communication between the subjects and the world, as well as between the subjects themselves.

Wilhelm von Humboldt described this process in terms of a linguistic relation, language being the active mediation. According to Humboldt, using language signifies in its deepest meaning establishing a relationship that takes the form of reciprocal action (*Wechselwirkung*). It allows subjects to express a force (*Kraft*) by referring to a form, which is shared by others – they "submit" to language, in order to act and speak, and they express themselves through this process of becoming a subject. Based on this expression, they create and recreate a common language, both stabilizing and renewing its common signifying forms. On the one hand, the speaking subject is formed as a subject and is constantly transformed, in its relationship to the world and to others. On the other hand, language is reproduced and transformed through its active expressions, while the world is made by language. This process enables therefore a double expression, of the subject and of the world, a dual constitution of subjectivity and objectivity, a mutual transformation of both sides. The establishment of such a dynamic relationship is made possible by virtue of linguistic mediation. This philosophy of language based on the idea of mediation constitutes the generic model of communication as well as the normative dialogical ideal that founds it. This dialogical model mediated by language also constitutes the modern ideal of culture as *Bildung*, as the practical ground on which the generic ideal of communication takes shape in the tradition of Critical Theory (Schiller, 2016).

A Critical Diagnosis

The mediation model of language forms the core of the critical theory of communication developed by Jürgen Habermas. Drawing from the new tools provided by the philosophy of language of the late twentieth century, as well as the theory of intersubjectivity (Mead) and socialization (Durkheim), he reshapes this mediation model – which is also, although often implicitly, an important philosophical background for the authors of the so-called "first generation" of Critical Theory (Adorno, Horkheimer, Marcuse, Benjamin, Kracauer). Nourished by the same philosophical legacy of German idealism, these authors focus mainly on critical reflections anchored in their own historical period with the help of new theoretical tools. They think and write in a historical and political situation where the social conditions to realize this ideal of communication have vanished. Driven by a revolutionary bourgeoisie committed to these ideals (freedom, reason, culture, etc.) (Marcuse, 1968), the liberal era has been replaced by an authoritarian one. As capitalism altered into a monopolistic system, and as the state went authoritarian under the growing pressure of "instrumental rationality" (Weber), the real conditions of communication according to the mediation model have been profoundly affected. Under the new social, economic and political situation of the twentieth century, the ideal that forms the core of the generic model of communication becomes a mere utopian reverie.

The originality of Critical Theory is to bring this process to light by closely associating philosophy with research practices inspired by sociology as well as social psychology. At the end of the 1920s, the members of the *Institut für Sozialforschung* (IfS) based in Frankfurt (until 1933) tried to understand the transformations of individuation under the new conditions of capitalism by focusing empirically on the changing forms of socialization within the family, considered as a "small society." In order to do so, they use theoretical tools from Marxism and psychoanalysis. The main idea of this research was that the young subject develops merely within the close family whose structures socialize him, and he develops into a full adult able to take his place – and play his role – within the "big" society. During this first socialization, young individuals develop a specific structure of the psyche – among various possibilities – linked to the family structure with its own educational model. Researchers conclude that the dominant form of family structure developing within monopoly capitalism – because of a specific oedipal constellation undermining the symbolic referent of the "father" – does not contribute to the growth of an autonomous individual possessing a "strong self" able to judge for himself and to exercise his free will autonomously.

On the contrary, the dominant structure encourages a type of individual whose ego lacks solidity. A "weak ego" has a strong tendency to rely on third-party instances helping him or her to navigate through life. Well before these subjects become mature, those authorities play an important symbolic role, promptly taking the place of the missing "father." Instead of referring to these authoritative instances dialogically and critically, "weak egos" tend to rely on their power. Instead of a dialogical relationship with such collective authorities and social institutions, they merely identify with them. They delegate to these instances their own missing agency and renounce autonomously acting in exercising critical judgment. Erich Fromm says that they have a "fear of freedom" (Fromm, 1941). Consequently, they tend to be docile toward authoritarian power structures. Far from being restricted to a psychological approach, the members of the IfS refer to various disciplines – in an interdisciplinary framework – in order to understand how key transformations of economy and society affect the ego's formation and the dialogic capacities of the subjects. The appearance of a "weak" ego, namely of individuals showing an "atrophied self," not only reveals a major change of the psyche during the twentieth century, but also a groundbreaking change in the socio-economic structures of capitalism.

The "Anthropology of the Bourgeois Era"

Max Horkheimer's research on the "anthropology of the bourgeois era" is part of all these concerns about the future of the individual under monopoly capitalism and authoritarian power (Horkheimer, 1982). His early philosophical studies revisit the birth of the bourgeois philosophy of history in light of its political and moral principles. According to Horkheimer, one of the traits of bourgeois morality is the rejection of selfishness and its prevalence of the "general interest" – in the liberal version of "public benefit" – as well as its collective version of the "political will" (Horkheimer, 1982). The moral principles of the bourgeoisie concern the equality of human beings and their freedom as individuals. The bourgeois "form of life" is based on economic exchange, a principle of performance, and self-realization through work, individual freedom, reason and responsibility. As a revolutionary class fighting against the powerful aristocracy in order to realize these principles in the actual social order, the bourgeoisie eventually succeeds after a long historical process of liberation in which it politically overthrows the "*Ancien Régime*" by rejecting its "values." In material terms, however, the new revolutionary class does not realize the universal promise of freedom and equality for all, as previously fought for. As soon as they start to exercise political power, the bourgeois political leaders restrict these principles to their own group, therefore "betraying" their own moral principles. Thereby, bourgeois ideals of justice, equality and reason turn to mere ideology (Marcuse, 1968, 1937).

Contradictions of the Bourgeois Era

Horkheimer identifies several *contradictions* in this historical "freedom movement." The first one concerns the relationship between morality and materiality: on the one hand, the bourgeoisie struggles for the moral principles of liberty, justice, equality and reason; on the other, however, it unfolds exchange relations that tend to isolate individuals and to generate abstract relationships, coldness and indifference between them. As a result, practical possibilities of putting into reality these moral principles are countered. A real contradiction between the moral principles of the bourgeoisie and material existence therefore takes place, the following marked by the antithesis of these values: alienation instead of liberty, inequality instead of equality, indifference instead of general interest.

Horkheimer identifies another contradiction: while the bourgeoisie makes a plea for reason, it actually brings about unreason. Such a contradictory process happens through the political discourse of the bourgeois leaders: in order to avoid speaking about social inequalities and injustices (therefore of real equality and freedom for all), bourgeois political leaders use a manipulative rhetoric filled with emotional effects – the precursors of fascist rhetorical technic analyzed in the 1940s (Löwenthal & Guterman, 1949). They gain adherence of the "masses" through identification to the political order at the high cost of dialogical reason. Because these rhetorical strategies reinforce unreason, they encourage immediate uncritical adherence to the world "as it is," rather than the dialogical process of communicative reason conceived by the philosophy of *Bildung* (Horkheimer, 1982). Communication thus becomes mere ideology, manipulative and irrational.

A third contradiction stressed by Horkheimer is inspired by Sigmund Freud's argument in *Civilization and its Discontents* (2002). According to Freud, whereas culture refers to the dealing of drives and instinctual life by consciousness or the "rational" faculties of the subject, it becomes repressive through the austere self-discipline of bourgeois morality. Inside a culture that so strongly rejects pleasure, the process of disciplinary self-formation develops through a self-aggressing ego provoking many frustrations. According to psychoanalysis, drives must be realized in a certain form, if not in the form of pleasure by "sublimation," then in the form of aggressiveness and violence. Culture allows one to become an adult and manage frustration through rationality. However, the bourgeoisie's mold of self-control drives the subject to aggressiveness as the only form capable of enduring the culture's constraints.

Fromm described this subject as being internally structured by a "sadomasochistic character" embodying the very type of psyche that emerged as the dominant figure in that time (Fromm, 1941). Constituted by self-repression, the subject is led to augment his potential to violence. He discharges this violent weight through several kinds of "compensation" – the "pleasure" taken in seeing others suffering. This kind of pathological psyche is the fundament of the "authoritarian personality" (Adorno et al., 1950; Fromm, 1984).

In this historical process, culture (*Bildung*) loses its dialogical and communicative dimension shifting toward a mere repressive form. Mirroring the dialogic process of subject formation, in its relation to oneself, to others and to the world, culture even becomes the source of the very process of aggression and unreason. Horkheimer comes to the same conclusion as Fromm, whose psycho-sociological approach is grounded by an empirical research on workers and employees in pre-Nazi Germany (Fromm, 1984): monopoly capitalism and industrial society of the early twentieth century encourage the rise of a kind of subject characterized by a damaged dialogical communication with himself, with others and with the world. In the form of self-discipline, mastery and control, in identifying with "what it is," this subject is no longer able to relate dialogically and sensitively to the social world surrounding him. In its reified form, culture highly contributes to this damaged communication, which tends to become authoritarian in modern capitalist societies.

Culture Industry

Theodor W. Adorno (1903–1969) takes up Horkheimer's study on the contradictions of the bourgeois era in an essay on Richard Wagner. In his monograph, Adorno broadly studies the transformations of music in the "bourgeois era" of the nineteenth century (Adorno, 2009). He sees Wagner as the figure corresponding to the descriptions of the bourgeois individual as having a strong sense of freedom, justice, equality and a strong reverence for an ideal of personal autonomy. Wagner has high ambitions, he thinks about himself as being a free composer in a situation where the former artistic dependencies and social chains limiting artistic creativity have been broken. As a free individual, he should be able to follow his aesthetic impulses, rather than submitting to canons imposed by discretionary masters.

However, as an artist in an era of the free "bourgeois entrepreneur," Wagner experiences on a social level dire uncertainty and precariousness. As a free subject, he has neither power nor property, and his artistic activity remains highly dependent. Deprived from material control over his life, he must accept relations of economic subordination and submit to his superiors. In the light of the moral principles of liberty and independence of his social class, his actual situation is underpinned by a rising tension between his objective status of dependence and his high self-esteem as an artist. Morally free, he is actually objectively constrained. He fails to be able to live up to his moral requirements. Such a contradiction induces an inner suffering and a frustration that arouses resentment.

Adorno's study of Wagner's music consists of subtly showing how this inner split linked to this contradiction finds expression in his compositions. Adorno does this through an immanent critique, which focuses on the musical material, its internal structure, inner organization, its rhythm and its instrumentation. He develops a sociology of artworks linked to musicology, showing how art should be analyzed according to its "objective" content. A material hermeneutic of musical contents also indicates something about the musical experiences that are made, the same way that it says something about its creation process. Musically, Wagner expresses the hopes and contradictions of the "bourgeois era," through the form of communication that the compositions established with the public. This happens indirectly, through the aesthetic form as well as through the mediation of musical language.

Adorno shows that a certain aggressiveness is to be found in Wagner's compositions, toward weak people, especially Jewish figures. Wagner's egalitarian bourgeois principles are present in his music through a moral concern to bring the people on stage and to give them a place in the musical composition. Adorno underlines the contradictions of such a gesture made "for the people," because it develops at the very price of the aesthetic relation itself – implicitly based on the modern ideal of communication. According to Adorno, in bringing people in front of the scene, Wagner must favor a relationship to the public by reaffirming its already existing expectations. A search for approval encourages standardized schemas based on what is "already here." As a result, Wagner appears as a forerunner of a model of cultural communication that will soon become a decisive feature of the culture industry with the rise of mass media in the twentieth century.

A Reified Culture

One of the main features of culture in the twentieth century is its departure from the *Aufklärung's* dialogical model of communication, because of its integration into the capitalist valorization process. Culture is transformed at its very core in order to fit into the logic of the capitalist circuit, being submitted to administration, industrial production and distribution. If economics always influenced cultural production, it has become an immediate goal of cultural production in the era of the culture industry. Cultural products are shaped from the very beginning according to industrial criteria, at the very cost of their aesthetic elements. In Marxist terms, the exchange value of cultural goods takes precedence over their use value (Adorno, 1991).

Partly because of its costs and the huge amount of economic investments it requires, in order to assure economic effectiveness, the culture industry has to deal with the concern of selling its industrial cultural products to large audiences understood as its market. In the logic of "exchange value," these markets appear as opposed to a public seen as a critical instance of aesthetical experience, similarly to « autonomous art », which is not directly submitted to similar pressures (Adorno, 1998). By seeking to "speak" to a common ordinary experience that is "already there," the culture industry constantly tries to activate layers of meaning that are previously shared by a huge number of "consumers." One of the most common experiences among large audience members is childhood – every member of an audience has previously been a child involved in narratives defined by highly emotional engagement as well as a mixture of fear and fascination. Anchored in a long human history such as popular stories adapted from the Hollywood industry, these stories are reshaped in cultural products that are standardized, structured in such a way that they have high emotional appeal. Being strongly repetitive, even elementary in their narrative form, such narrative products play a key role in the cultural production under the conditions of industrialization (Horkheimer & Adorno, 2002). According to Adorno, the very idea of culture is deeply redefined by this process: even when they are presented as "new" products ("the new film is coming"), cultural products are standardized and remain identical in their innermost structures. Advertisements of cultural products insist on "novelty," "originality," i.e., the supposed "individuality" of these products. Such an insistence reveals in fact their underlying sameness, an "identity" to each other, constantly repeating similar patterns (Adorno, "On popular music"). Such a stubborn insistence on their individuality reveals in fact their standardization.

An Identity Logic

Products of the culture industry open up a form of communication with their audiences that is structured by a strategy of producing immediate emotional involvement. Even when they appear as in being "individualized," such products address them in a way that they repeat what is "already there," "already known" and "already lived." They are addressed without endeavoring to produce any disrupture. They reproduce elementary forms of common experience, trivialized in their symbolic patterns. The inner quality of aesthetics is to engender troubling experiences about a given reality. The structural pressure for the economic valorization of cultural contents leads to cultural products that lack aesthetic content. According to Adorno, dramatic consequences arise from the cultural experience of modern capitalist societies characterized by immediate communication. It stimulates the reproduction of what is "already there," the existing, as well as it reinforces the "damaged life" within a "reified world." The relationship established by the culture industry with its recipients is restricted to this search for direct adherence of the audiences to "what it is."

Adorno considers that, in an industrial structure using a cultural process historically formed around the free principle of culture as self-education, such a cult of the "already there" radically undermines the very idea of culture (as *Bildung*) and the dialectical reason related to it. In opposition to the kind of communication shaped by the culture industry, the dialectical model of culture supposes an experience that engenders a critical distance from the existing sensory reality. It proceeds from a figurative gesture by which an artist tries to elaborate an adequate aesthetic form capable of expressing a real experience without abstraction.

Such an act of shaping neither replicates the real, nor abstracts from it; it finds the adequate language that figures reality relocating it into an abstract figuration – or into the identity principle of cultural commodities. Finding a fair form means to make reality exist at a second level of aesthetic expression. However, such a language cannot take form by submitting to the expectations of the recipients, or by merely reproducing "common sense." It takes shape by searching the features of an aesthetic form considered to be adequate because it signifies a reality without a violent subsumption. Since this aesthetic form doesn't correspond to the daily common language, it should be re-elaborated in connection with common practices, according to the terms of aesthetic praxis. According to Adorno, it has inevitably an "enigmatic character," which cannot be deciphered by an immediate experience (Adorno, 1998). Because its meaning is not immediately given, an artwork must be sensitively experienced and gradually reconstructed through reason, in order to decipher its inner "enigma." Applying pre-established schemas, abstract categories or general concepts doesn't help such a practice of deciphering an aesthetic enigma *toward* and *beyond* its conceptual-scientific understanding. It only happens by experiencing it, step by step, in several interpretative "trials." Only after a long "testing" process implying mimetic experience as well as reasoning, the "truth-content" of an artwork can progressively appear. It is not as a conceptual meaning but as a kind of "flash" or sudden "illumination." As soon as the enigmatic character of an artwork appears as "fixed," new "enigmas" appear once again – according to a hermeneutical circle than can never be closed.

An Aesthetic Praxis

A practical and sensible commitment of individuals is required in such an experience – on both sides, equally for the production of aesthetic form and the deciphering of artistic contents. Such a relation refers to an active practice of subjects in the aesthetic process (Adorno, 2002). In contrast to participation that is reduced to a psychic identification with non-enigmatic products, such a practical commitment requires a non-affirmative participation, as well as a hermeneutic effort. A mere "identification" to cultural products makes this experience impossible. Above all, what makes it impossible is the objective poverty of cultural contents, suggesting such an immediate identification rather than an exploratory process of deciphering. The culture industry reproduces common-sense categories and, at the very best, locks people into what they "already are," or what they "were" as infants, by using narrative patterns buried in childhood experiences. Adorno uses therefore the word "regression" or "infantilization" to describe the mode of communication initiated by the culture industry (Adorno, 1991).

The Refusal of the Non-Identical

Subjects get used to immediate approval, an easy adhesion to trivialized cultural contents that constantly reassure them by reinforcing what they already are and know. In the long run, subjects get so used to such a cultural universe that they even manifest hostile attitudes toward cultural contents that destabilize them by questioning their habits or demanding for an interpretative effort on their behalf – like contemporary art, for instance. Such a "rejection" is the by-product of the general "affirmative" culture generated by the culture industry (Adorno, 1998). Autonomous art is perceived as complex, inaccessible, impenetrable, requiring such intensive efforts that it is rejected, even aggressively. This aggressive rejection witnesses the fear of being troubled by aesthetic experiences, which would force an involvement based on a subjective sensory experience thus breaking the "identity logic" of the culture industry. Such a subject makes an experience of non-identity, interacting with otherness in a troubling process of deciphering. According to Adorno, even in negative terms, such a process is the fine last trace of the dialectical ideal of communication that can be experienced in modern capitalist industrial culture.

The "fear of otherness" in a culture, in which the traces of its internal demands, as a culture – corresponding to the dialogical ideal of communication – are still alive, testifies a real anxiety toward non-identity. On this point, the development of the culture under the pressure of the culture industry in the era of monopoly capitalism meets the social and political processes of rejection of otherness linked to authoritarianism. The "fear of otherness" on the aesthetic level responds to the fear of the social other: both testify a rejection of the "non-identical" in a mode of communication that is the opposite of the dialogical ideal stressed by the Enlightenment (Adorno, 1973). To this culture deprived from dialogical communication also corresponds an "identical" subject, cut off from otherness in his relationship to the world and to others – reflecting with anxiety on everything that does not reproduce the comfort of identity logic. In other words, the typical subject of the culture industry era is a "narcissistic subject" whose relation to himself is merely self-referential, because he is no longer able to develop a dialogical relationship with the world, with himself and with others. Authoritarian features are expressed in this fear from otherness that is expressed through aggressive rejection processes (Adorno et al., 1950). In an era of an industrialized culture, narcissistic subjects showing strong authoritarian tendencies have even become a real "archetype" of the human subjects – in opposition to the "dialogical" subject anchored in the classical ideal of communication.

Negative Communication

Adorno's writings on the culture industry (radio, film, television, etc.) may seem uncompromising. They show how much the generic ideal of dialogic communication expressed by the Enlightenment seems to be annihilated in late capitalist society – or reduced to a mere utopia – under the conditions of monopoly capitalism, reification and authoritarianism. However, dialogic communication does not completely disappear in Adorno's philosophy, it is indeed ascribed to his "negative dialectic." As such, it negatively suppresses real possibilities of communication. For him, the only way to think about communication is through a "negative dialectic" of the aesthetic relationship. A negative communication can only be thought through an aesthetic process, which rejects the identity logic between sign and thing, subject and object, subject and others. It is only possible in the negation of the identity logic, in its inner contradiction. It is not realized at all in an effective "true communication" supposed to carry out the dialogical ideal, nonetheless it may subsist as negation – an inner contradiction negatively turned toward the attempt to make it happen.

Affirmative communication appears as an ideological process within the identity logic that predominates in the "administered world" of modern capitalist society. Because it tends to exclude resistance as well as critical judgment against power structures, the affirmative form of communication is simultaneously a form of domination. On the other hand, negative communication consists in considering the possibility of a dialogical communication negatively. "Non-identity" escapes the identity logic that structures affirmative communication dialectically present in that logic – but only negatively. Such a negative formulation – philosophically thematized by Adorno in his *Negative Dialectic* (1973) – does not lead to an abandon of the dialogical model of communication. On the contrary, it stresses its possibility in negative ways, as a practical contradiction of the dominant affirmative communication produced by the culture industry – and it witnesses the possibility of considering communication in a non-ideological form. Far from being a sad prophet of a cultural decline, Adorno appears as a radical critical thinker who stresses the negation of this ideal of a communication model implied by the « administrative world » of late modern capitalism.

Other Versions in Critical Theory

Regarding the theory of communication, Critical Theory is broader and richer than the works of Max Horkheimer and Theodor W. Adorno as discussed until now. Herbert Marcuse (1898–1979) developed a concept of communication closely articulated to culture and language and based on the idea of the extension of expressive forms anchored in human sensibility and drives (Eros). Struggles against the "one-dimensional man" and the society linked to it are,

inevitably, struggles for expressive forms, poetic creation, language transformation and new forms of culture and communication (Marcuse, 1955). According to him, a creative extension of such a communication is at play within the
protests during the 1960s, thanks to the counterculture and political movements of that decade (Marcuse, 1972). These struggles developed sensible
forms of communication through creative linguistic expressions that translate
social experiences into a sublimated poetic language – a "new sensibility" that
deepens social meaning thanks to the "aesthetic dimension."

However, only radical opposition to the "one-dimensional society" can still
give a practical significance to the dialogical ideal model of communication
(Marcuse, 1964). Despite several differences, Marcuse rejoins Adorno in his
diagnosis of modern capitalist societies. Here, communication is subsumed into
capitalist valorization processes as well as the rationalization logic of industrial
structures. The linguistic mediation becomes merely functional, and language
is transformed into a "tool" that fulfills strategic functions – in advertising, it
serves economic goals which engender a desire for consumption. Poetic aspects
are cut off from this operational language, which has lost its inner negative
dimension – it just positively signifies what it denotes. In such a positive form,
language also cancels objective contradictions in hiding contradictory features
of the world. Expressions like "clean bomb" or "just war" are examples of
such ideological concealing (Marcuse, 1964, 92). As a result, communication
erases negativity and is used by institutional or economic powers in order to
play an affirmative role of ideological adherence to "what it is."

Walter Benjamin

Walter Benjamin develops an approach to communication in his early writings
based on a language theory highly inspired by theological motives. In a 1916
letter to Gershom Scholem on the question of language, Benjamin develops several motives of central importance in his thinking on communication. He distinguishes the "language of the name" from the "language of the word." The
former is a language that depicts a thing in a mimetic way, without abstraction.
A name is involved in the thing it describes, without separation; language is the
world and the world exists in language. On the opposite, the "language of the
word" results from a break with the world, marking an unavoidable distance
between sign and thing. It introduces a break between word and world, so that it
never can signify the world without abstraction – the world remains "alien" to
language. The gap between language and world designated through it condemns
the human beings to continuous desperate efforts in order to signify it
adequately – they multiply words and languages. Human beings are facing an
expressive disability through a spoiled language, they remain deeply affected in
their being by this split or "reification" (Benjamin, 1979). For example, traumatic
experiences of modern times, especially the extreme violence of the First World

War, have left human beings deprived of any language for expressing it. Deprived of language in their inner misery, they experience a new form of poverty related to the destruction of communication (Benjamin, 2005).

The principle of information carried out by modern media reinforces this split between language and world. Based on the idea of "truth," the modern press is no longer able to tell stories based on truthfulness. It is substituted by lived stories in which diffuse traces of the "language of the name" are still present. Modern media introduce a form of "communication" between human beings that breaks their links to the world, and between them: they desperately confront troubles by expressing their world-experience to make sense of it. A "loss of experience" due to a narrative pauperization in the modern world generates new forms of alienation (Benjamin, 1968).

Nevertheless, literary works still allow to revive traces of "language of the name" when they mimetically express the world-experience in its "truth-content" (Benjamin, 1996). An aesthetic relationship can re-open possibilities of having a world-experience that isn't totally submitted to an alienated reified language. In his writings of the late 1930s, Benjamin envisions possibilities – in opposition to Adorno – for such a mimetic relationship within modern media technologies, especially radio and cinema (Benjamin, 2008). A new technological mediation makes possible a perception of the world that widens visibility and generates experiences that are likely to recompose an emancipated political collective. Artworks are part of modern media and are experienced by wide audiences – collective experiences that are simultaneously canceled by the organization of modern societies. Benjamin's theses on modern media forms of communication have given rise to important debates within Critical Theory, especially with Adorno, who did not share his "optimism."

Siegfried Kracauer

Siegfried Kracauer (1889–1966) develops critical observations similar to those of his Frankfurt colleagues, especially in his Weimar writings of the 1920s, on the topic of the image, either fixed in the case of photography or in movement in the case of cinema. Kracauer stresses that modern media – especially magazines – mobilize photography, but don't contribute to a better or increased knowledge of the world. Photography contributes to a loss of meaning, through an "ornament" process (Kracauer, 1995a, 75–88). "Ornament" refers to a mythical image construction of society in a modern industrialized period of massification. Modern media promise to widen the knowledge of the world as well as to "save phenomena" through their pictorial immortalization, but they actually fail to bring a better knowledge of the world. According to Kracauer, photographic fixation is a substitute for the practical memory of real situations (Kracauer, 1995b, 47–64), because memory proceeds from a narrative selection within a situation from which only what makes sense for a living experience is retained. Unlike memory, photography records all the slight details, even the meaningless ones, producing a plethora of signs that exceed all possibilities of sense-making. Recipients are subjected to such a number of signs that they cannot integrate them in a meaningful experience – the outcome consists of a world depicted in a meaningless way. An overflow of factual traces produced by media doesn't contribute to a better knowledge of the world, or to its pictorial "rescue" – it kills it in the very promise of saving it. Nonsense and amnesia are the consequences. A process of communication is established under a reifying modality, because of a superabundance of images deprived of meaningful stories able to make sense of the loss of meaning of the world inevitably occurs. Ornament is the nonsense of a society damaged by a dominant ratio – a rationality reduced to instrumentality in modern societies. Only in his post-war writings, especially his *Theory of Film* published in 1960, did Kracauer envision a dialogic communication based on the moving image of the cinematographic medium (Kracauer, 1997). The film medium allows a "redemption of physical reality" thanks to the life movement rendered by moving images. It resuscitates the dead, thereby making them visible. Film thus saves history from oblivion while rescuing justice; furthermore, it allows for the portrayal of the multiple horrors of the world, which have become intolerable to the naked eye. A sensory contact to the world is possible through mediation, to its past, to "others," and to ourselves as individuals and as a community. A medium of the senses – cinema – makes possible a form of communication in a modern media world where it almost disappeared.

Jürgen Habermas

Many traces of the communicative ideal formulated by the German Enlightenment can be found in the authors of Critical Theory discussed above, its negation by analysis of its « liquidation » in modern societies. Be it in its negation, or buried under layers of domination, communication is still thought about in similar terms. More than any authors discussed here, Jürgen Habermas (born in 1929) has built his thinking in reference to this dialogical ideal of communication. With Habermas, it even becomes the normative core of Critical Theory. Habermas comes back to this ideal by referring to the idea of a "public use of reason" supported by Kant in his writings on *Aufklärung* (Habermas, 1989). In Kant, public reasoning is closely articulated with private reasoning through language. In public reasoning through deliberation with others a "constraint-free communication" takes place. Far from being a utopian ideal, it is a common *modus operandi* in everyday life, which enables cooperative language use. For Habermas, language practice immanently excludes coercion and violence; in linguistic practices, individuals "suspend" their social identity, social status or power, in favor of the logic of a dialogical-practical agreement seeking.

A "constraint-free communication" that takes place is not governed by social domination, but by the logic of the "better argument" that establishes what is accepted as truth in social life. Habermas describes this logic of the "better argument" as a "constraint-free force" opposed to a force based on rhetorical technic or physical constraints (Habermas, 1984, 1990).

Habermas gives primacy to language as a practice that is immanently based on non-violent social cooperation between human beings – it also makes it possible to think beyond direct face-to-face communication by including large-scale "media communication" within modern democratic societies. Referring to the pragmatics of language, Habermas shows that language is immanently intersubjective and oriented toward mutual understanding. To speak is always to speak with others and to refer to a common medium. In an intersubjective conception, persons making statements dialogically build a practical "place" for other participants (Habermas, 2008, 12–13). At the very core of communication lie normatively structured practices that are immanently dialogical; they are focused on mutual understanding as well as the building of a common world. A dialogical model of social communication is inherent to language, activated by social subjects in their daily interactions. For Habermas they are simultaneously a practical reference allowing the reconstruction of theoretically already there features of a "good life" and a normative practical reference for a general critical theory of society. "Discourse ethic" developed jointly by Habermas and Karl-Otto Apel (1924–2017) is based on such a descriptive – practical as well as normative – principle of dialogical communication, that allows anchoring a procedural "ethics" in shared common practices. At the very roots of the Habermasian Critical Theory of society lies a practical model of communication that goes hand in hand with a deliberative theory of democracy and participatory politics in modern societies (Habermas, 1996).

In Habermas's work, reflections on damaged communication, or its "pathologies," are also present (Habermas, 1991). However, they are part of a more "optimistic" diagnosis than that of his former Frankfurt colleagues whose reflections on modern societies derive from a reflection on Nazi domination, authoritarianism and the culture industry. Habermas starts from the democratic reconstruction and moral re-education of post-war Germany (Habermas, 1996); in doing so, he develops a normative theory of society based on practice, whose philosophical roots are to be found in the dialogic model of communication developed by the *Aufklärung*. Nevertheless, Habermas doesn't stop stressing the threats to public reason and the public sphere which are still present in modern society, especially in terms of technocracy and "legitimation crisis." Habermas witnessed, early in the 1960s, the dramatic rise of technocracy, which he saw as threatening to replace practical-political problem-solving process within the public sphere through technical expert systems that systematically transform practical issues into technical ones – even into a technical rhetoric alien to public shared language (Habermas, 1970, 2015).

The practical as well as democratic "alienation" that results from such a process not only threatens societies in their ability to political self-determination, but also in their ability to reproduce – through communicative action – the symbolic meaning structures on which societies are based and on which they depend. In the *Theory of Communicative Action*, Habermas stresses the "lifeworld colonization" by "systems" (economy and power). Because functional "systems" are carried out by technical-functional processes that bypass social coordination based on language used between social subjects, they tend to invade spheres of language activity based on practical forms of coordination. When "subsystems" colonize the lifeworld of mutual understanding, they empty its practical content in a way that deeply damages social coordination based on language activity. A loss of meaning, as well as a growing inability to reproduce cultural references through communication are its fatal consequences (Habermas, 1987, 2008).

Axel Honneth

In the continuation of Habermas' philosophy, Axel Honneth (born in 1949) digs further into the "communication paradigm" of Critical Theory by showing that discursive processes are anchored in intersubjective practices of recognition – mutual as well as institutional – that constitute them. Intersubjective recognition appears as a practical precondition of linguistic interaction as well as social communication (Honneth, 1992). In his own way, Honneth revisits the dialogical model of communication by referring to the intersubjective philosophy of Hegel. He radicalizes the communicative model in order to integrate affective dimensions, neglected by discursive approaches of Critical Theory (Habermas and Apel). Honneth also considers practical forms of mutual appreciation anchored in prelinguistic interactions, focusing especially on moral injuries experienced by subjects that alter their ability to participate in the public sphere. He also shows that these negative experiences generate struggles for recognition when they are socially and politically articulated (Honneth, 1992).

Honneth also insists on "pathologies" of intersubjective recognition, as a fragile precondition of social communication subjected to several forms of deterioration – disrespect, invisibility, reification (Voirol, 2015). These injuries to individual subjects as well as to social processes of moral recognition destroy the very conditions of social communication. According to Honneth, Critical Theory must be able to theoretically reconstruct the moral-normative content of social practices in order to contribute to the realization of "communicative freedom" – in affective relations, law, state, media, public sphere and political culture (Honneth, 2014). "Pathologies" of communication are to be found when institutional conditions for "social freedom" are undermined. According to the idea of "normative reconstruction" supported by Honneth, practical forms of communication, included in mutual recognition ensured by "communicational freedom," offer an immanent practical reference anchored in social practices in order to allow a non-idealized critique of pathological institutional arrangements.

Conclusion

With a varying degree and intensity, all authors of Critical Theory discussed here refer to the dialogical ideal of communication inherited from the Enlightenment, as a core principle for analyzing the possibilities of communication as well as its impossibilities – or "pathologies" – in modern capitalist societies. Critical Theory offers an original perspective on communication by making this dialogical ideal a key issue of its critical analysis. Especially in Habermas' "paradigm of communication," followed by Honneth, this dialogical model is seen as offering an immanent practical reference for "good life." However, for the first generation of critical theorists, the dialogical model of communication appears as a background that functions as a critical motive. It is a key element of a Critical Theory of modern capitalist societies and their institutions. The modalities of reification of modern capitalist societies are subject to radical criticism with reference to such a dialogical model. Authors of the first generation underlined the fatal degeneracy of the dialogical model, and its transformation in mere utopian ideals – if not ideology – within a "false life." As an implicit or explicit reference, the dialogic model of communication makes it possible to criticize forms of pathological or non-communication, therefore structuring a critical social diagnosis.

The importance given by these authors to one or multiple of the aforementioned aspects may be very different, according to the critical model they use. The dialogical model of communication can function as a normative reference for a reconstructive critique that reveals the "pathologies" of modern societies (Habermas, Honneth). On the other hand, it can be – like in Adorno's perspective – the reference of its decay, in order to stress its negation. It is precisely the type of diagnosis made about modern societies by the different members of Critical Theory that differentiates their accounts on communication. Diagnoses made by Benjamin, Kracauer and Marcuse, agree broadly with this second critic. For Adorno, the transformation of modern capitalist societies such as the concentration of power, commodification, and reification of social relations as well as culture, has deeply damaged these societies that communication can only be thought in negative terms. The post-Fascist era did not get rid of all the authoritarian generating processes that have developed within the "damaged life" of modern capitalist societies. As a result, its political threat remains as long as this mutilation persists. As for Habermas and the authors who are part of the "paradigm of communication" of Critical Theory, a bet is made about the possible democratic reconstruction in post-war societies, about the possibilities of a deliberative democracy, and therefore the practical effectiveness – even partial – of the dialogical ideal of communication.

References

Adorno, T. W. (1973, 1966). *Negative Dialectics*. New York: Seabury Press.

Adorno, T. W. (1991, 1938). On the Fetish Character in Music and the Regression of Listening. In *The Culture Industry: Selected Essays on Mass Culture*. Edited by J. M. Bernstein. London: Routledge (pp. 29–60).

Adorno, T. W. (1998). *Aesthetic Theory*. Edited by Robert Hullot-Kentor. Minneapolis, MN: University of Minnesota Press.

Adorno, T. W. (2002, 1932). On the Social Situation of Music. In *Essays on Music*. Edited by Richard Leppert and Trans. by Susan H. Gillespie. Berkeley, CA: University of California Press (pp. 391–436).

Adorno, T. W. (2009). *Current of Music Elements of a Radio Theory*. Edited and Trans. by Robert Hullot-Kentor. Cambridge, UK and Malden, MA: Polity Press.

Adorno, T. W., Frenkel-Brunswik, E., Levinson, D. J., & Sanford, R. N. (1950). *The Authoritarian Personality*. Harpers.

Benjamin, W. (1968, 1936). The Storyteller. Reflections on the Works of Nikolai Leskov. In *Illuminations*. Edited by Hannah Arendt and Trans. by Harry Zohn. New York: Harcourt Brace Jovanovich (pp. 83–109).

Benjamin, W. (1979, 1916). On Language as Such and on the Language of Man. In *One-way Street, and other Writings*. Edited by Peter Demetz and Trans. by Edmund Jephcott & Kingsley Shorter. London: NLB (pp. 107–123).

Benjamin, W. (1996) The Concept of Art-criticism in German Romanticism. In *Walter Benjamin: Selected Writings, Volume 1: 1913–1926*. Edited by Marcus Bullock & Michael W. Jennings and Trans. by David Lachterman, Howard Eiland & Ian Balfour. Cambridge, MA: Harvard, Belknap Press of Harvard University Press (pp. 116–219).

Benjamin, W. (2005, 1933). Experience and Poverty. In *Walter Benjamin: Selected Writings, Volume 2: 1931–1934*. Edited by Michael W. Jennings, Howard Eiland, Gary Smith. Cambridge, MA: Harvard University Press (pp. 731–736).

Benjamin, W. (2008, 1937). *The Work of Art in the Age of Its Technological Reproducibility and Other Writings on Media*. Cambridge, MA: Harvard University Press.

Fromm, E. (1941). *Escape from Freedom*. New York: Farrar & Rinehart.

Fromm, E. (1984). *The Working Class in Weimar Germany: A Psychological and Sociological Study*. Cambridge, MA: Harvard University Press.

Habermas, J. (1970, 1968). Technology and Science as Ideology. In *Toward a Rational Society: Student Protest, Science, and Politics*. Boston: Beacon Press (pp. 81–122).

Habermas, J. (1984). *Reason and the Rationalization of Society*. Trans. by T. MacCarthy. Boston, MA: Beacon.

Habermas, J. (1987). *The Theory of Communicative Action, Vol. II: Lifeworld and System*. Cambridge: Polity Press.

Habermas, J. (1987, 1981). *Theory of Communicative Action. Vol. 2: Lifeworld and System: A Critique of Functionalist Reason*. Trans. by Thomas A. McCarthy. Boston: Beacon Press.

Habermas, J. (1989, 1962). *The Structural Transformation of the Public Sphere: An Inquiry into a Category of Bourgeois Society*. Trans. by Thomas Burger and Frederick Lawrence. Cambridge: Polity Press.

Habermas, J. (1990, 1983). Discourse Ethics: Notes on a Program of Philosophical Justification. In *Moral Consciousness and Communicative Action*. Cambridge, MA: MIT Press (pp. 43–115).

Habermas, J. (1996, 1992). *Between Facts and Norms: Contributions to a Discourse Theory of Law and Democracy*. Trans. by William Rehg. Cambridge, MA: The MIT Press.

Habermas, J. (1997). *Between Facts and Norms: Contributions to a Discourse Theory of Law and Democracy*. Cambridge: Polity Press.

Habermas, J. (2001, 1974). Reflections on Communicative Pathology. In *On the Pragmatics of Social Interaction: Preliminary Studies in the Theory of Communicative Action*. Trans. by Barbara Fultner. Cambridge, MA: The MIT Press (pp. 129–170).

Habermas, J. (2008, 2004). Public Space and Political Public Sphere – The Biographical Roots of Two Motifs in My Thought. In *Between Naturalism and Religion: Philosophical Essays*. Cambridge: Polity Press (pp. 11–23).

Habermas, J. (2015). *The Lure of Technocracy*. Trans. by Ciaran Cronin. London: Polity Press.

Honneth, A. (1995, 1992). *The Struggle for Recognition: The Moral Grammar of Social Conflicts*. Cambridge, MA: Polity Press.

Honneth, A. (2014). *Freedom's Right: The Social Foundations of Democratic Life*. Hoboken, NJ: John Wiley & Sons.

Horkheimer, M. (1982, 1937). Egoism and the Freedom Movement: On the Anthropology of the Bourgeois Era. *Telos, 21*(54), 10–60.

Horkheimer, M. & Adorno, T. W. (2002, 1947). *Dialectic of Enlightenment*. Redwood, CA: Stanford University Press.

Humboldt, W. (1960–1981). *Werke in fünf Bänder*. Edited by A. Flitner & K. Giel. Darmstadt: Wissenschaftliche Buchgesellschaft.

Kracauer, S. (1995a). Mass Ornament. In *The Mass Ornament: Weimar Essays*. Edited and Trans. by Thomas Y. Levin. Cambridge, MA: Harvard University Press (pp. 75–88).

Kracauer, S. (1995b). Photography. In *The Mass Ornament: Weimar Essays*. Edited and Trans. by Thomas Y. Levin. Cambridge, MA: Harvard University Press (pp. 47–64).

Kracauer, S. (1997, 1960). *Theory of Film: The Redemption of Physical Reality*. Princeton, NJ: Princeton University Press.

Löwenthal, L. & Guterman, N. (1949). *Prophets of Deceit. A Study of the Techniques of the American Agitator*. New York: Harper & Brothers.

Marcuse, H. (1955). *Eros and Civilization: A Philosophical Inquiry into Freud*. Boston: Beacon Press.

Marcuse, H. (1964). *One-Dimensional Man: Studies in the Ideology of Advanced Industrial Society*. Boston: Beacon Press.

Marcuse, H. (1968, 1937). The Affirmative Character of Culture. In *Negations*. Boston: Beacon Press (pp. 88–133).

Marcuse, H. (1972). *Counterrevolution and Revolt*. Beacon Press.

Schiller, F. (2016, 1794). *On the Aesthetic Education of Man*. London: Penguin.

Voirol, O. (2015). Une critique immanente de la communication sociale. Sur quelques potentiels de l'approche honnéthienne. *Réseaux, 193*, 43–77.

6

REIMAGINING COMMUNICATION IN MEDIATED PARTICIPATORY CULTURE

An Emerging Framework

Usha Sundar Harris

Introduction

Our real-life experiences are increasingly being influenced and shaped by relationships forged in the virtual world. The networked world of social media draws people and cultures from far corners of the earth into a dynamic social and cultural milieu. Interactions may be brief over shared user-generated content (UGC) or formal and long-term conducted in a virtual professional sphere. This article reimagines the field of communication within the context of a mediated participatory culture in which cooperation and collaboration are valued. It considers the opportunities and challenges of communication in a digital environment; how it extends an individual's intercultural experience, builds new alliances in support of the environment and promotes sustainable future for a living earth.

From scholars to practitioners, there is growing recognition for the need to change the way in which we see the world and the concepts we use to understand it. The importance of transdisciplinarity, and a need to invite wider perspectives, are part of this process. One of the ways this transformation can be facilitated is through the triad of theory, experiential learning and reflexive research.

With this need to find new ways of thinking about and practicing communication, I proffer a framework that incorporates three interrelated elements: *Diversity, Network and Agency*. In my book, *Participatory Media in Environmental Communication* (Harris, 2019), these three concepts form the foundations of a conceptual framework for participatory environmental communication (PEC). In this chapter, I suggest that the DNA framework is just as useful in other communication contexts and provides valuable analytical tools by which to monitor and evaluate the effectiveness of communication for collaborative action. This will be demonstrated through examples of participatory media projects undertaken in the Asia and Pacific regions, and students' experiences of virtual communication in the higher education sector based on the author's teaching practice.

The Changing Context of Communication

Participatory media is defined here as a media culture that enables ordinary people to collectively identify problems, gather information, analyze, design and share media content which is of benefit to them and their network. The term "participatory media" is used to emphasize the pivotal role of participation of lay-people in making media content for their own purpose. The process of production invites collaboration, dialog and mutual problem-solving, whereby people gain a critical understanding of the problem and find appropriate solutions (Kheerajit & Flor, 2013). Proactive communication between members empowers endogenous networks and uncovers unique solutions to vexing problems arising from a particular context – be it historical, social, cultural, economic, political or geographic – of the community. In doing so it embeds local knowledge and finds solutions from within. Here, the boundary between producer and consumer of content disappears.

Digital Media

While participatory communication and the culture of community media has been around for more than half a century, technological affordances in the digital age have enabled participation by diverse populations around the world. Easy access to digital media has resulted in a paradigm shift from producer-oriented, top-down, capital-intensive content creation to user-generated content using peer-to-peer networks. Digital media resulting from the convergence of telecommunications, broadcasting and internet technologies eventually gave rise to web 2.0 interactivity. Web 2.0 services focus more on socially connected web applications such as wikis, blogs, photo-video sharing, community media, participatory video, social networking and podcasts where the online presence of the users is important (Andersen, 2007). User-friendly web 2.0 applications thus allow active participation of end-users as media creators and require very limited technical knowledge compared to traditional media (Harrison & Barthel, 2009). People form online communities to exchange content using social media networks such as Facebook, Twitter, YouTube, Google+, Instagram and other emerging platforms.

Participatory Culture

Digital media technologies and social media platforms have created significant transformations in the participatory culture. Henry Jenkins and his colleagues at Massachusetts Institute of Technology (MIT) provided one of the first insights into digital participatory media culture, which they explained as having "relatively low barriers to artistic expression and civic engagement, strong support for creating and sharing one's creation and some type of information mentorship whereby what is known by the most experienced is passed along to novices" (Jenkins et al., 2006, 3). They add that a participatory culture "is also one in which members believe their contributions matter, and feel some degree of social connection with one another" (Jenkins et al., 2006, 3).

According to Jenkins (2006, 2–3), media convergence means "the flow of content across multiple media platforms, the cooperation between multiple media industries, and the migratory behavior of media audiences who will go almost anywhere in search of the kinds of entertainment experiences they want." Media convergence relates to participatory culture and collective intelligence, thus representing a cultural shift rather than just a technological shift where consumers willingly participate in searching out new data and making interactions among scattered media content (Jenkins, 2006). Convergence involves a change in the way media is produced and consumed. Such participatory culture, using web 2.0 applications and technological communication development, has transitioned the user from passive consumer to active producer of content in the social media platform.

Critical scholars question the participatory potential of web 2.0 because it entrenches commercial and marketing exploitation of users such as data mining and corporate dominance, and promotes simplistic notions of participation which contribute to a self-branding culture, or worse still, fundamentalism and fascism (see Fuchs, 2017). It also creates an echo chamber where people gravitate toward people and information reflecting views that are similar to their own (Clark & Van Slyke, 2010). Fuchs has advanced the theory that while web 2.0 is a "computer-based networked system of human communication", the next level of digital technology development is web 3.0, "a computer-based networked system of human cooperation" (2017, 46). An example of this is Wikipedia which, he says, is "a new way of producing, owning, consuming and distributing goods and a new way of collaborative decision-making" (2017, 319).

The term participatory media in this chapter aligns closely with the definition of "citizen's media" given by Clemencia Rodríguez:

> In sum, alternative and community media research should re-centre the agency of communicators over technologies and refocus on context, uses and needs … as communities experience shifting contexts because of climate change, health crises or political uprising and protests, communication and media ecologies, needs and uses shift and change.
>
> *(2016, 36–37)*

Fenton concurs and proposes "not beginning with the media at all, but rather starting with context" (2016, 10). Thus participatory media is not solely about accessing technology. It is about ways in which participatory platforms enable people to have a voice, to organize and to demand change. Ordinary people who are excluded from participating in top-down hierarchical communication systems, be it village meetings or mainstream media, where they are only passive receivers of information, become active agents of change as they gain power over their own storytelling. The process of dialog which engages diverse networks in finding new solutions to old problems enables transformative thinking by their active engagement in media production.

The DNA Framework

The framework offered here incorporates three interrelated strands: *Diversity, Network and Agency* – the essential building blocks of a resilient society. I refer to these as the DNA of participatory environmental communication because each of these strands is important to the survival of both natural and social systems. The DNA model brings people together in a process of dialog and collaboration which inspires collaborative action in the face of many environmental challenges. *Diversity* enables innovative and transformative thinking. The term here means both difference and inclusion of a broad range of factors – different knowledge systems, socio-cultural values and beliefs, abilities, talents, demographic variables (age, gender, class, ethnicity) and the non-human world (ecosystems, technologies, texts). *Networks* are a complex system of relationships connecting both human and non-human worlds. They include human networks and non-human networks. The understanding that humans are but one part of this intricate web of creation, and not separate from it, is essential in the action-reflection-action cycle of the PEC process. *Agency* is an action or a doing of human and more-than-human entities, which leads to an effect or outcome. Agency in the natural world contributes to the efficient working of an ecosystem. Human agency results from a realization of our own potential through dialogic encounters that act as the catalyst for change.

Embedded in these elements are the core principles of participatory communication which White and Patel so eloquently describe as "equalitarian, transactive and dialogic" in nature (White et al., 1994, 363). The process of dialog is central to this framework because it engages diverse networks in finding new solutions to old problems thereby enabling transformative thinking. As a dialogical approach that engages people in meaningful conversation this change requires a shift from an individualistic win/lose mindset toward a collectivist approach. The participatory environmental communication model which underpins this framework is a process-oriented method whereby users can share knowledge, create awareness and take action on issues through their unique perspectives and intimate lived experiences.

FIGURE 6.1 The DNA framework
Source: Adapted from Harris (2019, 36)

Finding Connections

The DNA framework (Figure 6.1) identifies important attributes of each element – *Diversity, Network and Agency* – which enable collaboration and dialog
among different actors, inviting them into a dynamic web of relationship. It is
an integrated approach that can be replicated for different purposes, much like
the DNA in living organisms, and should be integrated thoughtfully in participatory communication processes. Each thread of the DNA is not separate, but
is intertwined in a holistic relationship, influencing and shaping the other in
a communicative process, as illustrated below:

diversity + network stimulates cross-fertilization and cross-pollination of ideas by
   proactively linking disparate networks that previously may not have been
   linked or had weak links. These manifest as communication between different networks (as identified in Table 6.1) such as intergenerational,

TABLE 6.1 DNA essentials

Elements	Diversity	Network	Agency
Definition	*Diversity* describes both difference and inclusion of a range of entities, abilities and ways of knowing.	*Networks* are complex systems of relationships and interactions (links) between human and non-human worlds (nodes) vital for exchange of information and resources.	*Agency* is the capacity of an actor to have the power to take action in a given context.
Attributes	knowledge systems socio-cultural values and beliefs demographic variables (age, gender, race, identities, education, place disabilities etc.) interface non-human world biosphere holistic	human non-human relational (form and content) reciprocity social capital social trust social cohesion	dialog participation empowerment transformation capabilities and functionings voice and storytelling active listening agents of change participatory action research
Manifestations	Communication cross-fertilization new connection innovation intergenerational intercultural interdisciplinary inter-faith inter-agency cross-sector	Communication local or endogenous network heterogeneous network trusted network top-down horizontal bridging and bonding collaboration cooperation	Communication catalyst for change mutual learning problem-solving gather information, analyze, design, knowledge sharing transformative thinking awareness-raising active participation marginalized voices action-reflection-action cycle

Source: Harris (2019, 37)

intercultural, interdisciplinary, inter-faith and inter-agency dialog, leading to innovation and collaboration previously untapped.

network + agency enables new relational possibilities. As actors in networks participate in reciprocal acts they begin to understand the importance of their relational possibilities in expanding networks. The process of communication brings new insights into old problems. The relational aspects bring to light both positive and negative interactions. As such power, privilege and inequality should be studied alongside empowerment.

agency + *diversity* facilitates problem-solving. Agency is a critical component in solving problems, both personal and societal. When people and organizations with different abilities, beliefs and knowledge systems interact proactively, transformational learning results.

Diversity + *Network* + *Agency* opens up a dynamic space for dialog and collaboration. This is based on the belief that the cornerstone of human progress is collective enterprise instead of conflict and competition.

The DNA framework actively seeks to integrate diversity in all forms, and emphasizes the act of building networks at different levels – national, local and individual. DNA is a horizontal model of communication that gives voice to those on the margins including women, indigenous groups and people with disabilities. By telling their own stories in their own way, people are able to communicate compelling accounts of how different, but interrelated, environmental, political and economic issues converge and impact at a local level. These local stories, told through personal narratives of adversity and triumph, bring a lived human dimension to environmental changes occurring on a global scale.

The DNA framework also places special emphasis on non-human networks such as local ecosystems (inclusive of animals, plants, river systems, etc.) as important nodes of communication. Their presence helps a community to observe and learn how their own interactions with these nodes impact on the resilience of the whole ecosystem. Indigenous traditional knowledge is an important node in the wider network of knowledge holders. Participatory mapping of informational networks assists individuals and organizations to uncover the relational links (e.g., strong, weak, bonding, bridging), and the types of actors, resources and information that are available to them. It can be applied at the micro-level and scaled up depending on the context and needs of different users.

Using DNA in Intercultural Contexts

An aim of the DNA framework is to reveal connections between people and organizations and how well these connections enable information flows, i.e., how, where and with whom people exchange information, and the quality of these exchanges within a given context. These below-the-surface information flows are also vital in the early hours and days of emergency response and aid distribution. The intrinsic information exchanges may be quite obvious to individuals in the local community, but not so obvious to outsiders such as first responders and aid workers. Community participants and aid agencies can use this tool to identify the community information flows in the local area.

A diversity mapping exercise at an early stage will ensure that a cross-section of the community is included. The diversity mapping creates a visual representation of the forms of diversity one would find in different contexts. These can be expanded further to uncover related factors. Creating a diversity matrix for a community, no matter how big or small, assists both community members and outsiders to understand the complexity and information needs, especially during disasters and for resilience building.

The map would also identify information exchanges specific to different ethno-cultural groups in a society. Unequal distribution of resources is often a cause of conflict. Neglected communities perceive aid disparity during emergencies as favoritism for one group over another or as racial discrimination, instead of a possible oversight on the part of agencies who are unaware of the appropriate channels through which information is exchanged in specific groups.

Table 6.2 provides an example of a diversity matrix that can be considered when creating resilience-building tools. By undertaking a diversity mapping exercise, people are forced to consider their actions on the environment, including connections with non-human entities such as river systems, forests, oceans and animals that have either direct or indirect impact on a community's livelihood. Participants would identify resources, skills and knowledge already present within the group and those they would have to source from outside. For example, the community may have a strong base in local knowledge, but require current scientific knowledge to help them understand soil salinity or reasons for a decrease in soil quality that is affecting their crop output.

In designing disaster prevention and management plans, each element in the column is identified and mapped according to the local context. The elements

TABLE 6.2 Diversity matrix

Knowledge systems	Cultural factors	Social factors	Non-human	Modes of communication	Services
Scientific	Values/ beliefs	Education/ literacy	River systems	Interpersonal	Emergency/ DRR
Indigenous/ Traditional	Race/ ethnic Tribe/clan	Children/ Youth	Coastal ecosystems	Print	Medical
Local	Place-based	Gender/ LGBTI	Ocean	Broadcast	Community
Experiential	City/rural	Disabilities/ abilities	Animal population	Internet	Private/ commercial
Faith-based	Religion	Elderly	Plant varieties	Transport	Faith-based
Ecological	Language	Wealth	Soil biome	Non-human	Ecosystem

Source: Harris (2019, 69)

can be deleted or new ones added, based on different contexts and changing needs. Once the diversity matrix has been created, community members undertake a network mapping exercise to identify the relational links between nodes and their value in either strengthening or weakening individual, family or community resilience in a particular context. Once identified, communities can then engage constructively in the participatory mapping process of vulnerability assessment and adaptation planning. Diversity mapping initiates a culture of integrated problem-solving and a move away from thinking in silos. The identification of these actors and entities is the first step toward building networks.

An example of this is found in a sustainable agriculture project in Kiribati, a nation that is consistently portrayed in international media as an extreme example of climate change impact on a population that is destined to lose its islands to the rising seas. The I-Kiribati decision-makers emphasize that adaptation is about giving people choices. This means assisting those who want to migrate to gain new skills so they can migrate with dignity, and helping those who want to stay in Kiribati to live at home enjoying the rights to clean water, food security and safe housing. This requires the affected communities to gain the skills to grow food and build strong structures and safe homes, which would withstand increasingly severe weather events. But to do this Kiribati needs the support of the international community in the form of technical expertise and climate funds.

Agricultural specialist, Keboua John Collins, is an inspiring young I-Kiribati who has returned to Abaiang after gaining his education overseas to seek solutions which will allow the people to continue living in their homeland. He is focused on finding sustainable ways of turning to productive use the resources of the ocean and the thin sliver of the atoll which is home to more than 110,000 people. Keboua received part of his training in South Korea, where he learned about the New Village Movement known as Saemul Undong, a community-led development plan introduced by President Park Chung-Hee to rebuild the society after the Korean War (see The Movement for Community-Led Development, 2017). Influenced by this Korean cultural practice of self-sustainable community building, Keboua sought help from village elders to teach unemployed youth about sustainable building practices that were being lost. This has aided intergenerational knowledge transfer and capacity-building for unemployed youth.

Keboua also teaches communities about making compost using local organic material to build up the soil content. This allows villagers to grow vegetables and reduce reliance on imported foods. One of the challenges is access to good quality seeds. He describes his way of working with the Abaiang community as the "less approach – how to work with the available resources to engage or develop the community." His relationship with the community is based on trust and fair exchange known as *Bubuti*, a form of reciprocity.

We find all the attributes of the DNA framework in this story. Keboua uses scientific knowledge that he has acquired as an agricultural specialist to teach others on the island about composting. In an intercultural exchange, he has combined the cultural knowledge of his own people with the South Korean community-building practice of Saemul Undong to find a sustainable development model for his island community. He uses his kinship network to enact intergenerational knowledge exchange, at the same time catalyzing the agency of others in the community through group dialog.

DNA Framework in Participatory Media

The participatory production process engenders diversity through cross-fertilization of ideas between previously disparate groups. Vertical networks initiate dialog with decision-makers on policy reform, and facilitate knowledge transfer between experts and communities. Horizontal information exchange improves information flow and networking with heterogeneous groups on a range of social, political, and environmental issues to build consensus and resolve conflict.

The process of content creation empowers individuals through their new-found agency. Experiential learning and the action-reflection cycle improves participants' capabilities for self-reliance and self-organization. As participants engage in the production process and begin to understand visual grammar, they also become critically aware of the way in which storytelling can be manipulated by the choice of shots and the interview process. They begin to critically reflect on mainstream media's role in influencing public understanding of issues by giving greater voice to political and economic elites. Through a feedback loop of viewing and reviewing the videos, the participants also identify strengths and weaknesses in their own practice and recognize the stock of knowledge and skills available within the group.

Social action researchers who have facilitated participatory media (PM) activities demonstrate that agency, voice, empowerment and transformation are more than mere words on a page when media spaces are created for marginalized groups who have been silent in the past. These words become living representations of people who provide the concrete meaning of these concepts through action.

Two examples of community media use are shared below to illustrate its use in building networks and enabling agency. The first is the use of films and film-making in West Papua for empowerment and advocacy. The second is the use of community radio to assist networking and information flow within communities in Aceh, Indonesia after the 2004 Tsunami.

The conflict between West Papuans and the Indonesian Government remains largely hidden from the world. Despite decades of hostility and violence, demands for justice have received little global attention. Papuan Voices is a collective movement that aims to bring the life stories of West Papuans to a wider audience by training Papuan filmmakers both for empowerment and for film production.

During the last six years, young West Papuans have been using video cameras to tell their own story from their own perspective. Wensislaus Fatubun, filmmaker and founder of Papuan Voices describes this process:

> From my experience, film-making is able to connect people. It is not only communication, but more of an emotional connection. People who are at different places can have an emotional connection to each other because they watch the film. The process of film-making for advocacy and memory is an intersubjective experience and dialogue on story, history and culture. This is the first process when we produce a film. I visit the villages, explain what a camera is and what purpose it serves and convince the villagers of its usefulness as a tool. Once a plan for the sequence is established to everyone's satisfaction, we can film. The interest is not just in making films. Each sequence is an opportunity to talk with the villagers about human rights, the rights of Indigenous peoples and their socio-economic rights. Making a film is an occasion for us to settle on a strategy with the villagers for them to claim their rights.
>
> *(Harris, 2019, 153–154)*

On December 26, 2004, a tsunami devastated Nangroe Aceh Darussalam (NAD), the westernmost region of Indonesia. As many as 164,000 people died or were reported missing and more than 400,000 people became homeless. The infrastructure collapsed, including communication networks. With poor communication infrastructure, coordination and information exchange became very difficult among the organizations and within communities. After assessing the situation, the availability of trusted and equitable information was the biggest and most valuable form of aid for the affected communities. A network of community radio stations was set up in the affected regions of Aceh and Nias. Iwan Awaluddin Yusuf, Independent Project Reviewer, explains the effect of the community radio:

> The majority of the stations were located in *gampông* (villages) affected by the tsunami. The stations then took on the characteristics of *gampông*-based community radio. Hence the characteristics of each community radio in Aceh varied according to its village. After the initial stage, the local citizens took control of the improvement of their community. Generally, the people of Aceh eagerly welcomed radio as it could accommodate their need for information and entertainment in order to release anxiety and uncertainty after the tsunami. It also empowered them because they could engage actively in the process of rehabilitation and reconstruction through community radio programs.
>
> *(Harris, 2019, 84–85)*

When facilitating a participatory media project, the facilitator should embed practices (discussed below) that ensure community voice thereby guarding against top-down initiatives that pretend to be participatory.

Whose Knowledge?

Both participants and project planners should establish what reciprocal arrangements exist between parties for the exchange of knowledge and resources. Has the community developed the idea? Has someone else approached the community? It is important to recognize the "tensions, contradictions, dilemmas, and power imbalances inherent in all forms of knowledge production and communication" when "scientists and citizens engage in mutual learning on the basis of the different knowledge forms that they bring with them" (Phillips, 2011; Phillips et al., 2012, 4).

Who Will Participate?

Establishing trust with the community is vital to the success of participatory projects. Finding a community leader who has the trust of all sections of the community and the authority to engage with them is essential. This may include the local social worker, faith-based leader or president of a social club.

A diversity mapping exercise (see Table 6.2) at an early stage will ensure that a cross-section of the community is included. The diversity mapping creates a visual representation of the forms of diversity one would find in different contexts. These can be expanded further to uncover related factors. Creating a diversity matrix for a community, no matter how big or small, assists both community members and outsiders to understand the complexity and information needs, especially during disasters and for resilience building.

The map would also identify information exchanges specific to different groups in a society made up of diverse ethno-cultural groupings. Unequal distribution of resources is often a cause of conflict. Neglected communities perceive aid disparity during emergencies as favoritism for one group over another or racial discrimination, instead of a possible oversight on the part of agencies who are unaware of the appropriate channels through which information is exchanged in specific groups. The intrinsic information exchanges may be quite obvious to individuals in the local community, but not so obvious to outsiders when they arrive in a community. Creating a map of the community information flow with key stakeholders is the necessary first step.

Choosing the Right Fit

Decisions about the forms of participatory activity to use should be made in consultation with the community. Young people may prefer to use video and social media, while older members may prefer radio or traditional forms of face-to-face communication. Consider the audience and final dissemination of the end product. Will it be distributed globally or shared for local information only?

Mediated Intercultural Communication in Higher Education

Previously, I have identified the growing importance of computer-mediated communication (CMC) in the educational sector to prepare students for professional work online (see Harris, 2014, 2017). In this section I incorporate my research on e-service learning to illustrate the importance of collaboration and cooperation between the higher education sector and social action at a community level through the discussion of an undergraduate unit which I have convened. Macquarie University's Professional and Community Engagement (PACE) initiative is "based on the principle of reciprocity" which provides undergraduate students a range of experiential and practice-based opportunities to "collaborate with a vast network of partner organizations from the business, government, non-government, and not-for-profit community sectors" (PACE, Macquarie University, n.d. https://staff.mq.edu.au/teach/teaching-at-macquarie/PACE). Students are able to engage in learning through participation in both real and virtual environments. In ICOM204 International Communication Campaigns (formerly ICOZ202), students work in the virtual environment with international partners using computer-mediated communication.

Computer-mediated communication has played an important role in the development of global partnerships with multinational and multicultural virtual teams sharing knowledge and skills without the restrictions of time and space in various fields of commerce, medicine and civil engagement (Ratcheva & Vyakarnam, 2001). CMC has provided new forms of virtual partnerships that "build rich shared learning and knowledge cultures" (Starke-Meyerring & Wilson, 2008, 7). It has afforded new opportunities for educators to develop curricula that enable students to collaborate with partners in different parts of the world, while gaining important life skills as an extension of their classroom learning while empowering communities as collaborators in knowledge production and social action (Crabtree, 2008).

The virtual environment offers various options for continuous dialog, through the use of such tools as asynchronous e-mail or threaded discussion forums, and synchronous real-time conferencing platforms such as Skype or basic chat and instant messaging functions (Barab et al., 2001; Meyers, 2008). The Internet becomes an educational tool that expands working in a local community to working in a global community. An important benefit of virtual placement for the partner is that they are relieved of the responsibility of managing student activities and having a duty of care toward them, which are special considerations for resource-poor NGOs. It also opens up opportunities for students who are unable to travel overseas because of disability or cost factors. Also, as we become conscious of our carbon footprint, online modes of communication are favored.

A study with students and partners revealed that the communication exchanges provided important lessons in intercultural dialog, both in real and virtual spaces (see Harris, 2017). The remote engagement allowed the students to reflect on the contextual validity of theory, develop cross-cultural insights and acquire knowledge and skills in problem-solving which have implications for their future careers. The experiential learning came from interaction with NGO staff and responses to the project brief in campaign design. Significantly, it gave students exposure to a transnational working environment in an e-service-learning context, thereby developing their identity as global citizens. They learned how to use ICT and media within a social justice framework and recognized that the cultural diversity of their own team was beneficial. By researching the assigned topic and designing campaigns, students were able to translate classroom knowledge and skills into the work environment, as a student noted in her self-reflection (2018): "I have learnt to focus solely on teamwork. It has enhanced my emotional intelligence as I have learnt to focus on the outcomes this project will have for the partner rather than myself to produce the best campaign possible."

The partnership is of mutual benefit to both the students and the organization. Students have the benefit of working with a partner who has significant experience in community development and who provides valuable feedback on various design elements as well as insights into the socio-cultural context of the society in which it will be implemented. In exchange, the partner receives a completed project which meets its specific objectives and can be used immediately. Students gain unique intercultural experience, be it mediated, in the context of a professional setting while working remotely on issues facing culturally divergent communities. The resulting sense of connectedness to a community with which they have had no prior links gives young people an emerging sense of what it means to be a global citizen in a digitally networked world as observed by this student:

> I have found this exercise extremely useful in pushing me outside my comfort zone and exploring cultures and issues I have never even contemplated before. It has also made me aware of the power individuals and groups have to cause a positive effect, and makes me feel more connected to the world as a global citizen.

Conclusion

The DNA framework discussed in this chapter, offers a systems approach to problem-solving. The approach integrates interdisciplinary knowledge, inspires collaboration and dialog at all levels, and utilizes information networks to catalyze the agency of citizens toward collective action. It also includes systems of knowledge that have been silenced. These are the use of folk songs and stories, or the powerful acts of protest that activists and indigenous groups deploy.

An aim of the DNA framework is to reveal connections between people and organizations and how well these connections enable information flows, i.e., how, where and with whom people exchange information, and the quality of these exchanges within a given context. It improves people's ability to identify the networks and resources available to them and helps them to make that critical link between vulnerability and the necessity to adapt and change.

Through the use of participatory media ordinary citizens open up shared spaces for modern-day storytelling, enabling a shift from binary frameworks toward a more holistic and networked communication model. It improves collective action and knowledge exchange between diverse stakeholders across different networks; intercultural, intergenerational, interdisciplinary, interagency and cross-sectoral. Collaboration encourages transactional communication and reciprocity through the production process. Enabling agency of a community means providing opportunities for members to engage in collective dialog to find solutions that improve their community and consequently improve their own lives. It asks participants to imagine a different reality. To achieve this, people need access to reliable information and the ability to act upon it collectively. Participants' active engagement in communicative processes increases knowledge exchange and improves adaptive capacities.

Finally, the DNA framework offers a theoretical tool that fills an important gap in communication between disparate and divergent groups. *Diversity*, in the form of collaboration among various entities, enables innovative and transformative thinking; *Networks* form complex webs of relationships vital for information exchange and survival of an ecosystem; *Agency* is the capacity of all beings to act upon their environment in order to bring about change.

Acknowledgment

This chapter is based on the author's recent book, *Participatory Media in Environmental Communication: Engaging Communities in the Periphery* (Routledge, 2019).

References

Andersen, P. (2007). *What is Web 2.0?: Ideas, technologies and implications for education.* Retrieved from Bristol, UK: http://21stcenturywalton.pbworks.com/f/What%20is%20Web%202.0.pdf

Barab, S., Thomas, M. & Merrill, H. (2001). Online learning: From information dissemination to fostering collaboration. *Journal of Interactive Learning Research, 12*(1), 105–143.

Clark, J. & Van Slyke, T. (2010). *Beyond the echo chamber: Reshaping politics through networked progressive media*. New York, NY: The New Press.

Crabtree, R. D. (2008). Theoretical foundations for international service-learning. *Michigan Journal of Community Service Learning, 15*(1), 18–36.

Fenton, N. (2016). Alternative media and the question of power. *Journal of Alternative and Community Media, 1*, 10–11.

Fuchs, C. (2017). *Social media: A critical introduction*. London, UK: Sage.

Harris, U. (2014). Virtual partnerships: Implications for mediated intercultural dialogue in a student-led online project. In S. H. Culver & P. Kerr (MILID Yearbook) (Eds.), *Global citizenship in a digital world* (pp. 177–190). Göteborg, Sweden: NORDICOM, University of Gothenburg.

Harris, U. S. (2017). Virtual partnerships: Engaging students in E-service learning using computer-mediated communication. *Asia Pacific Media Educator, 27*(1), 103–117. doi:10.1177/1326365X17701792

Harris, U. S. (2019). *Participatory media in environmental communication: Engaging communities in the periphery*. London, UK: Routledge.

Harrison, T. M. & Barthel, B. (2009). Wielding new media in Web 2.0: Exploring the history of engagement with the collaborative construction of media products. *New Media & Society, 11*(1–2), 155–178.

Jenkins, H. (2006). *Convergence culture: Where old and new media collide*. New York, NY: New York University Press.

Jenkins, H., Clinton, K., Purushotma, R., Robison, A. J. & Weigel, M. (2006). *Confronting the challenges of participatory culture: Media education for the 21st century*. Retrieved from Chicago, IL: www.macfound.org/media/article_pdfs/JENKINS_WHITE_PAPER.PDF

Kheerajit, C. & Flor, A. G. (2013). Participatory development communication for natural resources management in Ratchaburi province, Thailand. *Procedia – Social and Behavioral Sciences, 103*, 703–709. doi:10.1016/j.sbspro.2013.10.390

Meyers, S. (2008). Using Transformative pedagogy when teaching online. *College Teaching, 56*(4), 219–224.

Phillips, L., Carvalho, A. & Doyle, J. (2012). *Citizen voices: Performing public participation in science and environment communication*. Bristol, UK: Intellect Bristol.

Phillips, L. J. (2011). *The promise of dialogue: The dialogic turn in the production and communication of knowledge* (Vol. 12). Amsterdam, the Netherlands: John Benjamins Publishing Company.

Ratcheva, V. & Vyakarnam, S. (2001). The challenges of virtual partnerships: Critical success factors in the formation of inter-organisational teams. *AI & Society, 15*(1), 99–116.

Rodríguez, C. (2016). Human agency and media praxis: Re-centring alternative and community media research. *Journal of Alternative and Community Media, 1*, 36–38.

Starke-Meyerring, D. & Wilson, M. (2008). Learning environments for a globally networked world: Emerging visions. In D. Starke-Meyerring & M. Wilson (Eds.), *Designing globally networked learning environments: Visionary partnerships, policies, and pedagogies* (pp. 1–17). Rotterdam, the Netherlands: Sense Publishers.

The Movement for Community-Led Development. (2017). *Saemaul Undong – The Republic of Korea's New Village Movement, Part 1* [Online]. Washington, DC: The Movement for Community-Led Development. Retrieved from https://communityleddev.org/2016/03/29/saemaul-undong-the-republic-of-koreas-new-village-movement-part-1/

White, S. A., Nair, K. S. & Ascroft, J. (1994). *Participatory communication: Working for change and development*. New Delhi, India: Sage.

7

GLOBAL CULTURE

Tanner Mirrlees

Introduction: Globalization and Global Culture

We live in a world in motion. Each day, people make millions of large and small purchases across borders, cargo ships and aircrafts traffic commodities across oceans, and earth-orbiting satellites beam TV shows down to viewers wherever they live, in many different countries, simultaneously. Even though the planet is being networked together by capitalism's cross-border trade, transportation and communication, the earth's land, water and sky are still divided into 195 states, and each of these political entities attempts to exert sovereignty over its territory, and the culture supposedly contained by its borders. As states pursue and clash over economic and ideological interests, governments use nationalist rhetorics and sometimes even walls to secure their borders against supposed "threats." Meanwhile, cultural and linguistic differences divide certain people while uniting others. In this antagonistic world, is it possible to say that a "global culture" even exists?

Given the vast number of national states and cultures around the world, conceptualizing and concretely studying a "global culture" is a challenging prospect. What would even constitute a global culture – as distinct from a national culture? Could we define "global culture" as an individual's cosmopolitan feeling or affective identification with a world community, not just one country? Or would it be more fitting to define global culture as the particular social structures, characteristics and practices of powerful Empires that have been spread across borders? Or might we say that global culture is simply the popular entertainment products that people all around the world consume?

For almost three decades, social theorists have posed such questions about the definition of global culture, but clear answers remain hard to come by.

The concept of global culture itself became the interest of British and American scholars at the end of the US–Soviet Cold War, when a bevy of journal issues and books about the history, novelty, agents, processes, characteristics and effects of "globalization" and "global culture" appeared. A 1990 double issue of the journal *Theory, Culture and Society* put globalization and global culture on the map as a salient social science and humanities research issue and, in that same year, Featherstone's (1990) landmark book, *Global Culture: Nationalism, Globalization and Modernity* was published. Throughout the 1990s, key works on "globalization" and "global culture" were written by British and American-based social theorists, and circulated trans-nationally to academics around the world by prestigious Anglo-American publishers such as the University of Minnesota Press, Duke University Press, and University of Chicago Press. Appadurai's (1996) *Modernity at Large: Cultural Dimension of Globalization*, Wilson and Dissankayake's (1996) *Global/Local: Cultural Production and the Trans-National Imaginary*, King's (1997) *Culture, Globalization, and the World-System: Contemporary Conditions for the Representation of Identity*, Jameson and Miyoshi's (1998) *The Cultures of Globalization*, and Tomlinson's (1999) *Globalization and Culture* marked "globalization" and "global culture" as new and important areas of research.

By the turn of the millennium, the topics of "globalization" and "global culture" had become bywords in academic publishing, conferences and curricula. Outside of academe, news journalists, business writers and cultural policy-makers frequently took up these terms as well. While everyone from universities to news networks to neoliberal think-tanks to government bureaucracies produced an immense discourse about "globalization" and "global culture," consensus about what these terms actually referred to remained elusive, and they continue to be contested concepts that spark debate over their meanings. As Hafez (2007) notes, "attempts to systematize the field of globalization scholarship have shown a lack of empirical clarity and of a workable theoretical concept" (5). Jameson (1998) says global culture is "the modern or postmodern version of the proverbial elephant, described by its blind observers in so many diverse ways" (xi).

What, then, has "globalization" meant to theorists? Throughout the 1990s, academics regularly invoked globalization as a "periodizing" term and a synonym for new economic, political and cultural trends that emerged and began reshaping the modern world in 1991 as the Cold War drew to a close. The end of the Cold War marked Communism's demise as an ideology at the helm of nation-states and capitalism seemed triumphant; furthermore, the spread of neoliberal democracy seemed unstoppable and, for some, heralded the "end of history"; borders were criss-crossed by flows of media, technology, money, people and ideas; Hollywood and media corporations pushed a postmodern consumer culture while a recently corporatized internet and world wide web traversed states and loosened the fabric of national communities. Additionally, academics noted that globalization also described a set of processes that were *integrating* the world, and leading to greater *inter-dependency* and *inter-connectedness* between all peoples, cultures and countries.

Like globalization, "global culture" was a loaded term. From the early 1990s on, it has connoted: lifestyle identifications "de-territorialized" from the geographical and temporal constraints of place, especially places within nation-states (Featherstone, 1990, 2002); a new "culture which transcends national borders and exists in many different places around the world" but is "just as meaningful as the idea of national and local culture" (Robertson, 1992, 114); a homogenizing and assimilating corporate consumer culture (Sklair, 2001) marked by diversity (Hannerz, 1996) and a hybrid mélange of cultures (Pieterse, 2003); globalizing media corporations whose product imagery and messages glorify global capitalism, and promote the for-profit production, circulation and marketing of commodities everywhere (Herman and McChesney, 1997); "a diversity of cultural artefacts, products and services" rapidly "traversing national boundaries and becoming part of a new world [postmodern] culture" (Cvetkovich and Kellner, 1997, 7); and, "complex connections between societies, cultures and individuals worldwide" (Tomlinson, 1999, 170).

The many meanings of the terms "globalization" and "global culture" seemed to render them immune to a single definition: "Everyone agrees that we live in a more 'globalized' world, but views differ as to what this means and whether it is a trend for good or ill" (Christopherson, Garretsen and Martin, 2008, 343). Yet, it is crucial to recognize that these terms are not value-neutral, but fought over by a number of different actors and interest groups. The unionist who promotes the labor rights of Chinese workers gluing smartphones together for 12 hours a day at Foxconn, for instance, may hold a very different view of globalization and global culture than the Apple CEO who sub-contracts low waged tasks to those same workers. Critics on the Left warn that globalization is an ideological ruse for the rule of transnational corporations over sovereign nation-states, the erosion of democracy, and the deepening of class divisions (Klein, 2000, 2007). Proponents on the Right champion globalization as "flattening the world" by spreading free markets, liberal democracy, and more opportunities, to all (Friedman, 2000, 2005). Given the many meanings linked to globalization and global culture, perhaps the only thing that academics can agree on is that these terms are multifaceted.

Nevertheless, it may be possible to get a little closer to the meanings of these terms by interrogating their literal definitions. "Global" is a complicated word, but it tends to be an adjective defined broadly as "relating to, or involving the entire world" (e.g., "worldwide") or "relating to or applying to a whole" (e.g., "universal"). "Culture" is also a complex word, but according to Williams (1976), it frequently refers to: 1) "a particular way of life, whether of a people, a period or a group or humanity in general" (90); and, 2) "the works and practices of intellectual and especially artistic activity" (90). A portmanteau of these two concepts, "global culture" might be conceived as a whole way of life of the world's people, and also, cultural works that are produced and commonly consumed by people who live within and across many countries, not just one country. Taking this broad definition of global culture as a useful heuristic, this chapter aims to contextualize, summarize and critically assess *three* narrower meanings of "global culture." These articulations of global culture include: 1) mediated sociality as a whole way of life (e.g., the "global village"); 2) an Empire's universalization or trans-nationalization of a particular way of life (e.g., "cultural imperialism"); and, 3) cultural works that are financed, produced, circulated and consumed by people across the borders of nation-states (e.g., "global popular culture").

Mediated Sociality in the Global Village

One articulation of global culture pertains to the worldwide transformation of sociality through information and communication technologies (ICTs). For much of modern history, one's social interaction with others was tied to a territorial place, and socialization mostly referred to one's face-to-face inter-action with family members, friends, bosses, co-workers and acquaintances in place-based locales, such as households, workplaces, pubs, town squares and marketplaces. Although place still matters to social life, in the twenty-first cen-tury, it no longer constitutes the only space in which social interaction takes place. ICTs have enabled forms of mediated sociality between and among people located in many different places, and they facilitate new forms of social interaction that do not rely upon physical proximity. The networking of much of the planet with personal computers, smartphones, the internet and the world wide web has enabled millions of people who are separated by territorial distance to virtually connect. In effect, sociality has become mediatized and "de-territorialized." Family members in diaspora – some in New York City, others in Shanghai – interact each day through Facebook. Exchange students from India studying in Toronto use Skype to converse with family members in Mumbai. Managers for the local subsidiaries of global corporations interact with virtual teams using Google Hangouts on Air with YouTube Live. As ICTs connect together two or more people located different places together, *simultaneously*, this intertwining of absence and presence is called "distancia-tion" (Giddens, 1991). People are physically absent, yet audio-visually present, far away, yet virtually near.

ICT-afforded mediated sociality, however, is not entirely new. After all, each and every major ICT development – from the telegraph to the telephone to the radio to satellite TV to the personal computer to the internet to social media websites – seems to have shrunk the distance between two points on the world map. From the late nineteenth century to the early twenty-first, successive developments in ICTs have enabled people separated by territorial distance to connect and communicate with increasing rapidity, and finally, to experience "real-time" mediated sociality (or, in Apple's lingo, "Facetime"). As O'Neill (1993) says, "As the speed of communication rises" due to ICT developments, "social distance shrinks and ever-larger numbers of people, widely separated by place, are drawn together into common experiences" (24). As ICTs criss-cross borders, they blur perspectival and audio distinctions between here and there, near and far, national and international. The resulting feeling may be that the world is becoming one, that everyone is always already inter-connected with everyone else, and that the world itself is becoming smaller and smaller. As Harvey (1989) notes,

> as space appears to shrink to a 'global village' of telecommunications [...] and as time horizons shorten to the point where the present is all there is, we have to learn to cope with an overwhelming sense of compression of our spatial and temporal worlds.
>
> *(p. 240)*

This "global village" refers to a global culture of people linked and connected by ICTs.

Of course, the "global village" concept is indebted to the Canadian media theorist, Marshall McLuhan (1964), who in *Understanding Media*, declared that the world's many national "tribes" were converging into a "global village" because new ICTs were overcoming the longstanding spatial and temporal barriers to human connection, forming bridges between geographically separated individuals and groups, and instantaneously linking humans from many distinct locales together in a shared audio-visual experience. Enthused by the spread of radio broadcasting and satellite TV networks and programs, McLuhan (1964) quipped: "Today, after more than a century of electric technology, we have extended our central nervous system itself in a global embrace, abolishing both space and time as far as our planet is concerned" (19). For McLuhan, the technologies of radio and TV brought people living in many nation-states into faster contact with each other, and the transition from an individualist print-based culture to one submerged in electronic media heralded a new age of "electronic interdependence," a "global village" (McLuhan and Powers, 1989). Although McLuhan coined the "global village," ICTs have long been surrounded by optimistic projections that these time and space compressive tools and mediums would unite physically separated individuals and groups into a peaceful and positive world community.

In the late nineteenth century, the British Empire's ambassador Edward Thornton praised the telegraph as "the nerve of international life, transmitting knowledge of events, removing causes of misunderstanding, and promoting peace and harmony all over the world" (cited in Mosco, 2004, 90). In the

1920s, the telephone was lauded for sending voices across vast distances and helping people come together as an audience for the planet's best orchestras, orators and educators (Mosco, 2004, 126). The inventor Guglielmo Marconi declared wireless radio to be "the only force to which we can look with any degree of hope for the ultimate establishment of world peace" (cited in Mosco, 2004, 129). Throughout the 1960s and 1970s, the American prophets of post-industrialism linked McLuhan's notion of the global village with the exigencies of the US's globalizing post-Fordist economy, and projected that the "technological convergence of television, satellites, and computers into the Net would – at one and the same time – create a single social system for the whole of humanity" (Barbrook, 2007, 75). In the 1990s, *Wired* magazine investor and MIT media lab founder Nicholas Negroponte (1995) cast the internet as a "totally new, global social fabric" (183) that was breaking down national borders and uniting the world in harmony. In the early twenty-first century, the *New York Times* columnist and neoliberal globalizer Thomas Friedman (2005) argued that the internet, the personal computer and workflow software "flattened" the world, and described the effect of these technologies and free-market economics as global amity. Google executives Eric Schmidt and Jared Cohen (2013) likewise celebrated digital technologies for tearing down "age-old obstacles to human interaction, like geography, language and limited information" and lifting up "a new wave of human creativity and potential" (4).

As social media companies such as Facebook and Twitter spread social networking services and data-veillance business models around the world, global business journalists dreamed of a "global village" constituted by the instantaneous feedback loops between smartphones, platforms, users and content. Writing for the India-based *MoneyControl*, Krishna (2018), deemed Facebook and Twitter to be "revolutionary, bringing unprecedented change and truly creating the global village." In a story for the Nigeria-based *Vanguard*, Anthony (2015) opined that social media has transformed "the way individuals interact" and resultantly "made the world a global village." In *Forbes*, Fenech (2018) depicted "social networks" as helping us to become "a truly global village." In a report for Deloitte, Khoury and Maayeh (2015) contend that social media companies have "transformed [the world] into a global village with people just one click away" and this is good because they are "empowering youth, women, governments and entrepreneurs" (1).

The notion that the development, diffusion and use of new ICTs will eventuate a peaceful and emancipatory global village is a popular trope, and quite a technologically optimistic and deterministic one. Mediated sociality is in fact a growing way of life around the world, and many people located in many parts of the world are virtually coming together through screens, devices, platforms and applications. Yet, the trope that everyone is now living in a "global village" owes more to technophiles than to empirical reality. The trope is optimistic because it invites us to imagine a world of inclusion, peaceful one-ness and collective uplift without considering how the Westphalian system continues to exist, and with it, old conflicts between nation-states, social classes and cultural-linguistic identities. It is deterministic because it makes technology, not power relations between people, appear to be the primary "cause" or "agent" behind global social change. For a global village to exist, the whole of the world population would have to be included in it, but this is not the case. While old and new media connect people from many locales in shared spaces of mediated sociality, interlink communities and cultures across vast temporal and geographical divides, and rapidly deliver digital images and messages about the world to people living all around the world, much of the world's population is still not "in" the global village.

Consider the following.

To be part of the global village, one must first have access to the internet, but at present, only a little over half of the world's total population is online. The US's share of the world's total internet users is 8.8%, but its internet penetration rate is 88.5%. That means 60% of the world population (billions of people) and 12.5% of the American population (millions of people) cannot access the global village, even if it wanted to. The continent of Africa's share of the world's total internet users is 10.9%, but its internet penetration rate is 35.2% (Internet World Stats, 2018). On a continent of 1.2 billion people, more than half a billion people are thus barred from immersion in the global village. In some poor countries, only a minority of people – usually the most wealthy – have internet access: in the Democratic Republic of the Congo, 3.9% of the population uses the internet; in Somalia: 1.7%; and, in Eritrea: 1.1% (Internet Live Stats, 2018).

To access the global village, one must also possess a personal computer or smartphone of some kind. Yet, in 2017, a little under half of the world's total households owned a personal computer and a little over 32% of the global population owned a smartphone (Statista, 2018). Additionally, while access to the internet and a digital device may enable people to access the global village, to really experience it, one must already possess digital literacy skills. While digital literacy – the knowledge required to use digital media technologies to find, analyze, create, impart and receive information – is considered an essential skill, our world is one in which 750 million adults do not possess basic literacy skills (Montoya, 2017). Billions more have never operated a personal computer, sent a text message on a smartphone, produced a blog, or performed a Google search. Access to the internet and a smartphone are not very useful to the billions of people that have not yet learned how to use these digital tools, nor does it predict a person's identification with the world. In many countries, the daily grind of meeting subsistence needs take precedence over the pursuit of cosmopolitan knowledge about other peoples, places and cultures.

The reality of the digital divide – lack of access to the internet, a digital device and digital literacy – within and between countries deflates the ballooning hope that ICTs have brought about a "global village." Due to these divides, a genuinely "global" – that is, universally accessible – village, community and culture, do not seem forthcoming. Presently, the global village excludes the "4 billion people do not have any Internet access, nearly 2 billion do not use a mobile phone, and almost half a billion live outside areas with a mobile signal" (World Bank, 2016, 4). But even if the gap between the digital haves and have nots was effectively parsed and the barriers to accessing the global village were lowered, it is unlikely that access would lead to greater understanding between people, peace between warring states, and the emancipation of billions from abject poverty.

US Empire and Cultural Imperialism

A second articulation of global culture pertains to an Empire's universalization of a particular way of life, or, a way of life way that is trans-nationalized to become a global way of life at the expense of others.

Throughout the 1960s and 1970s, researchers from the global South developed theories of the world system, imperialism and dependency to challenge the idea that poor countries would automatically "develop" by embracing US and Western capitalist social and communication models (Frank, 1979; Wallerstein, 1974). These researchers scrutinized capitalism's uneven geographical development, the rise, fall and rivalries between different Empires, the international division of labor, and racial, sexual and class oppression and inequalities between and within countries. According to Wallerstein (1974), the world system consists of three zones: the core, the semi-periphery and the periphery. These zones are not static, but shifting. Nevertheless, in each major historical period, a core or imperial country emerges and tends to benefit economically and politically from an unequal and exploitative exchange relationship with semi-peripheral and peripheral countries. As core states try to organize the world system to serve their interests, this comes at the expense of others. The "development" of the world system's centers of power relies upon the "underdevelopment" of the still colonized and newly post-colonial peripheries.

Building upon these critical frameworks, early cultural imperialism researchers showed how in the post-World War II period, US and Western-led capitalist modernization and commercially-oriented communication and media development projects spread from the imperial centers of the world system outward, interlinking and integrating the post-colonial peripheries. Far from supporting the rapid economic and social "development" of post-colonial countries, these links exacerbated their "underdevelopment." Instead of being instruments of national independence for the post-colonial countries, the new communication and media arrangements, originating and expanding outward from the world system's power centers, were maintaining old and new relations of dependency. To escape the "development of underdevelopment," cultural imperialism scholars encouraged post-colonial states to de-link from the world system's power centers and pursue self-reliant development. With little success, some attempted to cut their ties with formerly colonialist and new neo-imperial states. But by the early 1990s, many peripheries had been integrated with the core, and many post-colonial countries were on the receiving end of US and Western media flows.

While earlier theories of world systems, imperialism and dependency explained global change with regard to the rise, maintenance and fall of Empires in a hierarchically organized and unevenly developed world system, leading globalization theorists often represented globalization as an all-together new agent driving processes that were changing the world as opposed to the outcome or effect of pre-existing processes (Appadurai, 1996; Tomlinson, 1991). Yet, in the 1990s, the agents most powerfully positioned to purposefully drive globalization were US corporate and state actors, and these had long been at the helm of the world's most powerful Empire. In this context, some scrutinized globalization as a concept with immense strategic utility to the Clinton Administration's pursuit of a neoliberal world order, and as a mask for the continuity of the US Empire (Bacevich, 2002). Observing an eerie convergence between the concept of globalization and the real exigencies of the US Empire, Bourdieu and Wacquant (1999) linked "the strongly poly-semic notion of globalization" to American cultural imperialism, which "rests on the power to universalize particularisms linked to a singular historical tradition by causing them to be misrecognized as such" (41). For these scholars, "globalization" was a particularistic concept produced and promoted by Anglo-American state agencies, think-tanks, universities and corporations, and then exported to the whole planet, and it "ha[d] the effect, if not the function, of submerging the effects of imperialism in cultural ecumenism or economic fatalism and making trans-national relationships of [economic, geopolitical and cultural] power appear as a neutral necessity" (42).

While it is reductive to suggest that all conceptualizations of globalization are little more than covert rationalizations for US foreign policy and tools of American cultural imperialism, the problematization of the concept encourages materialist research and analysis of how the world's real geopolitical economy shapes "global culture." In this regard, the cultural and media imperialism paradigm is invaluable, as it focuses on the role that Empires and their ICT and cultural industries have played and continue to play in universalizing national particularisms, emphasizes the longstanding and continuing power of specific countries to disproportionately shape "global culture," and explains ostensibly neutral expressions of global culture with regard to imperialist practices.

For hundreds of years, the world system has been structured to serve the geopolitical, economic, and cultural interests of powerful imperial countries at the expense of less powerful ones. As far back as 1901, people contemplated how American culture might one day become the globe's culture. In 1901, British journalist W. T. Stead published *The Americanization of the World*, and in it he described how US industrial production models, consumerism and liberal bourgeois values were going global and remaking the world (Rydell and Kroes, 2005, 9). During World War II, Henry Luce (1941) spoke of the "American Century" and described how "American jazz, Hollywood movies, American slang, American machines and patented products are in fact the only things that every community in the world, from Zanzibar to Hamburg, recognizes in common" (65). Having fled Nazi Germany to the US, the critical theorists Max Horkheimer and Theodore Adorno (1995) painted a less sanguine image of Americanizing global culture. They coined the term "culture industry" to highlight US monopoly capitalism's incorporation of culture into its circuits of accumulation and interrogate the for-profit production, distribution and marketing of all the world's cultural forms – high and low – as commodities. For them, cultural commodities seemed to help people cope with the alienating work routines they endured each day, concealed the class relations that divided society with ideals of consumerism, and closed minds to the badness of the present instead of opening them to alternative futures. Building upon these ideas, Herbert I. Schiller (1969, 1976) offered the most significant theorization of the centrality of the cultural industries to the US Empire's post-World War II expansion, or "cultural imperialism."

For over a century, the US Empire has indeed relied upon and utilized developments in the ICT and cultural industries to build, project and maintain its power, and much research supports this conclusion (Boyd-Barrett, 2015; Boyd-Barrett and Mirrlees, 2019; Jin, 2015; Mirrlees, 2016). Theorizations of cultural imperialism point to asymmetrical and unequal power relations between the US and other countries, and between the US-based globalizing ICT and cultural industries and those elsewhere. Generally, they make the following claims: the US is an Empire, and home to the world's largest, and most significant ICT and cultural industries; the US is the center from which most ICT and cultural products and services flow to the world, and there is little reciprocation or counter-flow; the US government supports the global business activities of US-based ICT and cultural industries; US ICT and cultural products carry stories, messages and images that extend American culture into other countries around the world, making American culture the predominant "global culture"; and, these processes buttress the US's structural power in the world, often to the detriment of non-US economies, ICT and cultural industries and cultures.

Cultural imperialism theory continues to be important because the US is still an Empire, and US-based ICT and cultural industries continue to be dominant around the world. The Forbes (2018) list of the world's largest companies identifies a total of 172 corporations across the many "communication" and "media" sectors that constitute the largest global ICT and cultural industries: broadcasting, telecommunications, advertising, computer electronics retailers, computer hardware, computer software, computer services, diversified media, internet and catalog retail, printing and publishing, and semiconductors. Of this total, the United States is home to 76 of these corporations (44%), the BRICS (Brazil, Russia, India, China and South Africa) together are home to 25 (14.5%), and China singularly is home to 16 (9.3%). Clearly, the BRICS combined and China alone do not come close to matching the United States' privileged position as the world system's headquarters to the globe's biggest "communication" and "media" corporations. In fact, the United States has three times the world's largest of these corporations as the BRICS, and over four times the number of the world's largest corporations in these industries as China. And Hollywood rules the world entertainment market, annually producing most of the highest globally grossing films (Mojo, 2018).

Furthermore, US-based social media companies are the world's most powerful and global. To bring the cultural imperialism paradigm up to speed with the platform era, Jin (2015) recently coined "platform imperialism" to conceptualize the economic, technological cultural dominance of US-based platform corporations around the world. Currently, the US is home to three of the world's top five most visited sites: Google.com (#1), YouTube.com (#2), and Facebook.com (#4) (Alexa 2020). As Jin (2015) observes, the US "seems to dominate the world with platforms, benefitting from these platforms in terms of both capital accumulation and spreading symbolic ideologies and cultures" (7). Jin shows how US-based platform corporations rule the lion's share of the global digital economy's hardware and software, intellectual property rights – copyrights, patents and trademarks – and user data and highlights how this corporate platform dominance is shored up by the US state's foreign policy of digital free trade and democracy promotion. This platform imperialism is also cultural, as it exerts influence upon the global norms and values of emerging digital cultures by pushing "the spiritual hegemony of American-based entrepreneurship" (Jin, 2015, 185). In sum, the US's ICT and cultural industries are globally dominant, as they are the most significant owners of the world system's technological infrastructure, the means to service and access this infrastructure, and the lion's share of the means of producing, distributing, platforming and exhibiting the commercialized informational and cultural goods pulsing through it each day.

While US-based ICT and cultural industries have economic power around the world, that power does not always or easily translate into the power to turn American culture into global culture, or destroy other national cultures. US media corporations produce and sell commercial cultural goods that represent and may influence national culture, but these products are not in themselves identical to culture, or a direct threat to it. Moreover, nation-states have not lost sovereignty to globalizing media corporations, and the political and economic elites of each state are significant gatekeepers of cross-border media and cultural flows, and they use national cultural policy to protect national cultural industries from a takeover by US firms. At the same, time, they promote the internationalization of the business interests of these industries and their products to the US, and to other countries.

Moreover, people on the receiving end of US-owned media and cultural goods may select, adapt, indigenize, mix and redeploy the texts of these works in complex and contradictory ways (Ang, 1985; Appadurai, 1996; Liebes and Katz, 1990). Around the world, most people's media diets are comprised of a mix of globalizing American media fare and national or culturally proximate cultural goods that express their own identity and language (Straubhaar, 1991). Additionally, the notion that American national culture can dominate or eliminate another national culture rests upon the simplistic notion that cultures begin and end at a state's territorial borders and neglects to consider how all cultures are hybrids, dynamic mixtures of many elements (Kraidy, 2005; Pieterse, 2003). Importantly, American culture is not an ethno-monolith reducible to "blood and soil," but rather, the terrain of political struggle between US citizens over the meaning of their nation. If US cultural imperialism really was effective at turning the particular characteristics of American culture into a new global culture, we might see signs of this in a rapid decline in nation-states and a deterioration of national cultures. Yet, nationalist parties and ethno-nationalism is surging across Europe, and around the world (Rachman, 2018). Cultural imperialism exists in state and media-corporate practices, but these may be less and less effective.

Global Popular Culture

A third articulation of global culture refers to media and cultural works that are common to people all over the world, or, globally popular cultural forms that are produced, circulated and consumed by a global – as opposed to distinctly national – audience. "Globally popular culture" refers to entertainment media that travels the world and cultural works that are consumed by many people in many countries (Mirrlees, 2013). In the early twenty-first century, globalizing media corporations are designing globally popular cultural forms such as blockbuster films, global–local TV formats and brands, and global media events.

Hollywood's power over transnational production, distribution and exhibition chains supports the global reach of its films (Miller et al., 2005). People all over the world frequently choose to watch Hollywood films and often enjoy them, sometimes even more than those cultural products released by "national" film industries based in their respective countries. This is due in part to Hollywood's success at designing globally popular blockbuster films as opposed to distinctly "American" films. Heeding the "cultural discount" – the notion that a film that exclusively represents one national culture and which is originally intended for one national audience will have diminished appeal elsewhere – Hollywood creates blockbuster films intended to address a multinational as opposed to national audience. For example, in the first decade of the twenty-first century, the highest-grossing Hollywood films worldwide were not explicitly about American people, places and cultures: *Star Wars: The Force Awakens* (2015), *Avatar* (2009), *Jurassic World* (2015), *The Avengers* (2012), *The Dark Knight* (2008), *Rogue One: A Star Wars Story* (2016), *Beauty and the Beast* (2017), *Finding Dory* (2016) and, *Avengers: Age of Ultron* (2015). When creating global blockbuster films like these and distributing them to cinemas worldwide, Hollywood appeals to and integrates a world audience, not just a national one.

While global Hollywood designs globally popular blockbusters to avoid the cultural discount, globalizing TV corporations are sensitive to the tastes and preferences of culturally proximate audiences (Straubhaar, 1991). As such, they abide by the business strategy of glocalization – or "think globally, act locally" – when designing global TV formats and brands that are intended to criss-cross many borders. TV formats are "program concepts that can be re-packaged to suit particular [national] markets and tastes" (Freedman, 2003, 33), and worldwide audiences are watching nationally adapted versions of global TV formats, and *Top Model* is a salient example. Created by Tyra Banks, and produced by 10 x 10 Entertainment and Bankable Productions, *America's Next Top Model* was initially made for broadcast in the US, but soon after, *Top Model* was licensed for globally flexible adaptation by national TV networks in at least 46 countries, from Cambodia to Russia to Mexico. The globally popular TV format is coupled with global–local brands, such as MTV. Viacom Media Networks currently owns a large number of regional and nation-specific MTV channels in North America, and across Africa, Asia Pacific, Europe, the Middle East and Latin America. While people continue to be divided by borders and cultures, the form, script and aesthetics of global–local TV formats and brands are a referent for millions of people worldwide.

Another form of globally popular culture is the "global media event." In the early 1980s, Meyrowitz (1986) argued that satellite TV turned viewers into "audiences to performances that happen in other places" (7) and transformed presidential speeches, military invasions and sporting spectacles into "dramas that can be played on the stage of almost anyone's living room" (118). A familiar global popular cultural form, a "global media event" is a happening that becomes the site of routinized coverage by media corporations, and repeat viewing by audiences, all around the world. Global media events convey topics of collective importance or intrigue such as war and peace, miracles and disasters, and sport, celebrity and royalty (Couldry, Hepp and Krotz, 2009). Although the occurrence may happen in one place, it is viewed around the world, on TV sets, computers and smartphones. A few significant twenty-first-century global media events were the September 11, 2001, terrorist attack on the US; the 2003 US "shock and awe" bombing of Baghdad; the 2006 FIFA World Cup Final; the

Beijing 2008 Summer Olympics opening ceremony; and the 2011 British Royal Wedding. Global media events like these are watched by millions, sometimes billions of people, within and across national borders. They constitute a globally common referent for people from many different countries, and they unite people in a shared ritual of consuming and experiencing the same event, often at a great distance from where it actually happens.

As the US ICT and cultural industries go global, they create and sell global blockbusters, global–local TV formats and brands, and global media events to audiences across borders. The old cultural commodity forms of cultural imperialism – US nationalist films and TV shows that promote one-dimensional or homogeneous images of "the American Way of Life" to the US and to the wider world – may be giving way to or co-exist with emerging globally popular cultural commodity forms.

Conclusion

This chapter has contextualized, described and assessed three articulations of a "global culture." While developments in ICTs are accompanied by great hope (sometimes verging on hype) for a peaceful and empowering "global village," and the experience of mediated sociality is more common for more people worldwide, ICTs do not automatically produce a "global culture" because nation-states and digital divides exist. The US is an Empire, and cultural imperialism theory is correct in emphasizing how US-based globalizing ICT and cultural industries and their cultural products continue to be dominant around the world. Yet, US cultural imperialism has not universalized the particulars of the American Way of Life: US state and media-corporate actors have not *effectively* Americanized national cultures, nor have they created a global culture that liquidates the world's many non-US national cultures (however constructed). Some of the world's most powerful media and entertainment corporations – many of which are headquartered in the US – are trying to design globally popular culture and watered-down versions of national or local culture.

While growing worldwide access to and certain uses of ICTs and popular culture may support a "structure of feeling" that "we are one," unfortunately, place-based divisions rooted in national chauvinism, classism, racism, sexism and forms of religious zealotry continue to be significant parts of the world we live in too. Globalizing ICTs and popular culture offer the possibility of making a new global culture, but the old geopolitical and economic structure of the world system stymies this potentiality. After all, ICTs and global popular culture may enable post-national, cosmopolitan and global identifications, but they also remain instruments for the spread of nationalist, nativist and localized hate ideology. For the majority of people still living in local, regional and national villages, global culture is a distant horizon as there are not yet any global equivalents to the narratives, myths and affective investments in and attachments to, the national community. But there ought to be given the real interdependence of the human species, and the ever-looming threat posed to our collective existence by global climate change. For the near future, though, most people will likely continue to identify as part of local and national cultures as opposed to global ones.

References

Alexa. (2020). "The Top 500 Sites on the Web." www.alexa.com/topsites.

Ang, I. (1985). *Watching Dallas: Soap Opera and the Melodramatic Imagination*. London: Methuen.

Anthony, F. (2015, September 1). "How Social Media Makes the World a 'Global Village'." *Vanguard*, www.vanguardngr.com/2015/09/how-social-media-makes-the-world-global-village/.

Appadurai, A. (1996). *Modernity at Large: Cultural Dimensions of Globalization*. Minneapolis, MN: University of Minnesota Press.

Bacevich, A. (2002). *American Empire: The Consequences and Realities of U.S. Diplomacy*. Cambridge, MA: Harvard UP.

Barbrook, R. (2007). *Imaginary Futures: From Thinking Machines to the Global Village*. London: Pluto Press.

Bourdieu, P. and Wacquant, L. (1999). "On the Cunning of Imperialist Reason." *Theory, Culture and Society*, 16 (1): 41–58.

Boyd-Barrett, O. (2015). *Media Imperialism*. Thousand Oaks, CA: Sage Publications.

Boyd-Barrett, O. and Mirrlees, T. (Ed.). (2019). *Media Imperialism: Continuity and Change*. New York, NY: Roman & Littlefield.

Christopherson, S., Garretsen, H., and Martin, R. (2008). "The World is Not Flat: Putting Globalization in its Place." *Cambridge Journal of Regions, Economy and Society*, 1 (1): 343–349.

Couldry, N., Hepp, A., and Krotz, F. (Eds). (2009). *Media Events in a Global Age*. New York, NY: Routledge.

Cvetkovich, A. and Kellner, D. (1997). *Articulating the Global and the Local: Globalization and Cultural Studies*. Boulder, CO: Westview Press.

Featherstone, M. (Ed.). (1990). *Global Culture: Nationalism, Globalization, and Modernity*. New York, NY: Sage.

Featherstone, M. (2002). "Post-national Flows, Identity Formation and Cultural Space." In E. Ben-Rafael and Y. Sternberg, (Eds.), *Identity, Culture and Globalization* (pp. 482–526). Boston: Brill.

Fenech, G. (2018, November 18). "Leap in the Blockchain Development: What Will be in Demand on the Decentralized Market in 2019?" Forbes, www.forbes.com/sites/ger aldfenech/2018/11/18/leap-in-the-blockchain-development-what-will-be-in-demand-on-the-decentralized-market-in-2019/#7c3271b3045d

Freedman, D. 2003. "Who wants to be a millionaire? The politics of Television Exports. Information." *Communication and Society*, 6 (1): 24–41.

Forbes. (2018, December 15). "Global 2000: The World's Largest Public Companies." *Forbes*. www.forbes.com/global2000/list/.

Frank, A. G. (1979). *Dependent Accumulation*. New York, NY: Monthly Review Press.

Friedman, T. (2000). *The Lexus and the Olive Tree: Understanding Globalization.* New York, NY: Random House.

Friedman, T. (2005). *The World is Flat: A Brief History of the Twenty-First Century.* New York, NY: Farrar, Straus, and Giroux.

Giddens, A. (1991). *The Consequences of Modernity.* Cambridge, MA: Polity Press.

Hafez, K. (2007). *The Myth of Media Globalization.* Cambridge, MA: Polity Press.

Hannerz, U. (1996). *Transnational Connections: Culture, People, Places.* New York, NY: Routledge.

Harvey, D. (1989). *The Condition of Postmodernity: An Enquiry into the Origins of Cultural Change.* Cambridge, MA: Blackwell.

Herman, E. and McChesney, R. (1997). *The Global Media: The New Missionaries of Corporate Capitalism.* London: Continuum Press.

Horkheimer, M. and Adorno, T. (1995). *The Dialectic of the Enlightenment.* New York, NY: Continuum.

Internet World Stats. (2018). "Internet World Stats." www.internetworldstats.com/stats.htm

Jameson, F. and Miyoshi, M. (Eds.). (1998). *The Cultures of Globalization.* Durham, NC: Duke University Press.

Jameson, F. (1998). "Preface." In F. Jameson and M. Miyoshi (Eds.), *The Cultures of Globalization* (pp. xi–xvi). Durham, NC: Duke University Press.

Jin, D. Y. (2015). *Digital Platforms, Imperialism and Political Culture.* New York, NY: Routledge.

Khoury, H. and Maayeh, E. (2015). "The Time is Now." *Deloitte,* www2.deloitte.com/content/dam/Deloitte/xe/Documents/About-Deloitte/mepovdocuments/mepov17/the-time-is-now-mepov17.pdf

King, A. (Ed.). (1997). *Culture, Globalization, and the World-System: Contemporary Conditions for the Representation of Identity.* Minneapolis, MN: University of Minnesota Press.

Klein, N. (2000). *No Logo: Taking Aim at the Brand Name Bullies.* Toronto: Vintage.

Klein, N. (2007). *The Shock Doctrine: The Rise of Disaster Capitalism.* Toronto: Alfred A. Knopf Canada.

Kraidy, M. (2005). *Hybridity or the Cultural Logic of Globalization.* Philadelphia, PA: Temple University Press.

Krishna, M. (2018, November 26). "As the Jack Dorsey Incident Shows, it's No Longer About the Technology." *MoneyControl,* www.moneycontrol.com/news/technology/opinion-as-the-jack-dorsey-incident-shows-its-no-longer-about-the-technology-3217601.html

Liebes, T. and Katz, E. (1990). *The Export of Meaning: Cross-Cultural Readings of Dallas.* New York, NY: Oxford University Press.

Luce, H. (1941, February 17). "The American Century." *Life* (Chicago, IL), 61–65. www.informationclearinghouse.info/article6139.htm

McLuhan, M. (1964). *Understanding Media: The Extensions of Man.* New York, NY: McGraw Hill.

McLuhan, M. and Powers, B. (1989). *The Global Village: Transformations in World Life and Media in The 21st Century.* New York, NY: Oxford University Press.

Meyrowitz, J. (1986). *No Sense of Place: The Impact of Electronic Media on Social Behavior.* New York, NY: Oxford University Press.

Miller, T., Govil, N., McMurria, J., Maxwell, R., and Wang, T. (2005). *Global Hollywood 2*. London: British Film Institute.

Mirrlees, T. (2013). *Global Entertainment Media: Between Cultural Imperialism and Cultural Globalization*. New York, NY: Routledge.

Mirrlees, T. (2016). *Hearts and Mines: The US Empire's Culture Industry*. Vancouver, BC: University of British Columbia Press.

Mojo, B. O. (2018). "2018 Worldwide Grosses." *Box Office Mojo*, December 15. www.boxofficemojo.com/yearly/chart/?view2=worldwide&yr=2018&sort=wwgross&order=DESC&p=.htm.

Montoya, S. (2017, May 9). "Tracking Literacy in an Increasingly Digital World." *UNESCO*, http://uis.unesco.org/en/blog/tracking-literacy-increasingly-digital-world-0

Mosco, V. (2004). *The Digital Sublime*. Cambridge, MA: MIT Press.

Negroponte, N. (1995). *Being Digital*. New York, NY: Knopf.

O'Neill, M. J. (1993). *The Roar of The Crowd: How Television and People Power are Changing the World*. New York, NY: Times Books Randomhouse.

Pieterse, J. N. (2003). *Globalization and Culture: Global Mélange*. New York, NY: Rowman & Littlefield.

Rachman, G. (2018, June 25). "Donald Trump Leads a Global Revival of Nationalism." *Financial Times*, www.ft.com/content/59a37a38-7857-11e8-8e67-1e1a0846c475

Robertson, R. (1992). *Globalization: Social Theory and Global Culture*. London: Sage.

Rydell, R. and Kroes, R. (2005). *Buffalo Bill in Bologna. The Americanization of the World, 1869–1922*. Chicago, IL and London: Chicago University Press.

Schiller, H. (1969). *Mass Communications and American Empire*. Boston: Beacon Press.

Schiller, H. (1976). *Communication and Cultural Domination*. Armonk, NY: M.E. Sharpe.

Schmidt, E. and Cohen, J. (2013). *The New Digital Age: Reshaping the Future of People, Nations and Business*. New York, NY: Knopf.

Sklair, L. (2001). *The Transnational Capitalist Class*. Oxford: Blackwell Publishers.

Statista. (2018). "Global Smartphone Penetration per Capita since 2005." www.statista.com/statistics/203734/global-smartphone-penetration-per-capita-since-2005/

Stats, I. L. (2018). "Users by Country." www.internetlivestats.com/internet-users-by-country/

Straubhaar, J. (1991). "Beyond Media Imperialism: Asymmetrical Interdependence and Cultural Proximity." *Critical Studies in Mass Communication* 8 (1): 39–59.

Tomlinson, J. (1991). *Cultural Imperialism*. Baltimore, MD: The Johns Hopkins University Press.

Tomlinson, J. (1999). *Globalization and Culture*. Chicago, IL: University of Chicago Press.

Wallerstein, I. (1974). *The Modern World System 1*. New York, NY: Academic Press.

Williams, R. (1976). *Keywords: A Vocabulary of Culture and Society*. London: Fontana.

Wilson, R. and Dissankayake, W. (Eds.). (1996). *Global/Local: Cultural Production and the Trans-National Imaginary*. Durham, NC and London: Duke University Press.

World Bank. (2016). "World Development Report 2016: Digital Dividends." Washington, DC: *World Bank*. http://documents.worldbank.org/curated/en/896971468194972881/pdf/102725-PUB-Replacement-PUBLIC.pdf

8

CULTURAL HYBRIDITY, OR HYPERREALITY IN K-POP FEMALE IDOLS?

Toward Critical, Explanatory Approaches to Cultural Assemblage in Neoliberal Culture Industry

Gooyong Kim

With the international success of K-pop since the mid-2000s, especially the recent global popularity of Psy, Girls' Generation, BTS and so on, there has been an increasing scholarly endeavor to understand the cultural phenomenon from cultural hybridity perspectives. However, the majority of academic discussion exclusively focuses on descriptive, functional and/or industrial dimensions of cultural assemblages: K-pop's cultural hybridity is believed to be a counter-argument against hegemonic globalization. Rather than Korea's entertainment market being dominated by American popular culture, the K-pop industry successfully practices a counter-flow of cultural production from a periphery country to metropolitan centers in its "indigenized and hybrid versions of *American* popular culture" (Joo, 2011, 496, emphasis added), not only for domestic consumption but also, more importantly, as an export item. In this respect, Choi and Maliangkay (2015) endorse that the decades-long issue of cultural domination by the West has been ameliorated, if not overcome, by "the presentational mode and content" of K-pop and its global popularity (13). This counter-flow of cultural production, or what they call "the role reversal in the global creative industry" (14), has been celebrated by current scholarship since Doobo Shim's (2006) foundational piece. Likewise, the literature is celebratory, merely focusing on microscopic analyses, ignoring larger socio-cultural and politico-economic contexts. However, I argue one has to pay critical attention to concrete conditions and modes that beget cultural hybridity in terms of its economic, historical, industrial, social and political contexts. Otherwise, one loses sight of the fetishism of locality in cultural production.

Contrary to a growing number of female idols and their increasing influence, there is a dearth of scholarly efforts to explicate how their personality, subjectivity and/or sexuality are formulated and promoted in K-pop. As an idiosyncratic feature of cultural hybridity in K-pop female idols, a contradictory personality that retains Korea's traditional feminine decency, passivity and innocence along with American hyper-sexualization has been rampant. To explicate this seemingly inconceivable character of K-pop female idols, I analyze the cultural and social backgrounds that led to the success of schizophrenic female subjectivity. Reconsidering the ethical dimension of cultural hybridity as a subaltern's transformative project, I in turn examine whether or not their increased presence in the industry has opened up new possibilities for women taking up new roles and careers in the nation. Specifically, I critically analyze whether K-pop female idols harbor an active alternative role model for female audiences in terms of female subjectivities and sexualities.

While the meaning of cultural hybridity is always in flux, I examine how it helped K-pop to enjoy popularity and profit in post-1997 Asian Financial Crisis neoliberal Korea, when American hegemony had permeated in every corner of society. Since hybridization is constituted and contested in a complex overdetermination of power, and local sensitivity should be the essence of hybridity with its "mutant result of fusion and intermixture" (Gilroy, 1993, 6), I investigate whether K-pop exercises a cultural command of locality that is reflective of Korean people's everyday lives in its creative expression. Furthermore, as a "meta-construction of social order" for a critical self-reflexivity (Werbner, 2015, 1), cultural hybridity is not just a creative fusion of different cultural artifacts, but more importantly the concrete result of a strategic, political action that emits a complex set of values, norms and meanings. In other words, as a cultural reorganization of power, cultural hybridity is a conscious embodiment of the society's ethos and language that manifests a certain world-view. With this self-conscious, continuous cultural rejuvenation, one's attention should be placed on the "*explanatory* power [of] the concept: studying processes of hybridization by locating these in structural relations of causality" (Canclini, 1995, xxix, emphasis original). In other words, since music is one of the most salient sites for hybridization as either cultural exchange or commodification (Hebdige, 1987; Lazarus, 1999), hybridity in K-pop must be discussed within concrete cultural, economic, political and social backgrounds of production and by its actual content. By understanding how K-pop female idols enable female audiences to place themselves in "imaginative cultural narratives and, as such, they help [...] to construct and provide insights into that wider experience" (Whiteley, 2013, 9), I recontextualize the phenomenal success of the female idols back into their role in contemporary, patriarchal and neoliberal Korea.

While Korea has become a major non-Western country that since the 2000s has commanded exports of diverse cultural products (Jin, 2016), whether or not K-pop contains local sentiments and creative characteristics is an open question that warrants critical examination. As Canclini (2001) acutely argues that "the point of view of the oppressed or excluded can be helpful in the *discovery* stage, as a way of generating hypotheses or counterhypotheses that challenge established knowledges" (12), with K-pop's solid success with its stylistic distinctiveness and stability, the existence of cultural hybridity is not meaningful unless it provides something significant. Differently put, I examine whether or not K-pop generates autonomous cultural self-reflexivity through the critical appropriation of the dominant forms that can effectively carry their expected feelings and grounded knowledges.

As the emulation of an ideal model for socio-economic development, Korea's popular culture has evolved through an active adaptation or reapplication of that of Japan and the US. In this conscious cultural assembly process, it has always and already been hybridized albeit to different degrees of critical self-reflexivity. Undergoing the Korean War, poverty, industrialization and democratization, the Korean people have admired the US as a mythical utopia that becomes part and parcel of their collective imagination and desire (Kroes, 1999). In this respect, American culture is conceived as Koreans' recognition and expectation of a better world. Post-Korean-War popular music in Korea was oriented to American GIs as chief consumers who "avidly embraced formal and informal offerings of rest and relaxation, from sexual services to musical entertainment" (Lie, 2015, 31). In this respect, while local musicians/performers tried their best to live up to "American expectations" (32), Korean popular music, no matter how much it has been hybridized, has retained a fundamental asymmetric reliance on this American reference. For example, Motown's girl groups, as a cultural icon of American affluence and global success, were replicated in Korea in the 1950s and 1960s. While one could argue Korean girl groups could be a local, or hybridized, version of American cultural ingenuity, we should pay closer attention to whether or not Korea's local form addressed its innate needs and tastes while maintaining cultural autonomy or identity against the dominant cultural, economic and political hegemony.

Since US popular culture has commanded global hegemony, emulating American pop values and systems provides a better chance of success with less market risk. Leaders in the K-pop industry can be regarded as an example of the "dominated group's internalization" (Lee, 2010, 30) of transnational capitalism's business mantra. Practicing the hegemony of consumerism, commodification of culture and sexualization of femininity (Schiller, 1996), the K-pop industry complicates the evasive characteristic of cultural hybridization. This is opposite to the growing recognition of peripheral countries' competence to produce and market their indigenous culture globally (Sinclair & Harrison,

2004). K-pop delineates how hegemony employs an ideological double play in local cultural production. Even if it allows "counter-hegemonic" practices on a local level, it establishes the local culture industry as a cultural hegemon, while perpetuating predatory labor conditions and fetishizing local cultural production. By neglecting the historical realities of inequalities in resources and developments, hybridity in K-pop literature exaggerates the locality of cultural production.

In this regard, K-pop as Korea's apex of a neoliberal service industry has subsumed the cultural to the sum of the economic imperative of transnational capitalism, by poaching any part of global popular culture as far as it assumes a marketability. Then, since it does not prioritize autonomous creativity but gobbles up anything flashy and marketable, hybridity in K-pop has to be examined through its specific socio-cultural implications, which is what it projects and promotes. In terms of the neoliberal industry's marketing and survival strategies, I suggest that K-pop's ideal femininity is hybridized between Korea's traditional gender norms and the neoliberal hegemonic cliché of female empowerment through sexualization in order to address the society's changing needs, as Ortner and Whitehead (1981) indicate gender/sexuality is always already imbricated in society's political and economic conditions. Since there is a strong correlation between representations of K-pop female idols and discriminatory gender attitudes in labor and personal relations (Lin & Rudolf, 2017), I explain how K-pop's musical and performative imaginations have transformed and reconfigured women's roles, meanings and existential conditions, and in turn how this cultural reconstruction of women's reality discloses or conceals their asymmetric and exploitative relation in the patriarchal society.

K-pop is an exemplar of a neoliberal service economy that reflects how business demands have shifted from sweatshop manual workforce to service, immaterial labor: K-pop female idols are the most salient example of the elite's strategic appropriation and application of neoliberal culture industry. Retaining Korea's traditional culture of patriarchal sexism, female idols are employed to meet the growing demand of the culture industry as a service sector economy, effectively reinterpreting the traditional gender-based code of ethics, diluting its strict areas of application and disseminating a new set of entitlements to inclusion and exclusion. As Kim and Choi (1998) acutely indicate contemporary Korea is a "palimpsest of multiple layers of Japanese colonialism and neo-imperial domination, especially by U.S. hegemony" (3), I contend that the recent development of K-pop has been as a result of hybridization with the experiences of the country's turbulent industrial capitalist nation-building. Like its predecessors decades ago, the contemporary K-pop industry takes advantage of a hegemonic model that quickly produces profitable, homogenized, disposable commodities from a highly concentrated, hierarchal production system. In this respect, the idols severely lack creative autonomy or authenticity to the extent that they "execute what has been conceived for them; they wear what they are told to wear; they sing what they are told to sing; and they move and behave as they are told to move and behave" (Lie, 2015, 141).

In this respect, on the contrary to a tautological limit of cultural hybridity, I reconsider K-pop's cultural mixing practices through Baudrillard's notion of hyperreality as an alternative heuristic. By taking various bits from different types of desirable femininities, K-pop female idols have been manufactured as a hybridity of images and signals, and in turn simultaneously become an object and a subject of simulation. By doing so, I better explain how K-pop appropriates different cultural artifacts and practices that entail broader socio-political ramifications. In other words, I not only re-assess the industry's strategic appropriation of various cultural elements as a commercial venture but also argue that its broader governmental practices condition audiences' subjectivities by promoting an imagined gender ideal to better guarantee its future profit in the neoliberal culture industry. In other words, with its governmentality power (Kim, 2017, 2019), I examine what kind of feminine ideal the K-pop industry has produced, which is more real than the actual one, that is, hyper-real femininity. In that regard, I shed critical light on the cultural politics behind the idols' split personalities, which can "train the broadest mass of people in order to create a pattern of undreamed-of dimensions" into conformative social behaviors (Kracauer, 1995, 77).

Schizophrenic Personality of Suzy: Between Innocence and Sexualization

As one of the most sought-after K-pop female idols, Suzy made her debut as a member of MissA in 2010, whose appeal is their aggressive sexuality at the age of 15, as the youngest female idols. Regarded as the Queen of Commercial Film for her popularity and influence, Suzy has been active in endorsements for various commodities like cosmetics, clothes, beverages and so on. Contrary to MissA's group identity, Suzy is famous for her wholesome image and an appealing appearance. While most K-pop female idols market a dual character of pure, innocent and cute femininity along with explicit sexuality, Suzy is most successful in this hybridized personality.

MissA deftly mixes different materials from various cultural and social backgrounds in its performative activities. Donned in all black or black/white clothes with natural make-up to defy an expectation of traditional femininity, they assume an aggressive attitude toward male counterparts, dancing in a school setting in their 2010 "Bad Girl, Good Girl" music video. The choice of color in their outfits indicates their desire for power, authority or dominance. While female affects used to be channeled to addressing male emotional demands, they do not show any facial expressions, showing an indifference to the male gaze. Rather, claiming how they look does not represent what they are, the idols declare their own (sexual) subjectivity. Dancing in a provocative manner in a ballet studio where "good girls" normally practice the feminine arabesque in classical music, MissA appropriates the conventionally perceived womanly behaviors and proclaim they are "bad girls" who do not conform to traditional femininity. Making direct eye contact with the camera, the idols look stern and strong while emitting a sense of rebellion. Sexualized, the idols' powerful dance routines seem to convey a visual message of female empowerment.

However, cautious analyses of the idols' outfits and choreography reveal a different layer of meaning behind their audio-visual seduction. They promote a *dubious* image of female liberation from the patriarchy: they promote an unexplained aggression to a fellow male student, which falls back onto a traditional notion of masculinity. The well-synchronized choreography format indicates the dominant, patriarchal rationality of control and manipulation. While there are occasions of individualized dance moves, all the group members eventually come to conform to a pre-determined, collective theme of corporeal arrangement and exhibition. To be more specific, when an individual member is spotlighted in a given shot, she is highly sexualized with more provocative outfits and even more provocative dance moves. While their abrupt, dramatized dance moves could symbolize their desire for liberation, these moves give way to sexually charged actions, such as groping their own bodies, dropping down, rolling on the floor and thrusting and gyrating their bottoms, and in turn perpetuate male-oriented female sexual conduct. Certain body parts rather than the whole body, especially over-emphasis on their bottoms and bosoms that shake, swing and gyrate, are emphasized and reify the idols' physicality. In this respect, their unconventional choreography is not to liberate their sexuality but updates and re-affirms male-oriented dominant sexuality. Either individually or collectively, they re-produce the dominant patriarchal norms through sexualization and group *conformity*.

In the 2013 "Hush" music video, MissA stage much more carnal visual and lyrical messages, unapologetically manifesting they are "bad girls." The song is all about a "secret party," filled with sexual activities. Opening with a flooded room insinuating that the idols are fully mature to the extent that they are "wet," the video visually lavishes sexual allusions and references. In risqué sartorial materials such as body chains, tight leather pants and high heels, the idols wear drastic make-up with eye-catching red lipstick on their game-faces. Through many occasions of touching themselves, bouncing, pumping and crouching in graphic poses, the choreography is filled with explicit sexual messages. The idols claim a direct and demanding sexuality: "Kiss kiss kiss baby. Hush hush hush baby. Hot hot make it hot, and melt me. Give it to me, give it to me oh." Their bodily and lyrical messages further complicate and perpetuate the hegemonic sexuality that reifies females as sex objects.

In K-pop, what the idols can claim for female empowerment is to push the envelope further to the extent that they shock the audience and create an illusion of power. In other words, the idols promote the patriarchal hegemony of female empowerment, staying within the dominant gender hierarchy. While it is obvious that MissA presents a rather aggressive femininity in their lyrical, physical and visual measures challenging conventionally quiet, obedient, passive and chaste femininity, they are still within the safe boundary of the patriarchal discourse of femininity and sexuality. Despite a hint of individuality with varied hair dyes and color of clothes, visual conformity to femininity is evident to the extent that all four women have a slender body with a similar height and fair skin. The idols do not have their own sexual subjectivity, and they can only be fulfilled by male sexual desire: "Hurry hurry boy. I want you." The idols' aggressive sexuality feeds the male sexual ego, and the lyrics manifest female passivity in the guise of aggressive sexuality: "I can't stand it, I can't take it, my heart palpitates, I can't keep straight." It is in this very ambiguity that the video further perpetuates hegemonic patriarchy while giving a false sense of female subjectivity and empowerment.

In "Good Girl, Bad Girl," Suzy wears a tutu-like skirt that seems to be less revealing than the clothes of the other members. However, since she has elongated legs, the skirt does not cover her body well, but attracts more attention to her body whenever it waves and exposes her thigh. Especially, when combined with her milky, porcelain skin, the black dress accentuates Suzy's physical attractiveness. This visual contradiction gets more salient when her ponytail hair with bangs, which is typical for teenage girls, is incorporated in the sexualized visuals, especially with explicit sexual moves, such as holding buttocks and swinging them sideways, and pelvic thrust on the floor. This is how her slippery image gets solidified when her wholesome physical attributes retain sexual connotations. Later, in a similar manner, Suzy harks back to a sense of purity and innocence in the hyper-sexualized theme of "Hush." Especially in the subway car scene of "Hush," Suzy holds a giant lollipop candy while wearing a white long-sleeve turtleneck sweater, which gives another layer of reified innocence and childhood purity, and this fetishism is revealing when Suzy sits upright and stiff with no facial expression like a china doll. While Suzy bites the candy in time to the lyric, "I can't hold it in anymore," she becomes fully sexualized, inviting the male sexual fantasy of violating virginity. In "Good Girl, Bad Girl," Suzy is sexualized in a sophisticated and teasing way; however, she is the most innocent and wholesome in "Hush" utilizing the exact same elusive sexuality.

Since every representation is a reconstruction of a political fantasy that arises from actual and potential social relations (Jameson, 1995), more specifically, femininity is configured in the "deployment of standardized visual images" (Bordo, 1992, 17), the idols' split personality can be understood a hegemonic mode of femininity that is hybridized between the patriarchal gender hierarchy and the neoliberal commodification of female sexuality. A schizophrenic structure of female identity thus exhibits the social condition of female lives under the two distinct worldviews at the same time: they are required to be innocent, cute and submissive patriarchal women, and an active, hyper-sexualized practitioner of neoliberalism. In turn, this rigid, double confinement of female identity does not grant any room for Korean women to construct their own autonomous subjectivity: at one point, failing to achieve individual development, a woman behaves obedient, submissive and subjectless like a little girl, and at another, she acts as a temptress and preaches neoliberal sexualization. This dual, contradictory demand has intensified patriarchy's "totally other-oriented emotional economy" (Bordo, 1992, 18) to satisfy male affective and sexual needs. By doing so, the patriarchal affective economy exhausts and appropriates the counter-hegemonic potential of women's subjectivity and sexuality into the service of the *status quo*. In its double hybridized constraints, which is traditionally Confucian and contemporarily neoliberal, pure/innocent and sexual, as a marketing strategy for commodified differences in the industry, the schizophrenic female personality further constrains the possibility of alternative thoughts and behaviors in women. In this "schizoid double-pull" (Braidotti, 2006) of femininity, K-pop female idols exercise a new set of sexual norms and values, which renew the traditional ones as the popular agent of daily gender experiences in the country.

Hybridity: Cultural Strategy of Transnational Capitalism

All culture is always already concurrent in cross-breeding between different social groups. For Bhabha (1994), hybridity comes from the in-betweenness of elite emigrants' cultural identities, by constantly engaging in a mutual, simultaneous reconstruction and destruction, a process that nullifies an essentialist notion of cultural authenticity. However, concrete hybridity stems from local agency's dialectic interaction with hegemonic power of transnational forces by "mitigating social tensions, expressing the polyvalence of human creativity, and providing a context of empowerment in individuals and communities are *agents in their own destiny*" (Kraidy, 2005, 161, emphasis added). Since cultural practices "develop and emerge as types of implicit (i.e., nonpropositional or nonverbal) knowledge created in response to lived experiences in a particular social location," not paying due attention to dominant institutional structures of cultural production results in "epistemic violence" (James, 2016, n.p.). Put differently, to correctly understand how cultural hybridity is rendered within global structures of various hegemonies, one has to scrutinize a concrete set of cultural production activities that engender different qualities of hybridity and in turn whether a concrete hybridity reflects or diverts the dominant hegemony. In this respect, rather than the mere existence of hybridity, a manifestation of critical agency in hybridization is the most important qualification (Brah & Coombes, 2000).

However, I indicate an embryo of theoretical complicity in consumerist capitalism from Bhabha's (1994) work that romanticizes the semiotic practice of cultural consumption. Like Radway (1991) celebrates the symbolic and individual pleasure of resistance from reading romance novels, Bhabha exults the subversive power of the subaltern's cultural practices against the imperial domination of cultural, economic and political ideologies. Later, Bhabha (2015) indicates hybridity is helpful to promote an "aesthetics of cultural difference and the politics of minorities" (ix); however, he is not able to acknowledge asymmetric power relations shaped by intricate and multiple layers of constraint during cultural production processes. Even though he acknowledges that cultural difference has been "reconfigured as spontaneous discrimination or systematic inequality," Bhabha rejects to examine cultural hybridity in its diagnostic relationship to local contexts of power relations by saying that it is "neither historically synchronic not ethically and politically equivalent" (ix).

Facing various criticisms, Bhabha (2015) admits that hybridity has been coopted by neoliberalism as a cultural strategy of transnational capitalism, turning what he envisioned as the subaltern's subversive cultural politics into the dominant cultural hegemony. As an epistemic improvement, he applies Antonio Gramsci's dialectics of agency that is always already conjunctured by the dominant social structure of constraints and the subaltern aspiration of social change. In turn, Bhabha accentuates that empowerment is what cultural hybridity aims for by "achievement of agency and authority" (xii). However, rather than paying due attention to the objective and structural conditions of cultural production that is constrained by a larger configuration of power, Bhabha (2015) once again returns to his myopic perspective that "hybridity derives its agency by activating liminal and ambivalent positions *in-between* forms of identification" (xii, emphasis original). Even if Bhabha's act of "enunciation [which] is *at the same time* an act of renunciation: a passionate ambivalence, a subaltern rejection of sovereignty" may be viable as an idyllic bourgeois individual cultural practice (xiii, emphasis original), it is a deadly impotent strategy of achieving agency in the current, neoliberal culture industry. What is worse is that his perspective of achieving a "hybrid voice" as the concrete result of the subaltern's enunciation and renunciation is rather reactionary or detrimental to the minorities, since he believes hybridity "can only accrue authority by questioning its *a priori* security, its first-person privilege, both grammatical and geopolitical" (xiii). A subaltern group that needs to claim cultural agency as a precondition of political agency does not have much privilege, if any at all. If this is the case in Bhabha's argument, his hybridity is a convenient and sure way for the hegemonic group to control the subaltern since the latter voluntarily gives any privilege or power to the sum of the former. This is exactly what has happened since cultural hybridity became vogue in the early 1980s, as it has became coopted to the service of the cultural hegemony of global capitalism.

Furthermore, as not every minority is progressive in society, hybridized local cultural text does not automatically represent critical reflexivity and creativity that leads to an autonomous subjectivity. When cultural alterity is the result of cultural hybridization, it is structured by various degrees of power and discourse in the local hierarchies of the global system, where an asymmetric relation of power and resources cannot be overcome by a cultural exchange or mix (Canclini, 1995). Thus, there must be something ethically transformative that re-kindles the socio-political imagination in one's everyday life for a more egalitarian and democratic society. However, decision-makers in the K-pop industry play the local hegemon who internalizes the industrial and managerial logics of the neoliberal culture industry, and in turn asymmetric relations in the local culture production get increasingly complex and nuanced. Differently put, leaders in the K-pop industry should be considered as a re-territorialized, semi-global center of transnational cultural production, who struggle to control local cultural capital that begets financial and social power by maximizing the benefits of making strategic alliances with the traditional metropolitan centers for cultural mixing and selling. As cultural hybridity has become co-opted by the dominant hegemony of transnational capitalism, as a commodification of difference, it reflects neoliberal mantras through the culture industry's increasing capacity to create and circulate seductive images and sounds (Yúdice, 2003). Or, hybridity and difference have become the bare minimum for access to the market. In this respect, a mere celebration of local cultural production is a blind-celebratory knee-jerk reaction to romanticized hybridity, which regards "subaltern cultural practices ensuing from a given social situation are necessarily subversive" (Yúdice, 2001, xv). And it does not explain the interwoven relationship between the global dissemination of the centers, the multiplicity of transnational capitalist hegemons and the omnipresence of market logic in hybridized local culture production. In sum, unless it leads to a broader transformative project that recognizes that "the margin is hierarchized, appropriated, tokenized or fetishized in order to serve the interests and maintain the order constructed by the center" (Papastergiadis, 2015, 272), cultural hybridity is a conniving, yet alluring justification of global exploitation in a cultural guise.

However, without paying any attention to critical power configurations in local culture production, the current K-pop (Korean popular culture in general) scholarship celebrates the mere existence of hybridized cultural texts as a successful commodity. Claiming that there are multidirectional cultural productions from conventional peripheries, Ryoo (2009) boldly maintains that the phenomenon is a "clear indication of new global, and regional, and transformation in the cultural arena" (147) as a sign of overcoming the American cultural hegemony. Furthermore, while neglecting the politico-economics of K-pop production that has been disproportionately conditioned by American cultural and technical criteria, Ryoo inadvertently attributes the K-pop industry's implementation of the American standard of media liberalization and culture industry to K-pop's success. In this respect, Ryoo's dramatization of local production should be regarded as what Appadurai (1990) criticizes as "production fetishism," an illusion of local cultural productive power in contemporary transnational capitalism, disguising the fundamental asymmetric global power structure. For Shim (2006), K-pop's hybridity was epitomized by the emergence of Seo Taiji and Boys, who mixed various Western music genres and invented a unique Korean flavor. Appropriating American genre formulas, the band successfully exemplified how to exert local agency's active and creative capacity to express local sentiments, issues and traditions and in turn engendered a broad practical transformation in Korea's soundscape. However, over artistic innovation, Shim focuses on industrial transformation: expanding Korea's music market scale, boosting album sales, fortifying record company's roles and, most importantly, heralding the birth of Korea's talent agencies and manufactured K-pop idols. Thus, K-pop is a new economic model that procures a faster and bigger profit margin than the traditional manufacturing industry as a "distinct spatiotemporal configuration" (Sassen, 2001, 268) of Korea's neoliberal economy.

As examined, the current K-pop scholarship is celebratory of the K-pop industry's commercial implementation of hybridization, and an exemplar of transnational capitalism's strategic rhetoric that "systematically seeks to capitalize on cultural fusion" (Kraidy, 2005, 90). While the center appropriates various cultural elements of the peripheries without having to be a part of them, despite various degrees and extents, the periphery nations have to incorporate the hegemonic system of cultural, economic and social production in "the global local articulations of the world system" (Friedman, 1995, 80). Thus, to avoid this empty celebration of cultural consumption, one must critically dissect the political economy of local culture production structured or motivated by transnational capitalism so that we do not fall into "endorsing the cultural claims of transnational capital itself" (Ahmad, 1995, 12). In this respect, Pieterse (1995) maintains that relationships of "power and hegemony are inscribed and reproduced within hybridity … [and] hegemony is not merely reproduced but refigured in the process of hybridization" (57). Therefore, the current hybridity literature retains what Ahmad (1995) criticizes postcolonial intelligentsias for: a "characteristic loss of historical depth and perspective" to "rapid realignments of political [economic] hegemony on the global scale" (16). This blind celebration only confers the "unlimited freedom of a globalized marketplace … [where] commodified cultures are equal only to the extent of their commodification" (17). In this respect, disregarding Korean people's common aspirations, experiences, feelings and lives, K-pop reduces them to the lowest common denominators, that is, addictive beats and rhythms, and explicit sexualization of female bodies for a universal neoliberal market transaction.

In this respect, as I previously investigated (Kim, 2017, 2019), cultural hybridity in K-pop is rather a local embellishment or application of the cultural hegemony of the US popular music, which follows a specific recipe for commercial success: hybridity's market "potency is secured by the implanting of the white seed in the nurturing indigenous womb" (Papastergiadis, 2015, 261). For Canclini (1995), while the scope and a speed of cultural hybridization have accelerated in the globalized society, local cultural production is increasingly "conditioned by a coercive *heteronomous hybridization*" in that only a few people in the headquarters of the neoliberal culture industry dominate initiatives, purposes and applications of new symbolic and semantic creations (xl, emphasis original). In turn, though there are ostensibly different cultural artifacts through the ceaseless inter-cultural bricolage, they become inevitably homogenized by a market imperative of profit-making, which does not provide local cultural practitioners with an opportunity for self-expressive and self-regulated creative production. Far from a creative embodiment of critical self-reflexivity, K-pop is an audio-visual manifestation of consumerism with a seductive visuality and an ephemeral innovation and obsolescence, which is a competition for the survival of the most sexualized, or commodified, overwhelms anything artistic or ideational. It is the K-pop version of the immediacy of frenzied consumption that dramatizes the fleeting nature of what one wants to buy, which is fabricated by the commercial media to increase sales and profit. In this commercial logic of the K-pop industry, the ethical dimension and the political potential of cultural hybridity are subjugated to the neoliberal imperative of profit-making. Thus, according to Pieterse's (1995) continuum of hybridities, K-pop is pervasive with an "assimilationist hybridity that leans over toward the center, adopts the canon and mimics the hegemony" (56), which in terms of cultural mixture is largely oriented to maximizing commercial profit rather than reconfiguring a traditional relationship of power and hegemony by cultural self-reflection.

Hyperreality: Neoliberal Cultural Assembly and Its Effects

In this section, I propose a better heuristic tool to understand K-pop's practices of cultural hybridity in its relationship with transnational capitalism and the commercial imperatives of the local neoliberal industry. Based on the political economy of the K-pop industry, I analyze the socio-cultural implications of K-pop female idols in Korea's growing neoliberalization that is characterized by the intensifying power of media spectacles, consumerism and affective service industries. By doing so, I provide a stronger explanatory approach to cultural assemblage practices in the global system of capitalism.

In the late capitalist economy, images and spectacles seduce individuals to adopt manufactured needs, fantasies and behaviors. As a fundamental source of expressive and interpretative communities (Featherstone, 2007; Kellner, 1995), media spectacles provide essential tools for social identification, outweighing traditional human bonds and experiences. For Baudrillard (1994), while representation works through the "equivalence of the sign and the real" (6), simulation is the generation of the real based on models of a real, defying a referential function of a sign. In turn, it ushers in a broader socio-cultural transformation "from the society of the commodity to the society of the spectacle to the society of the simulacrum, paralleled by increasing commodification and massification" of the image (Best & Kellner, 1997, 80). In this hyperreality, the real is artificially reproduced as "real," by being retouched and refurbished in a "hallucinatory resemblance" with itself (Baudrillard, 1983, 23). While a model precedes the real, simulation conditions the real, and a boundary between hyperreality and individuals' everyday lives disappear: "simulations come to constitute reality … [and] the reality of simulation becomes the criterion of the real itself" (Best & Kellner, 1991, 120). With this dramatic dominance of the visual, simulations structure individuals' emotions, experiences, and value systems, erasing the boundary between the imaginary and the real.

Pushing this argument further, signs and images become a powerful control mechanism over one's life. In this transformation of a strategic idea into reality by simulation, the social in hyperreality, or "hyperreal sociality," aims to transform individuals in the image of the model so that they can be models themselves (Baudrillard, 1994, 29). Through the "orbital recurrence of models" in hyperreality (3), there is "no more center or periphery [but] pure flexion or circular inflexion," to the extent that a seeming difference is an effect of simulation (29). From this perspective, with the meticulous deployment of different signs and commodities, the hyperreality of K-pop female idols is their own simulacrum of an idealized and fantasized femininity that is deployed to maximize the industry's interest. As a caveat, while his deterministic view, like an implosion or an evaporation of the social, does not exactly operate in the real world, Baudrillard's notion is helpful to understand how the idols, who embody an imagined fantasy of male desire, are hyperreal; that is fabricated from their images and visuality that transforms into reality without a real reference to a profound reality. Furthermore, as they become pervasive and influential, there is a continuous emergence of new realities that are initiated or motivated by the idols. In other words, while a simulation becomes real by various technological interventions such as plastic surgery, an imagined potential or a desired outcome becomes a new reality for a community of fans. By seductive visuality and sexualized bodily movements as a conjunction of desire, value, capital and power, K-pop idols, as an effective medium and a powerful message simultaneously, are the main premise of power that is effective in audience's mimetic desire, and in turn an active governmentality agent that shapes individuals' public and private lives. In this regard, unless critically aware of the political economy of the culture industry, the audience may lose their cognitive capacity to distinguish what is real from what is imaginary

(Jameson, 1991), and furthermore be turned into masses through the overwhelming power of the visual (Baudrillard, 1994).

As a fantasy simulated by mythical images of glamour, freedom and success, consumption is a social dynamic to condition individuals' behaviors, that is, to seek out different things to buy. The idols' corporeal and emotional signs take on meanings, and cultivate the audience with various desires for the consumer products they are promoting. Despite a mere consumption of images, individuals are to *"conjure away the real with the signs of the real"* by a false sense of liberation that they do not have in the realm of the social and the political (Baudrillard, 2004, 33, emphasis original). Weaved into a heteroglossia of commodities and services from fashion accessories to plastic surgery, the idols are simultaneously an image and a product, that is, what they promote and the object they are promoting. Like the process through which hyperreality becomes reality, while consuming a glossy heteroglossia of fantastic images, individuals become conformative since they want to live up to an "abstract model, to a combinational pattern of fashion, and therefore relinquish any real difference" (88). Through the market fetishism of differentiation, the K-pop industry manufactures an array of different feminine concepts so that a bigger audience can adopt a favorite personality and life-style through consumption. Furthermore, the female idols are particularly effective in perpetuating neoliberal commercial culture since capitalism exploits an *"extension of the feminine model to the whole field of consumption"* since they are more susceptible to the socio-cultural needs of conformity (98, emphasis original). K-pop female idols as a hyperreality of the neoliberal culture industry replace the harsh reality of women's lives with a superabundance of fantasy images that magically satisfy their desire for economic, political and social mobility.

Guy Debord's (1994) society of spectacle captures how K-pop idols contribute to promoting consumerism with a hyperreality of images and media spectacles. They are cultural linchpins that teach individuals how to utilize commodities as a means of self-transformation into someone better like the idols, and mobilize them to be a steady force of neoliberal consumerism. In this respect, hybridity in K-pop female idols is a result of the industry's simulation that intends to maximize commercial profit by assembling various cultural components to offer the audience aspired to lives, experiences and self-images for an imagined and transformative experience. An overwhelming visuality along with a positive emotionality helps audiences get immersed in the simulated fantasy world of K-pop, and buy what the industry promotes to keep this imaginary experience maintained. The ever-changing visual and emotional theme of K-pop female idols, which allows them and the audiences to enjoy image-switching or personal transformation, is a powerful marketing tool to create an ever-increasing demand for insatiable consumers in a market that sells a plethora of commodities for different images, personalities, styles and experiences. Thus, K-pop's relentless repetition of fantasy enforces a feedback loop upon its audiences in the eternal return of always-wanting-more.

Thus, the idols, and in turn their fans, are a commodity that is to be consumed for marketing commodities, which, as hyperreality, subsequently become reality in the fantasy world of K-pop. In turn, with their simulation power, K-pop idols legitimatize the commercial agendas of the neoliberal culture industry by setting a new set of fragmented, discontinuous trends, values and norms in one's everyday life. Specifically, since female bodies and sexualities are utilized to sell commodities and services as well as being an object of consumption themselves, young female audiences are targeted to assume a socio-cultural identity as consumers by various market entities (Cook & Kaiser, 2004). In this respect, women have been "inscribed as both consumer and commodity, purchaser and purchase, buyer and bought" to support the market economy (Roberts, 1998, 818). Rather than a liberating tool against Korea's centuries-long patriarchy, K-pop female idols' sexualities are commodified and deployed to control the transformative power of the eros, by being offered for mere consumption.

K-pop female idols sport a rhetoric of feminist empowerment, mainly by incorporating active sexuality that is a canonical element in US popular music. Conceiving the idols as an example of the unconstrained possibility of empowerment, the individuals try to emulate or live up to the simulated media figure whose mediated image is more real. In this respect, fandom becomes a hyper-real condition of individuals' desires and fantasies (Sandvoss, 2005), which is fulfilled through the combination of commodities that the idols promote: a "circulation, purchase, sale, appropriation of differentiated goods and signs/objects today constitute our language, our code, the code by which the entire society *communicates* and converses" (Baudrillard, 2004, 79–80, emphasis original). However, as "girl power" has been an effective marketing and branding tool in late capitalism (Klein, 2000), without a self-reflective articulation that carries a critical sensibility, it is a mere hyper-real strategy in an "instrumentalization of feminism as a source of innovation and dynamism for consumer culture" (McRobbie, 2008, 584). In this respect, K-pop idols embody a fantasized version of female empowerment that does not deal with everyday life, strife, prejudice, discrimination, exploitation and other problems, which in turn hides them by distracting individuals' attention away from them.

Therefore, since late capitalism runs on an economy of signs (Harvey, 1992), K-pop idols should be understood from the politics of a hyper-real visuality that imbues audiences with a stylized fantasy, audio-visual illusion of the appealing, beautiful and fashionable. In other words, divorced from any real referent in society, they simulate woman-as-image for the visual consumption of an imagined femininity as fantasy rather than image-of-woman. In this respect, with their governmentality function (Kim, 2019) and by an incessant practice of cultural hybridity, K-pop idols exert an influential role in maintaining and perpetuating gender-specific neoliberal conformative "behaviors, circuits of operationalization that frame thought and action globally" (Luke, 1995, 103).

Concluding Remarks

As examined, the female idols are simultaneously empowered and disempowered in the K-pop industry. As much as they are enabled by their presence, they are still subject to the traditional gender norms, updated and constrained by neoliberal market imperatives. The schizophrenic personality is a salient example that materializes how their subjectivity has been hybridized to satisfy the patriarchal gender values and expectations and to market an ephemeral taste of consumers in neoliberal hyper-capitalism. By doing so, it perpetuates the patriarchal value system which demands women be kind, gentle, decent, delicate and in need of care as well as sexually available. With neoliberal market imperatives as a strange bed-fellow, the old formula conspire to a marketization/commodification of femininity. As to the ethical dimension of cultural assemblage in the idols, far from disrupting or providing a moment of shock to change the dominant system of patriarchal capitalism, the schizophrenic female personality intensifies a scope and a degree of exploitation. "Hybridity and difference sell; the market remains intact" (Hutnyk, 2015, 122).

In this seemingly promising perspective on female success in the K-pop industry, women voluntarily contribute to renewing and perpetuating centuries-long gender oppression, by abandoning any sense of unfairness or oppression in social relations but equipping themselves with an extra amount of agency and effort. In this model of the voluntary internalization of exploitative social relations between genders, women, especially any K-pop female wannabes, trainees, and idols, become and exercise an ideal neoliberal subject, *homo economicus*, who capitalizes on their efforts in an already exploitative capitalist society. Differently put, with a powerful interplay of these two dominant ideologies that configures power relations, the idols' split personality endorses "terms of their subordination and are willing, even enthusiastic, partners in that subordination" (Scott, 1990, 4). As a concrete manifestation of the different effects on a physical body, specific behaviors and corresponding social relations, the idols "enthusiastically perform patriarchal stereotypes of sexual servility on the name of [female] empowerment" (Tasker & Negra, 2007, 3).

Thus, the idols' schizophrenic personality is a neoliberal hyperreality that aims to mobilize women to spark an affective economy, which monetizes an imaginary feeling of male superiority by watching girls playing innocent and cute and indulging sexual fantasy on the one hand, and updates the gender-based asymmetric development of capitalism on the other. As much as female bodies and sexuality have been mobilized to attract foreign capital for national development since the Korean War (Jager, 2003; Moon, 1997), they are still manipulated fashionably to legitimate masculinist development. The more the idols' cute and innocent behavior is highlighted, the more their sexualization is visually salient: the further the female bodies are displayed as a neoliberal commodity, the further they need to validate the traditional gender norms and expectations to maintain the dominant cultural and social *status quo* of the nation. In sum, as de Lauretis (1987) indicates, a female subject is formulated through a "multiplicity of discourses, positions and meanings which are often in conflict with one another" (x), the split personality of K-pop female idols is a carnal imprint of a cacophonous hybridization between the traditional mode of gender oppression and the current neoliberal hegemony.

References

Ahmad, A. (1995). The politics of literary postcoloniality. *Race & Class, 36*(3), 1–20.

Appadurai, A. (1990). Disjuncture and difference in the global culture economy. In M. Featherstone (Ed.), *Global culture: Nationalism, globalization and modernity* (pp. 295–310). London: SAGE Publications.

Baudrillard, J. (1983). *Simulations*. New York: Semiotext(e).

Baudrillard, J. (1994). *Simulacra and simulation*. Ann Arbor, MI: The University of Michigan Press.

Baudrillard, J. (2004). *The consumer society: Myths and structures*. London: Sage.

Best, S. & Kellner, D. (1991). *Postmodern theory: Critical interrogations*. New York: Guilford Press.

Best, S. & Kellner, D. (1997). *The postmodern turn*. New York: Guilford Press.

Bhabha, H. (1994). *The location of culture*. London: Routledge.

Bhabha, H. (2015). Foreword. In P. Werbner & T. Modood (Eds.), *Debating cultural hybridity: Multi-cultural identities and the politics of anti-racism* (pp. ix–xiii). London: Zed Books.

Bordo, S. (1992). The body and the reproduction of femininity: A feminist appropriation of Foucault. In A. Jagger & S. Bordo (Eds.), *Gender/body/knowledge: Feminist reconstructions of being and knowing* (pp. 13–33). New Brunswick, NJ: Rutgers University Press.

Brah, A. & Coombes, A. (Eds.). (2000). *Hybridity and its discontents: Politics, science, culture*. London: Routledge.

Braidotti, R. (2006). *Transpositions: On nomadic ethics*. Cambridge, UK and Malden, MA: Polity Press.

Canclini, N. G. (1995). *Hybrid cultures: Strategies for entering and leaving modernity*. Minneapolis, MN: University of Minnesota Press.

Canclini, N. G. (2001). *Consumers and citizens: Globalization and multicultural conflicts*. Minneapolis, MN: University of Minnesota Press.

Choi, J. & Maliangkay, R. (Eds.). (2015). *K-pop–the international rise of the Korean music industry* (Vol. 40). New York: Routledge.

Cook, D. T. & Kaiser, S. B. (2004). Betwixt and be tween: Age ambiguity and the sexualization of the female consuming subject. *Journal of Consumer Culture, 4*(2), 203–227.

de Lauretis, T. (1987). *Technologies of gender: Essays on theory, film and fiction*. Bloomington and Indianapolis, IN: Indiana University Press.

Debord, G. (1994). *The Society of the spectacle*. New York: Zone Books.

Featherstone, M. (2007). *Consumer culture and postmodernism*. London: Sage.

Friedman, J. (1995). Globalization system, globalization and the parameters of modernity. In M. Featherstone, S. Lash & R. Robertson (Eds.), *Global modernities* (pp. 69–90). Newbury Park, CA: SAGE Publications.

Gilroy, P. (1993). *Small acts: Thoughts on politics of black cultures*. London: Serpent's Tail.

Harvey, D. (1992). *The condition of postmodernity: An enquiry into the origins of cultural change*. Malden, MA: Wiley-Blackwell.

Hebdige, D. (1987). *Cut'n'mix: Culture, identity and Caribbean music*. London: Methuen.

Hutnyk, J. (2015). Adorno at Womad: South Asian crossovers and the limits of hybridity-talk. In P. Werbner & T. Modood (Eds.), *Debating cultural hybridity: Multi-cultural identities and the politics of anti-racism* (pp. 106–136). London: Zed Books.

Jager, S. M. (2003). *Narratives of nation building in Korea: A genealogy of patriotism*. Armonk, NY and London: M.E. Sharpe.

James, R. (2016, May 16). How not to listen to *Lemonade*: Music criticism and epistemic violence. *Sounding Out*. Retrieved from https://soundstudiesblog.com/2016/05/16/how-not-to-listen-to-lemonade-music-criticism-and-epistemic-violence/

Jameson, F. (1983). Postmodernism and consumer culture. In H. Foster (Ed.), *The anti-aesthetic: Essays on postmodern culture* (pp. 111–125). Port Townsend, WA: Bay Press.

Jameson, F. (1991). *Postmodernism, or, the cultural logic of late capitalism*. Durham, NC: Duke University Press.

Jameson, F. (1995.). Politics of utopia. *New Left Review, 25*, 35–56. January.

Jin, D. Y. (2016). *New Korean wave: Transnational cultural power in the age of social media*. Chicago, IL: University of Illinois Press.

Joo, J. (2011). Transnationalization of Korean popular culture and the rise of "pop nationalism" in Korea. *Journal of Popular Culture, 44*(3), 489–504.

Kellner, D. (1995). *Media culture: Cultural studies, identity and politics between the modern and the postmodern*. London: Routledge.

Kim, E. H. & Choi, C. (Eds.). (1998). *Dangerous women: Gender and Korean nationalism*. New York and London: Routledge.

Kim, G. (2017). Between hybridity and hegemony in K-pop's global popularity: A Case Of Girls' Generation's American debut. *International Journal of Communication, 11*, 2367–2386.

Kim, G. (2019). *From factory girls to K-pop idol girls: Cultural politics of developmentalism, patriarchy, and neoliberalism in South Korea's popular music industry*. Lanham, MD: Lexington Books.

Klein, N. (2000). *No logo: Taking aim at the brand bullies*. New York: Picador.

Kracauer, S. (1995). *The mass ornament: Weimar essays*. Cambridge, MA and London: Harvard University Press.

Kraidy, M. M. (2005). *Hybridity, or the cultural logic of globalization*. Philadelphia, PA: Temple University Press.

Kroes, R. (1999). American empire and cultural imperialism: A view from the receiving end. *Diplomatic History, 23*(3), 463–477.

Lazarus, N. (1999). *Nationalism and cultural practice in the post-colonial world*. Cambridge, UK: Cambridge University Press.

Lee, J. (2010). *Service economies: Militarism, sex work, and migrant labor in South Korea*. Minnesota, MN: University of Minnesota Press.

Lie, J. (2015). *K-pop: Popular music, cultural amnesia, and economic innovation in South Korea*. Berkeley, CA: University of California Press.

Lin, X. & Rudolf, R. (2017). Does K-pop reinforce gender inequalities? Empirical evidence from a new data set. *Asian Women, 33*(4), 27–54.

Luke, T. (1995). New word order or neo-world orders: Power, politics and ideology in informationalizing glocalities. In M. Featherstone, S. Lash & R. Robertson (Eds.), *Global modernities* (pp. 91–107). Newbury Park, CA: SAGE Publications.

McRobbie, A. (2008). Young women and consumer culture: An intervention. *Cultural Studies, 22*(5), 531–550.

Moon, K. (1997). *Sex among allies: Military prostitution in U.S.-Korea relations.* New York: Columbia University Press.

Ortner, S. B. & Whitehead, H. (Eds.). (1981). *Sexual meanings: The cultural construction of gender and sexuality.* Cambridge: Cambridge University Press.

Papastergiadis, N. (2015). Tracing hybridity in theory. In P. Werbner & T. Modood (Eds.), *Debating cultural hybridity: Multi-cultural identities and the politics of anti-racism* (pp. 257–281). London: Zed Books.

Pieterse, J. N. (1995). Globalization as hybridization. In M. Featherstone, S. Lash & R. Robertson (Eds.), *Global modernities* (pp. 45–68). Newbury Park, CA: SAGE Publications.

Radway, J. (1991). *Reading the romance: Women, patriarchy, and popular literature.* Chapel Hill, NC: University of North Carolina Press.

Roberts, M. L. (1998). Gender, consumption, and commodity culture. *The American Historical Review, 103*(3), 817–844.

Ryoo, W. (2009). Globalization, or the logic of cultural hybridization: The case of the korean wave. *Asian Journal of Communication, 19*(2), 137–151.

Sandvoss, C. (2005). *Fans: The mirror of consumption.* Cambridge, UK: Polity.

Sassen, S. (2001). Spatialities and temporalities of the global: Elements for a theorization. In A. Appadurai (Ed.), *Globalization* (pp. 260–278). Durham, NC: Duke University Press.

Schiller, D. (1996). *Theorizing communication: A history.* New York: Oxford University Press.

Scott, J. C. (1990). *Domination and the arts of resistance: Hidden transcripts.* London: Yale University Press.

Shim, D. (2006). Hybridity and the rise of Korean popular culture in Asia. *Media, Culture & Society, 28*(1), 25–44.

Sinclair, J. & Harrison, M. (2004). Globalization, nation, and television in Asia: The cases of India and China. *Television & New Media, 5*(1), 41–54.

Tasker, Y. & Negra, D. (Eds.). (2007). *Interrogating post-feminism: Gender and the politics of popular culture.* Durham, NC: Duke University Press.

Werbner, P. (2015). Introduction: The dialectics of cultural hybridity. In P. Werbner & T. Modood (Eds.), *Debating cultural hybridity: Multi-cultural identities and the politics of anti-racism* (pp. 1–26). London: Zed Books.

Whiteley, S. (2013). *Women and popular music: Sexuality, identity and subjectivity.* London and New York: Routledge.

Yúdice, G. (2001). Translator's introduction. In N. G. Canclini (Ed.), *Consumers and citizens: Globalization and multicultural conflicts* (pp. ix–xxxviii). Minneapolis, MN: University of Minnesota Press.

Yúdice, G. (2003). *The expediency of culture: Uses of culture in the global era.* Durham, NC: Duke University Press.

9

POSTCOLONIAL SCHOLARSHIP AND COMMUNICATION

Applications for Understanding Conceptions of the Immigrant Today

Adina Schneeweis

Introduction

An increasing number of institutions and organizations have begun implementing training and conversation on diversity, equity and inclusion; popular discourses have progressively taken on more diverse representations and have more explicitly tackled misrepresentation and injustices; while political discourses have exposed a sharp divide in approaches and actions accompanying a global society increasingly aware of its diversity. In the proclaimed age of globalization, as the world is increasingly challenged to live in closer proximity and interaction, one can no longer ignore minorities, diasporas and migrant, immigrant and displaced populations – we are all migrants (Nail, 2015). The closer we live to one another, the more observable the differences are, the more they overwhelm and demand a reaction, whereas the larger the physical distance, the more the moral responsibility for the *other* diminishes (Bauman, 2000).

At the center of these issues lie a historic preoccupation with *otherness*, a deep need to make sense of one's relationship to *otherness*, and belief systems about the superiority of some groups over others, which have motivated hatred, violence, segregation and extermination, on the one hand, and efforts toward assimilation of the *other*. Attempts to examine *otherness* and ways of misrepresentation and subjugation of the *other* are incomplete when approached as mere acts of stereotyping and when they are relegated to ingroup–outgroup relationship issues. In this chapter I present a noteworthy alternative offered by postcolonial scholarship that links contemporary discriminatory, excluding and racist practices with the construction, and perpetual discursive justification, of *otherness* and the fear of difference. For a collection on communication, I will narrow the discussion to a few specific examples – communicating about (im) migration, to be specific – yet readers will have to understand the difficult task of summarizing the work of a broad range of theorists in one chapter; the interested audience member will hopefully follow up by engaging with postcolonial ideas to explore the numerous avenues of examining communication, representation, geopolitics, difference, identity and knowledge production.

It is further difficult – if not a pointless exercise – to find a starting point for the conceptualization of an *other*. Psychological approaches, for example, have assessed practices of stereotyping to be justifiable, born out of arbitrary, rudimentary categorizations that are useful for diminishing everyday uncertainty. They position constructions of the *self* as "ingroup" or "we-group," with which the individual is familiar and comfortable, in contrast to the "outgroup" consisting of outsiders (Schneider, 2004; Tajfel, 1969). Scholars of postcolonialism have amply demonstrated that fear of the *other* and of difference in general continues to be entrenched in ideological systems of thought, with roots going back to Enlightenment concepts of rationality and civilization/civility, and evident in modern concepts like statehood, nationality, citizenship, economy, capitalism, science and deviance, to name a few (Gregory, 2004; Hart, 2002; Inayatullah & Blaney, 2004; Mehta, 1999; Mignolo, 2000; Said, 2003; Todorov, 1999). What is at stake in the absence of serious acknowledgment of difference is perpetuating a global culture of fear. In what follows, I begin with an outline of concepts that have informed the key debates and ideas around difference and *self–other* dynamics. I apply some of the theoretical concepts to contemporary representations of (im)migrant *others*.

The *Other* as Subaltern

In broad strokes, the *other* has been conceptualized as the indisputable half of a dyad of *self* and *other*, a pair that has historically been structured hierarchically with the *self* asserting itself above the *other*, which in turn must be controlled, subjugated and/or eradicated. While broadly adopted in contemporary scholarship, the concept has its roots in the work of Antonio Gramsci (1971), so I must begin here. To contextualize, in Gramsci's worldview, the state as central actor encompasses the political domain of government and institutions; it regulates and dominates the civil society, ruling through control and consent in a hegemonic process; while the space of civil society is one of struggle, contestation and negotiation (Anderson, 1977; Barrett, 1991). Gramsci's conception of the elite is relevant to the emergence of the concept of the subaltern, as its purpose is to create the conditions most favorable to its domination, stability and expansion (Gramsci, 1971). Hierarchy is necessarily embedded in this conceptualization of the elite's control of the masses via (1) dominance and (2) utilization of the intellectuals' noetic power. Even though there is fragility and contestation built into this system, in a Gramscian approach, the masses are always at the mercy of intellectuals who mediate all forms of resistance and counter-hegemony. At the same time that Gramsci was interested in the role of culture, often organized and logical, to produce revolutionary thought, the subordinate group must be mobilized from the outside – in Gramsci's focus, the Italian Northern revolutionary urban workers had to mobilize or "win over" the peasants of the South (Shome, 2018). It is precisely the conception of such dissent and counter-hegemony that has been one the most critiqued; no doubt an inheritance of Marxist thought, the overemphasis on the subaltern classes as powerless and void of autonomy and political voice has raised concerns (Chatterjee, 2004; Hall, 1996; Urbinati, 1998).

The Subaltern Studies group, originally from South-East Asia, appropriate and reconfigure Gramsci's masses, and explore the resistance of the *other* through examining the voice of the subaltern. Even though Gramsci did not explicitly address slavery or colonialism, many have shown how the Italian thinker's analysis of North and South can be taken to the analysis of the "third world" *other* (e.g., Burawoy, 2003; Hall, 1996). Subaltern Studies emerged in the early 1980s as a South-East Asian initiative, in reaction to concerns over "a depressing half-century of decolonization" in which "struggles against colonialism were fought on the unseen marshes of Western ideology" and conceptualized as a liberation from above (Burawoy, 2003, 247). Subaltern Studies historians led by Ranajit Guha (1988), Partha Chatterjee (2004) and Dipesh Chakrabarty (2000), shifted analyses from the study of colonizers to call for "immersion" in the world of the colonized in order to uncover the alternative visions that have existed there all along. They engage directly with the position of incapability – unless and until intellectuals translate reality to and for the subalterns. The initial mission of the Subaltern Studies was not to *give* voice to the subalterns, but rather to dig up and expose resistance that has been silenced, documenting and rewriting historical narratives in which the subalterns are subjects of their own histories (Burawoy, 2003). Through this lens, the subaltern's contribution to political life, for instance, is understood as different than that of the elite or of intellectuals – yet active/subjective/political in its own right (Chakrabarty, 2000; Cooper, 2003; Guha, 1988). Whereas originally deemed non-autonomous subjects, without political autonomy and voice, without the possibility of revolution and without even being able to imagine the state, scholars of the Subaltern Studies project assess peasants in the Indian context (the original locus of analysis) as a critical part of the enterprise of modernity, acutely aware of societal structures, and active negotiators of meaning in all moments of political life. Critical of nationalist histories of India as elitist and of the cannon of historiography in general, the subaltern (rural peasants in India, in the original project) voice was brought to light, through uncovering modes of resistance, for lack of a better word, through idioms and materials left out of archives. The focus was to recover the subaltern consciousness. Chakrabarty writes, "Once the subaltern could imagine/think the state, he transcended, theoretically speaking, the condition of subalternity" (Chakrabarty, 2002, 34). It is to this desire to investigate the question of consciousness, that scholars of postcolonialism, anticolonialism and deconstruction have responded (Shome, 2018).

Postcolonial Scholarship and Difference

Following the first wave of enthusiastic embrace of the refreshing approach of Subaltern Studies, concerns emerged over whether representation and voice can ever be abstracted from embedded ideology. It became evident that subaltern autonomy is not – and cannot – be conceived outside conjuncture. This systemic interconnection raises "conceptual difficulties" as one attempts "to recover consciousness and memory outside of a literate elite" (Cooper, 2003, 3). Postcolonial scholarship emerged to examine the complex interaction between colonial, imperial culture and moments of indigenous, local resistance, the "phenomena, and effects and affects, of colonialism that accompanied, or formed the underside of, the logic of the modern, and its varied manifestations in historical and contemporary times" (Shome & Hegde, 2002, 258). Its initial purpose was to expose the failings of colonial scholarship – written around Western and imperial ideology, explicitly or implicitly – and to react against the institutionalization and perpetuation of colonial discourses in modern-day phenomena (for instance, in development and modernization projects). Edward Said (2003), Gayatri Chakravorty Spivak (2003) and Homi Bhabha (1994) are recognized as groundbreakers in exposing colonial discursive practices in the construction of difference. Said has been regularly praised for his contribution to discourse as an ideological, alternative construction of reality that has little to do with anything "real," and to uncover the "complicity between politics and knowledge" (Young, 2001, 74). Bhabha has shifted the analytical focus from reading *otherness* in other, Oriental, spaces to listening to the voice of the *other*, while Spivak's celebrated intervention in her "Can the subaltern speak?" and other works questions the notion of autonomous subjectivity outside the politics of representation. Her deconstructive reading points out that subalternity is lost at the moment of its utterance, as the experience of alterity (or resistance) cannot be completely known within the bounds of (discursive, ideological) language. Spivak points out the paradoxical project of attempting to articulate subaltern consciousness, which cannot be uttered without and apart from the thought of the elite – i.e., intellectuals, including present-day scholarship (Shome, 2018; Spivak, 1994, 2003). It follows that attention to power and knowledge production have come more directly under scrutiny. More recently, postcolonial scholars have cautioned against contemporary writing that neglects to consider the way the past is written (as strategically colonial) and have called for a serious interrogation of power struggles embedded in social constructs in their particular historical location (Schwarz, 2005).

Postcolonial analysis, it is significant to note, is not relegated only to questions after colonialism. Rather, the attention is focused on the "continuing process of resistance and reconstruction ... of imperial suppressions and exchanges" in diverse societies in their institutions and interconnected discursive practices (Ashcroft et al., 2006, 2–3). The present-day scope of postcolonialism extends well beyond any critique of colonial projects in ex-colonial societies to global developments as after-effects of empire and colonialism. Such concepts as nation, migration, diaspora, first vs. third worlds, suppression, resistance, difference, ethnicity, identity, race, gender, place, subject, agency, empire, terrorism and violence, to name but a few, have come under the close scrutiny of the postcolonial lens (Quayson, 2005; Shome & Hegde, 2002). Diligent work applying a postcolonial lens to present-day phenomena cannot depart entirely from the context of colonialism, because it would lose its particular critical and contextual stance, focus solely on confrontation and "reproduce a dangerous acontextualism," in Shome and Hegde's words, who add that, "such acontextualism can flatten the story of modernities by implicitly denying any change in its relations from one time to another, from one context to another" (2002, 256–257; also see Cooper, 2003).

Nonetheless, many have examined colonial narratives and other similar phenomena in which a certain group is in an oppressed position and another in one of dominance, in order to grapple with the mentality of the privileged and of the oppressed alike. Important work by Franz Fanon (1963, 1967) has introduced the notion of internal colonization to explain the processes ensuing under oppression, as the oppressed come to largely perceive themselves as they are portrayed or understood to be by the oppressing group. For others, this "parallels (inter)cultural colonization" (Charnon-Deutsch, 2002, 22). Further attention is also required not to essentialize the concept of the *other* and any method of investigation associated with it – for example, colonial discourse analysis that may not operate identically across space and time (Young, 2001). Methodologically, a researcher must allow for alternative forms of "truth" while theoretically exploring moments of similarity and positioning them in broader and inter-cultural frameworks. Otherwise, "we run the risk of imposing our own categories and politics upon the past without noticing its difference, turning the otherness of the past into the sameness of the today," in Young's words (90). Spivak further complicates historical and archival research by casting doubt on the archives themselves, always incomplete; in a sense the very question of knowability of subaltern spaces and voices is "contaminated," as we learn from postcolonial work, and therefore must answer the call for "a critical reflexivity in our work as intellectuals, but not in a way that lapses into paralysis" (Shome, 2018, 21).

Postcolonial theory scholars have contributed important, essential and thoughtful insights into the social construction of the *other*, adding the nuance of difference to the concept of a subaltern. In the words of Bhabha, *otherness* is "an articulation of difference contained within the fantasy of origin and identity" (1994, 67). Some of the first conceptualizations of difference can be traced to Jacques Derrida's (1976) notion of *différance*, to psychoanalytic approaches (Lacan, 1978; Lévi-Strauss, 1983) and to Said's (2003) conceptualization of the *other* in his analysis of discourses of Orientalism (also see Bhabha, 1994; Hall, 1996). Yet practices of enacting difference can be traced further back; for example, many have shown how Enlightenment concepts underscored the envisioning of the white Western world as rational, homogeneous, and a standard against which all other forms of difference were contrasted. As such, the *other* was constructed as backward, uncivilized, and, significantly, needing the intervention of the *self* (Mehta, 1999; Mignolo, 2000, 2003; Todorov, 1999). Conceiving difference has been understood therefore as a dehumanizing process used to further justify discrimination, prejudice and phobias, as well as conflicts and wars, imperial projects to colonize, "liberation" missions in the recent past, and the present "War on Terror" alike (see Fanon, 1963; Gregory, 2004).

While the distinctions are subjective and constructed, the created categories of *self* and *other* become, and are treated as, fixed. Let us examine an example. The concept of integration of minorities has been increasingly popular in Europe, spearheaded by the fall of the Communist regimes in the early 1990s, waves of immigration, opening of borders and the growth of the European Union (EU henceforth). At the same time, communities of Roma,[1] which constitute the largest minority in Europe, have presented a consistent challenge for all countries on the continent. Responding to historic discrimination of the minorities groups that culminated with Nazi-era genocide and post-Cold War segregation, sterilization and eugenics policies, the EU and other intergovernmental bodies have increasingly paid attention to the Roma, also prompted in part by international media reports of ethnic and hate crimes post-1989 and the fall of Communist regimes (see Bancroft, 2005; Fekete, 2008; McVeigh, 1997; Penninx et al., 2008; Schneeweis, 2012; van Noije, 2010). Despite somewhat recent lobbying and activism by human rights groups and the emerging transnational social movement for Roma rights (coming up on roughly 30 years of experience), news accounts such as the following continue to pepper mainstream discourses:

> [I]f travellers really wanted jobs, rather than victimhood, they'd make one simple lifestyle change. They'd stop travelling. If they did this it would mean accepting the houses that the local authorities provide for them, maintaining them to a habitable standard, and making an effort to get on with the neighbours. They'd stay in one place long enough for the next generation to get the education that is available to them. They'd abandon their anachronistic, indefensible and ludicrous existence

in order to give their kids the chance to enjoy the rights, services and respect they demand for themselves. ... A shiftless, transitory way of life is incompatible with the sort of work that offers prospects and potential.

(Power, 2006, 18)

The definite finality and the unapologetic disregard for the Roma as *other* marks such reporting. This passage from *The Times* demonstrates the stability of the discourse of difference, and the implications that *all Travellers* are backward and different than us, and the possibility of inter-cultural harmony or cohabitation is rendered impossible, also because of the *other's otherness*. Any option toward integration emerges, in fact, as assimilation – using the representation of the *other* (justifying racism and xenophobia in the process), the writing calls for a solution that can only mean transforming the *other* into a non-*other*, into someone more like one*self*, as the only way to be accepted by the majority.

Other disciplinary approaches that emphasize urges to estrange or protect oneself from another as natural and a universal characteristic of relations are incomplete or unequipped to help combat inequity or understand the complexities of discrimination and xenophobia (Charnon-Deutsch, 2002). Postcolonial work has exemplified how the construction and enactment of difference as unfamiliar and as the foreign *other* is not a useful representation in itself for the *other*; rather, difference is constructed to represent otherness for the *self* (Said, 2003). The discursive distinction creates, on the one hand, the Westerner, powerful, "rational, peaceful, liberal, logical, capable of holding real values, without natural suspicion," "virtuous, mature, 'normal'" *self* (10, 49), and, on the other hand, the Oriental, the Indian, the Native, nearly always the non-white, and often the immigrant. The *other* is conceptualized as the opposite and backward versions of the Westerner, and therefore "irrational, depraved (fallen), childlike, 'different'" (40), an object, frequently disempowered and historically physically "out there," in "other worlds." The threat becomes real, usually climaxing into moral panics when that distance is bridged and the *other* is perceived to encroach on the space "normally" occupied by the *self* (see Shome & Hegde, 2002; Todorov, 1999). As the construction of difference is "set in motion and made meaningful through cultural practices" (Gregory, 2004, 18), it frequently becomes enacted through public discourse and communication strategies. Historically and contemporarily, the effect is the silencing of *the other* into a less-than-human(ized) state, so that the civilized Westerner has to either rule or come to the rescue of, often by deciding *the other's* history, past, traditions and beliefs (Mignolo, 2000; Spurr, 1999). Mechanisms of "knowing" the other – such as stereotypical knowledge, racialized theories of origin and the administrative experience of colonialism – have contributed to an institutionalization of knowledge about the subaltern *other* into "a range of political and cultural ideologies that are prejudicial, discriminatory, vestigial, archaic, 'mythical,' and, crucially, are recognized as being so," in

Bhabha's words (1994, 83). For instance, according to Bhabha, stereotypes impede the articulation of difference as anything other than as fixed entities, an insight that proves useful in contextualizing stereotyping practices within more complex mechanisms that keep them in place. The continued antagonism between *self* and *other* allows for cultures to interpret, represent and ultimately dominate and control one another (Spurr, 1999). A contemporary example is evident in the reporting on the migrant "caravan" traveling from Honduras toward the US through Mexico.

> Trashing the street. Doing drugs in public. Blocking traffic. Breaking into homes. … Our country's economy is becoming more automated and tech-centered by the day. It's obvious that we need more scientists and skilled engineers. But that's not what we're getting. Instead, we're getting waves of people with high school educations or less. Nice people, no one doubts that but as an economic matter, this is insane. It's indefensible, so nobody even tries to defend it. Instead, our leaders demand that you shut up and accept this. We have a moral obligation to admit the world's poor, they tell us, even if it makes our own country poorer and dirtier and more divided.
>
> *(Carlson, 2018)*

The discourse textually connects the criminality and immoral behavior assumed of all migrants with the need to intervene and control – the "moral obligation" to "defend" against the threat of difference.

Key to understanding postcolonial theorizing of the concept of difference is the ambivalence embedded in it. The separation between *self* and *other* can in actuality never be fully achieved since difference resides within the imagining of *the self* (Hall, 1997; Lévi-Strauss, 1983). The paradox is that *the other*, and thus difference, is necessary for *the self* – for domination, self-assertion and the articulation of identity – while *the self* continues to desire to eradicate it. "Diversity," writes Lévi-Strauss, "is less a function of the isolation of groups than of the relationships that unite them" (328). It can only be understood through dialog with *the other*, as *the other* is fundamental to the constitution and imagination of one*self* and of subjectivity (Bhabha, 1994; Hall, 1997; Inayatullah & Blaney, 2004; Ní Shuinéar, 1997). In the imperial context, the colonial *other* is a "mimicry" of *the self*, always "almost the same, but not quite," in a continual process of disavowal (Bhabha, 1984, 126). Herein lies the ambivalence of representing difference, wanting to reject or being repulsed by, *otherness* at the same time as needing it for one's own self-definition (Young, 2001). The contemporary global world of postcolonial difference continues to be characterized by the "not quite/not white" mimicry discourse (Bhabha, 1984, 133). Applied to the immigrant experience today, different others are both represented as apart, in their own world, and as appropriated, integrated or integrating into the mainstream world that defines its difference. After all, the political discourse is to accept those *others* that are no longer *others*, the "scientists and skilled engineers." In Lacan's words, "the effect of mimicry is camouflage, in the strictly technical sense. It is not a question of harmonizing with the background but, against a mottled background, of being mottled" (Lacan, 1978, 99; also see Bhabha, 1985). In this view, the migrant *other* must fit in with, and integrate within, dominant societies, not by becoming entirely the same, but by becoming sufficiently so in order to dampen their *too different* traits.

Some analyses of cultural practice have raised an epistemological critique to the theorization of the subaltern *other* in light of a continued focus on the object-position. Some object, in specific, to Guha's theoretical recognition of the peasant's intellectual freedom, while maintaining the peasant in the category of a subaltern. "Agency" in the postcolonial analysis is problematized in its relationship to the very ideological reproduction of subalternity, meaning that the *other* is conceptually trapped as a fixed category, exclusively defined through his/her subordination (Cooper, 2003; Robbins et al., 1994). Spivak (1994) also presents a critique – frequently ignored (Shome, 2018) – to Foucault and Deleuze for ignoring ideology in the production of the oppressed, for the notion that intellectuals (such as Western scholars) understand the masses, which they expect to arise to rebel outside the mechanism of ideology, and particularly for Foucault's ignorance of the violence of the colonial system that produced the same power and oppression structures he critiques in European modernity, but lacks the insight to examine internationally. Spivak (2003) further describes the possibility of a community where difference no longer fulfills a hegemonic function, yet she notes the impossibility of conceiving the community without embedding alterity into it. Likewise, Michelle Fine (1994) calls into question the dash between *self* and *other* as she critiques academic writing that merely reproduces the distinctions between *self* and *other* without interrogating the relations of power embedded in difference, and, I argue, the possibilities for discursive transformations. It is important to examine cultural processes with an eye on the need to conceptualize the subaltern as agent, at the same time recognizing the challenge (if not impossibility) of isolating the *other* from the *self* and from processes that (re)constitute difference. The struggle between these two aspects reflects, and is part of, the process of contradicting, while creating, consent; of social stability, while resisting it; of hegemony, while creating counter-hegemony.

A similar line of thinking is evident in anti-colonial theory critiques to what have been labeled present-day neocolonial practices; for example, capturing and problematizing potential continuities and discontinuities to make sense of the contemporary moment, asking what is "post" in the "postcolonial," examining the possibility of still colonized spaces, and investigating the relationships between colonialism, post and neocolonialism are central to the anticolonial project, significantly challenging Eurocentric approaches to such questions (Dei & Kempf,

2006). To return to Spivak (2012), she notes the new dynamics around the subaltern in global capitalism as a system of accessing the subaltern "one way" "as a source of trade-related intellectual property," which includes global humanitarian discourses (also see Shome, 2018, 21).

The Study of Communication and Difference

To a degree, communication studies and postcolonial research are concerned with similar issues – power relations and their creation, maintenance, transformation and establishment as dominant ideology and representation. While postcolonial studies have focused on uncovering mechanisms of constructing difference and their entanglement in contemporary hierarchies of power, communication scholars, particularly in the cultural and critical tradition, suggest that text and discourse play a significant part in the construction and negotiation of common sense. Communication analyses, particularly investigations of mediated forms of cultural practice, (can) benefit from the postcolonial lens to examine contemporary phenomena by focusing attention on the specific textual and discursive tools that (re)produce difference and objectify the *other*, and by problematizing and contextualizing social constructs often taken at face value.

Countless critical scholars are indebted to Gramsci's work; to name but one example, James Curran (1982) has highlighted the role of media institutions contributing to a process of negotiation of prestige, credibility and counter-hegemony. Many have by now noted the hegemonic workings of media communication, creating ways of seeing and legitimizing some representations and not others as dominant, giving "voice to some discursive positions while silencing others" (Jensen, 1999, 19; also see Barker, 1999; Hall, 1997; McRobbie, 2005; Scholle, 1988; Thompson, 1990). The postcolonial lens, however – although it has received substantial attention and use in disciplines such as English, anthropology, literary studies, history and women's studies – communication, journalism and mass communication research have been slower to recognize the intellectual developments introduced by postcolonial scholarship and analysis (with notable differences that include Raka Shome, Radha S. Hegde, Vamsee Juluri or more recently Radhika Parameswaran). Positioning communication processes (mediated or not) in their historical contexts, investigating them conjecturally, involved in multiple sites where power and ideology are enforced and challenged, and recognizing the "particular involvements and investments of communication in real historical social formations" (Grossberg, 1997, 84) becomes not only imperative, but an unavoidable task of postcolonial contextualization. To illustrate, in his postcolonial analysis of nationality, Stephen B. Crofts Wiley remarks that media research often takes for granted problematic concepts (e.g., nationality, in Wiley's focus), overlooking their historical developments and reliance on hegemonic processes.

Means of mass communication have been studied as having the role of creating, maintaining, challenging and changing the talk of the time. A rich body of literature has developed to contextualize and deplore media stereotyping, particularly since the socio-cultural changes following the 1960s. In exploring the role of communication, mediated or otherwise, in reproducing and interrogating difference, one must comprehensively place representation in specific historical, political, economic and social contexts, necessarily interrogating at the same time the hierarchical structures in place. "[P]ostcolonial theories of representation empower media critics to disrupt and denaturalize the subtle hegemony of the discursive myths that constitute the logic of globalization," writes Parameswaran (2002, 289). Postcolonial theory thus relocates problematic concepts and projects within geopolitical structures that shape those very concepts (Shome & Hegde, 2002; Wiley, 2004). Let me close this section by drawing from a fresh article by Shome (2018), who succinctly reminds of Spivak's contribution to communication studies – and, in large part, her comments can be generalized to the substantive work of postcolonial studies. She writes:

> [T]he cultural and historical contexts into which Spivak's work takes readers requires them to make huge leaps in their imaginations to enter worlds that are incomprehensible and unfamiliar to Western or Western-influenced frames of knowing. Spivak's work also requires the reader to suspend or challenge many of the theoretical logics or concepts through which they typically engage matters of social inequalities – for example, identity or reproductive rights – and universalize them to other national contexts …
>
> *(pp. 25–26)*

To this second point, it's worth remembering the dual challenge for academic research: to not ignore significant historical patterns of the construction of, and ensuing discrimination and violence against, *the other* at the same time as to not essentialize concepts that work in some environments yet are not directly translatable to others. Shome continues:

> Such recognition has implications for understanding global inequalities in knowledge production. They also have implications for thinking about our relationship with transnational "others." … Ultimately, they have implications for recognizing that the transnational other has a history, and it [is] our ethical responsibility to understand that history …
>
> *(p. 26)*

Shome's poignant and needed article on Spivak's relevance for communication studies recognizes the significance of thoughtful engagement with gender, asking:

> Can we imagine and build a "global" solidarity of women when so many women in the world fall out of our regimes of literacy and our models of solidarity in the (North Atlantic) West? The languages of such women or men may be charged with histories and traditions that we cannot comprehend, or their idioms might refuse or escape our attempts at translation.
>
> *(p. 26)*

The last sentence in this quote is my conclusion. The experiences of *others*, we learn from postcolonial scholarship, are infused with rich histories and contexts (often, though not universally, part of the fabric of colonialism and other forms of oppression and violence), which cannot be understood – easily, through Western-centric methods, or sometimes at all. None of this is to say that academia should stop trying, as an important voice against inequality, exclusion, persecution and brutality. It is hard to overestimate the influence of postcolonial scholarship on critical analyses of cultural processes and representation, yet it has been slower to penetrate communication studies. I close with a case in point to give a sense of the possibility of grappling with the construction of *otherness* in a communication context and through a postcolonial analytical lens.

Investigating and explaining discourses of (im)migration is a study of conjunctures, delving into (and deconstructing) intersecting discourses – popular culture and media, advocacy, politico-economic developments, socio-cultural traditions, health, education and religious tenets. All such sites are further complicated by histories of oppression (colonial, imperial or otherwise). Serving as a call for re-configuring both communication research about minority groups and analyses of colonial discourse, Bhabha (1994) suggests that "the point of intervention should shift from the ready recognition of images as positive or negative, to an understanding of the *process of subjectification* made possible (and plausible) through stereotypical discourse" (67; emphasis in the original). This is a useful reminder especially for existing research (in communication studies and otherwise) that tends to over-emphasize stereotypical representations, a point made anew by Parameswaran (2002):

> Applying the vocabularies of postcolonial critiques, which have previously dominated analyses of film and literary texts, to journalistic texts and images empowers media scholars to disrupt the hegemony of dominant discourses that shape conversations over key cultural and economic developments in the global public sphere.
>
> *(p. 312)*

As this chapter has demonstrated, approaches focused on identifying sexist and racist stereotypes are not equipped to deconstruct the subtle mechanisms of *othering* that structure neocolonial discursive practices in the age of globalization, and the conversation ends at pointing fingers (see Fine, 1994; Parameswaran, 2002). A similar outcome of incompletion and, to an extent, superficiality emerges in what are termed situations of reverse or positive discrimination and with many projects of intervention. For example, a business collaboration in Romania between an established architect and craftsmen of Roma ethnicity to create handmade products, and with the purpose of "revaluing traditional Roma craftsmanship" can be deemed as "positive"; the venture has enjoyed international success, with jewelry collections presented at important events and shows. Yet, beyond the (capitalist and positive) discourse, the project may be indicative of a different story. The business website lists its "designers," men and women from Austria and Sweden, visibly ranked apart and before listing its collections and, finally, its "craftsmen" – three men and a woman seemingly of Roma ethnicity. One of its projects, a jewelry collection, is framed as aiming to tell stories about Roma women, who "made a statement in the artistic field which reverberated in the social conscience of men." Yet the collection title "Enchant & Delight," and its summary explicitly invoke established ethnic and gendered tropes: "the *fascination for the gypsy woman* is both culture-specific and culture-independent. What is considered *forbidden* or regarded as *taboo* draws on cognitive schemes and emotional tensions and exceeds cultural specificities and temporal boundaries" (emphasis added). Such nods to familiar depictions in the literature, the arts and worldwide popular culture of bohemian, romantic, nomadic and artistic gypsies, flamenco dancers and dark-eyed fortune-tellers (e.g., Lemon, 2000) problematize what on the surface appears to be a "positive" enterprise meant to highlight and "revitalize" *otherness* to also be, at the same time, a project that builds on the same notions it seeks to reconfigure. Even though such insightful contributions are beyond the scope of this chapter, note that many have critiqued "positive discrimination" efforts that build on only visible markers of identity (problematic for its generalizing assumptions on its own) and ignore that positive or protective practices reconfigure, reproduce, and reify an ideology of primitivism, marginalization and backwardness (e.g., Chandra, 2013; Williams & Mawdsley, 2006).

To summarize, this chapter interrogates conceptualizations of difference and *self–other* dynamics through a postcolonial lens, grounded upon foundations within post-Marxist analyses of hegemony, within an understanding of the subaltern improved upon within Subaltern Studies, and within deconstructive critiques to hierarchy. Ultimately, this chapter hopes to contribute a call to attention to the role communication and representation play in the maintenance and transformation of difference as a social construct.

Note

1 The Roma are called many names in various countries. Some use tribal names (such as the Romanian Căldăraşi), others adopt wider group names (such as the British Travellers), and yet others opt for "Roma," a term also used in national and international discourse, and regarded to be the politically correct. "Roma" is most often used in reaction to the commonly used "Gypsy" and its linguistic variants (the German "Zigeuner," the Hungarian "Cigany," the French "tsigane," the Romanian "ţigan," etc.).

References

Anderson, P. (1977). The antinomies of Antonio Gramsci. *New Left Review, 100*, 5–78.

Ashcroft, B., Griffiths, G. & Tiffin, H. (2006). General introduction. In B. Ashcroft, G. Griffiths & H. Tiffin (Eds.), *The post-colonial studies reader* (2nd edn, pp. 1–4). New York, NY: Routledge. First published in 1995.

Bancroft, A. (2005). *Roma and Gypsy-Travellers in Europe: Modernity, race, space, and exclusion* (1st edn). Ashgate.

Barker, C. (1999). *Television, globalization and cultural identities.* Buckingham, PA: Open University Press.

Barrett, M. (1991). *The politics of truth: From Marx to Foucault.* Stanford, CA: Stanford University Press.

Bauman, Z. (2000). *Modernity and the Holocaust.* Ithaca, NY: Cornell University Press.

Bhabha, H. (1984). Of mimicry and man: The ambivalence of colonial discourse. *October, 28: Discipleship: A Special Issue on Psychoanalysis*, 125–133.

——. (1985). Signs taken for wonders: Questions of ambivalence and authority under a tree outside Delhi, May 1817. *Critical Inquiry, 12*(1), 144–165 ("Race," Writing, and Difference).

——. (1994). *The location of culture.* London: Routledge.

Burawoy, M. (2003). For a sociological Marxism: The complementary convergence of Antonio Gramsci and Karl Polanyi. *Politics & Society, 31*(2), 193–261.

Carlson, T. (2018, Dec. 18). Tucker Carlson responds to his immigration critics: We're not intimidated, we'll continue to tell the truth. *Fox News.* Retrieved from www.foxnews.com/

Chakrabarty, D. (2000). A small history of Subaltern Studies. In H. Schwartz & S. Ray (Eds.), *A companion to postcolonial studies* (pp. 467–485). Malden, MA: Blackwell Publishers.

——. (2002). *Habitations of modernity: Essays in the wake of Subaltern studies.* Chicago, IL: University of Chicago Press.

Chandra, U. (2013). Liberalism and its other: The politics of primitivism in colonial and postcolonial Indian law. *Law & Society Review, 47*(1), 135–168.

Charnon-Deutsch, L. (2002). Travels of the imaginary Spanish Gypsy. In J. Labanyi (Ed.), *Constructing identity in contemporary Spain* (pp. 22–40). New York, NY: Oxford University Press.

Chatterjee, P. (2004). *The politics of the governed: Reflections on popular politics in most of the world.* New York: Columbia University Press.

Cooper, F. (2003). Conflict and connection: Rethinking colonial African history. In J. D. Le Sueur (Ed.), *The decolonization reader* (pp. 23–44). New York, NY: Routledge.

Curran, J. (1982). Communications, power and social order. In M. Gurevitch, T. Bennett, J. Curran & J. Woollacott (Eds.), *Culture, society and the media* (pp. 202–235). London: Methuen.

Dei, G. J. S. & Kempf, A. (Eds.). (2006). *Anti-colonialism and education: The politics of resistance* (Vol. 7). Rotterdam, the Netherlands: Sense Publishers.

Derrida, J. (1976). *On grammatology.* Baltimore, MD: Johns Hopkins University Press (Translated by G. Spivak).

Fanon, F. (1963). *The wretched of the earth.* New York, NY: Grove Press.

———. (1967). *Black skins, white masks.* New York, NY: Grove Press.

Fekete, L. (2008). *Integration, islamophobia and civil rights in Europe.* London: Institute of Race Relations.

Fine, M. (1994). Working the hyphens: Reinventing self and other in qualitative research. In N. K. Denzin & Y. S. Lincoln (Eds.), *Handbook of qualitative research* (pp. 70–82). Thousand Oaks, CA: Sage Publications.

Gramsci, A. (1971). *Selections from prison notebooks.* New York, NY: International Publishers (Edited and translated by Q. Hoare & G. Nowell Smith).

Gregory, D. (2004). *The colonial present.* Malden, MA: Blackwell Publishing.

Grossberg, L. (1997). *Bringing it all back home: Essays on cultural studies.* Durham, NC: Duke University Press.

Guha, R. (1988). The prose of counter-insurgency. In R. Guha & G. C. Spivak (Eds.), *Selected Subaltern studies* (pp. 45–86). New York, NY: Oxford University Press.

Hall, S. (1996). Gramsci's relevance for the study of race and ethnicity. In D. Morley & K.-H. Chen (Eds.), *Stuart Hall: Critical dialogues in cultural studies* (pp. 411–440). London: Routledge.

———. (1997). The spectacle of the "other.". In S. Hall (Ed.), *Representation: Cultural representations and signifying practices* (pp. 223–290). London: Sage Publications in Association with The Open University.

Hart, G. (2002). *Disabling globalization: Places of power in post-apartheid South Africa.* Berkley, CA: University of California Press.

Inayatullah, N. & Blaney, D. L. (2004). *International relations and the problem of difference.* New York, NY: Routledge.

Jensen, K. B. (1999). Humanistic scholarship as qualitative science: Contributions to mass communication research. In K. B. Jensen & N. W. Jankowski (Eds.), *A handbook of qualitative methodologies for mass communication research* (pp. 17–43). London: Routledge. First published in 1991.

Lacan, J. (1978). *The four fundamental concepts of psycho-analysis.* New York, NY: Norton.

Lemon, A. (2000). *Between two fires: Gypsy performance and Romani memory from Pushkin to Postsocialism.* Durham, NC: Duke University Press.

Lévi-Strauss, C. (1983). *Structural anthropology* (Vol. 2). Chicago, IL: University of Chicago Press. First published in 1976.

McRobbie, A. (2005). *The uses of cultural studies: A textbook.* London: Sage.

McVeigh, R. (1997). Theorising sedentarism: The roots of anti-nomadism. In T. Acton (Ed.), *Gypsy politics and Traveller identity* (pp. 7–25). Hatfield: University of Hertfordshire Press.

Mehta, U. S. (1999). *Liberalism and empire: A study of nineteenth-century British liberal thought.* Chicago, IL: The University of Chicago Press.

Mignolo, W. D. (2000). *Local histories/Global designs: Coloniality, subaltern knowledges, and border thinking.* Princeton, NJ: Princeton University Press.

——. (2003). *The darker side of Renaissance: Literacy, territoriality and colonization*. Ann Arbor, MI: The University of Michigan Press.

Nail, T. (2015). *The figure of the migrant*. Stanford, CA: Stanford University Press.

Ní Shuinéar, S. (1997). Why do Gaujos hate Gypsies so much, anyway? A case study. In T. Acton (Ed.), *Gypsy politics and Traveller identity* (pp. 26–53). A companion volume to *Romani Culture and Gypsy Identity*. Hertfordshire: University of Hertfordshire Press.

Parameswaran, R. (2002). Local culture in global media: Excavating colonial and material discourses. In *National Geographic. Communication Theory, 12*(3), 287–315.

Penninx, R., Spencer, D. & Van Hear, N. (2008). *Migration and integration in Europe: The state of research*. Oxford: Economic and Social Research Council (ESRC) Centre on Migration, Policy and Society, University of Oxford.

Power, B. (2006, December 10). Travellers' ludicrous and selfish lifestyle. *The Sunday Times*. Retrieved from www.thetimes.co.uk/

Quayson, A. (2005). Postcolonialism and postmodernism. In H. Schwartz & S. Ray (Eds.), *A companion to postcolonial studies* (pp. 87–111). Oxford: Blackwell Publishing. First published in 2000.

Robbins, B., Pratt, M. L., Arac, J., Radhakrishnan, R. & Said, E. (1994). Edward Said's *Culture and imperialism*: A symposium. *Social Text, 40*, 1–24.

Said, E. (2003). *Orientalism*. New York, NY: Vintage Books. First published in 1978.

Schneeweis, A. (2012). If they really wanted to, they would: The press discourse of integration of the European Roma, 1990-2006. *International Communication Gazette, 74*(7), 673–689. doi: 10.1177/1748048512458561.

Schneider, D. J. (2004). *The psychology of stereotyping*. New York, NY: The Guilford Press.

Scholle, D. J. (1988). Critical studies: From the theory of ideology to power/knowledge. *Critical Studies in Mass Communication, 5*, 16–41.

Schwarz, H. (2005). Mission impossible: Introducing postcolonial studies in the U.S. academy. In H. Schwartz & S. Ray (Eds.), *A companion to postcolonial studies* (pp. 1–20). Oxford: Blackwell Publishing. First published in 2000.

——. (2018, September 26). Gayatri Chakravarty Spivak: Relevance for communication studies. *Oxford Research Encyclopedia of Communication*. Retrieved from http://oxfordre.com/communication/view/10.1093/acrefore/9780190228613.001.0001/acrefore-9780190228613-e-600

Shome, R. & Hegde, R. S. (2002). Postcolonial approaches to communication: Charting the terrain, engaging the intersections. *Communication Theory, 12*(3), 249–270.

Spivak, G. C. (1994). Can the subaltern speak? In P. Williams & L. Chrisman (Eds.), *Colonial discourse and postcolonial theory* (pp. 66–111). New York, NY: Columbia University Press.

——. (2003). *A critique of postcolonial reason: Toward a history of the vanishing present*. Cambridge, MA: Harvard University Press. First published in 1999.

——. (2012). The new subaltern: A silent interview. In V. Chaturvedi (Ed.), *Mapping subaltern studies and the postcolonial* (pp. 324–340). London: Verso.

Spurr, D. (1999). *The rhetoric of empire: Colonial discourse in journalism, travel writing, and imperial administration*. Durham, NC: Duke University Press. First published in 1993.

Tajfel, H. (1969). Cognitive aspects of prejudice. *Journal of Social Issues, 25*, 79–97.

Thompson, J. B. (1990). *Ideology and modern culture: Critical social theory in the era of mass communication*. Stanford, CA: Stanford University Press.

Todorov, T. (1999). *The conquest of America: The question of the other*. Norman, OK: University of Oklahoma Press. First published in 1982.

Urbinati, N. (1998). From the periphery of modernity: Antonio Gramsci's theory of subordination and hegemony. *Political Theory, 26*, 370–391.

van Noije, L. (2010). The European paradox: A communication deficit as long as European integration steals the headlines. *European Journal of Communication, 25*(3), 259–272.

Wiley, S. B. C. (2004). Rethinking nationality in the context of globalization. *Communication Theory, 14*(1), 78–96.

Williams, G. & Mawdsley, E. (2006). Postcolonial environmental justice: Government and governance in India. *Geoforum, 37*(5), 660–670.

Young, R. J. C. (2001). Colonialism and the desiring machine. In G. Castle (Ed.), *Postcolonial discourses: An anthology* (pp. 73–98). Oxford: Blackwell Publishers.

10

CYBERHATE, COMMUNICATION AND TRANSDISCIPLINARITY

Emma A. Jane and Nicole A Vincent

Language warning: This chapter contains uncensored examples of real-life gendered cyberhate including imagery of sexual violence, as well as racist, homophobic and transphobic slurs. As we explain, quoting such uncensored material is necessary for scholarly and ethical reasons.

Introduction

Imagine you're confronted in broad daylight by a menacing stranger who calls you a "stuck up cunt" and threatens to rape you with a combat knife unless you quit your job. Or a serial harasser sticks posters with your photo, phone number and home address up on street poles around your neighborhood, alongside a graphic suggestion that people who dislike your professional work show up at your door and teach you a lesson. Or your work colleagues tell you to "shut the fuck up" and stop being such an "attention whore" when you complain about the mountain of rape threats and transphobic letters piled up on your desk each morning. Or the police say there's nothing they can do about all the death threats you've been getting in the mail, because no-one really understands how the postal service works, and all those people who've been threatening to kill your children are probably just kidding.

In offline settings, such behavior – threats, breaches of privacy, reputational attacks, incitements to violence, coordinated bullying and vilification campaigns, economic vandalism, workplace harassment and institutional inaction – would not be tolerated. In addition to social norms rebuking such conduct, police, courts and various types of regulation exist to (at least in principle) protect victims and apprehend perpetrators. In modern, multicultural societies, the machinery of the state at least tries to maintain public order, to ensure safety, to protect citizens' physical and emotional wellbeing, and to keep things running smoothly. Guidelines are also in place for the design of safer and more inclusive environments, including adequate street lighting and wheelchair-friendly ramps into and around public buildings and spaces.

In stark contrast, identical online behavior often passes unnoticed, or else is framed as not *even* behavior – not even action – but *only* as communication, and either played down or just dismissed out-of-hand as unimportant. The technology responsible for the creation of the internet is nearly half a century old (iFactory, 2017), yet many spaces in the cybersphere still seem like the "wild west of the information age" (Tuovinen & Röning, 2007, 397) – under-policed frontiers where vigilantes, lynch mobs, sexual predators and violent abusers are not just tolerated, but even applauded by enthusiastic onlookers.[1] As we summarize below,[2] in recent years there have been increasingly frequent and urgent calls for intervention – including, for instance, from such authoritative organizations as the United Nations (UN), Amnesty International UK, YouGov UK, and the Australian Human Rights Commission (AHRC) – as a growing body of international research has increasingly documented both the prevalence and harm of online hate and harassment.

Unfortunately, despite overwhelming evidence of an increasingly urgent and serious problem – cited and taken seriously by the world's most reputable organizations – there remains an abundance of media and other commentary claiming that cyberhate is *not* a serious problem because it involves only words not actions, and only virtual not "real" life.[3] Cyberhate is apparently not a serious concern because it is, after all, "just words" and/or "just the internet."

As we have argued elsewhere,[4] part of the reason this gross mischaracterization of cyberhate is so difficult to dislodge is that it is in good measure self-perpetuating. This is a product of several interacting factors. First, in academic and lay contexts alike, there is a deficit of realistic examples and language in which to describe the phenomenon of cyberhate. Second, that makes it difficult for people and institutions to recognize cyberhate, to report it, to legislate about it and to study it. Third, numerous features of the technologies involved also present victims and onlookers with impenetrable barriers to collecting evidence of cyberhate. Fourth, even if victims and onlookers knew what to look for and had the means to record evidence of cyberhate, it is far from clear to whom they can even report it. Fifth, and finally, the implicated technologies' features also render authorities powerless to investigate cases of cyberhate, and related evidence, should any be brought to their attention.

The complex interplay between these factors – in which multiple stakeholders, institutions, technologies and norms play leading, support and extra roles – means that cyberhate is often not reported through official channels, and without official reports, the victims and impacts remain officially invisible. Thus, on our analysis, the lack of realistic, nuanced and finely-textured language, as well as uncensored examples of real-life cases of cyberhate, plays a key role in the self-perpetuating character of this entrenched mischaracterization of cyberhate.[5]

Given the theme of this edited volume, rather than recounting the details of the structural, institutional and technological factors which contribute to this state of affairs that we have detailed elsewhere, our focus will be on spotlighting the powerful role that communications scholars can play in remedying this situation. Specifically, on our account, a pervasive scholarly squeamishness around quoting unexpurgated examples of cyberhate in all of its offensive and affronting detail – combined with the deployment of unhelpfully narrow disciplinary lenses, and equally unhelpful *inter-* and *multi-*disciplinary lenses – fails to capture or convey the seriousness of this phenomenon, and at times even invisibilizes it completely. Critically, however, our point is that the limits of *mono-*, *inter-* and *multi-*disciplinarity are "limits not just of institutions, but also of *discovery*" (Mittelstrass, 2011, 330, original emphasis). In light of this, the ensuing discussion offers two quite independent reasons why scholars working in this area should consider including more vivid case studies and employing more transdisciplinary methods such as victim-inclusive stakeholder consultation: first, to support rather than hinder international efforts to combat cyberhate as well as misperceptions of cyberhate, and second to advance communications as a field of scholarship, by remaining receptive to new and potentially intellectually-fertile academic vistas. To reimagine communications, it helps to keep in mind that – borrowing from Kenneth Burke – our disciplinary lenses are just as much a way of seeing as they are a way of not seeing (1935, 49).

Definitions and Sources

In this chapter, we use the elastic term "gendered cyberhate" to refer to a range of phenomena occurring at the gender–technology–violence nexus. We intentionally use the word "phenomena," instead of "actions" or "discourses," to avoid begging the question in favor of either the claim that cyberhate is action or that it is communication, not the least because we take this to be a false dichotomy. By our account, gendered cyberhate includes sexually graphic invective, hyperbolic yet plausible rape and death threats, and/or persistent, unwanted sexual advances from senders who frequently lash out if they are ignored or rejected. It also includes: cyberstalking; rape blackmail videos; mob attacks (sometimes with the explicit purpose of causing job loss or career derailment); malicious impersonation; "sextortion" (the blackmailing of targets in order to extort them to perform sexual acts online); revenge porn (the uploading and circulation of sexually explicit material containing intimate graphic depictions of people without their consent); and "doxing" (the publishing of personally identifying information, often accompanied by encouragement for internet antagonists to hunt down targets in their offline lives).

This chapter draws on data from an ongoing series of research projects dedicated to mapping and studying the history, manifestations, nature, prevalence, etiology and consequences of misogyny online. Its sources include a 30-year archive of gendered cyberhate and an Australian government-funded study involving in-depth, semi-structured, qualitative interviews with 52 Australian targets of gendered cyberhate interviewed between 2015 to 2017.[6] This chapter also includes key insights gained from a transdisciplinary Cyberhate Symposium convened by the authors in Sydney in 2017 (Jane & Vincent, 2017b). While it is beyond the scope of this chapter to discuss the rationale for and methodologies involved in this symposium in detail, we note that it involved precisely the sort of transdisciplinary research approaches we advocate below.[7]

The Harms of Gendered Cyberhate

The first paragraph of the introduction to this chapter, in which we portrayed examples of online gendered cyberhate in forms it might take offline, is based on the experiences of the American games designer Brianna Wu, who received a large number of graphic rape and death threats during the global attacks on women in 2014 known as "GamerGate." We employ this commutation test[8] – a re-framing device – to illuminate the double standards that exist around what is considered unacceptable conduct offline by comparison to online (Jane, 2017b; Vincent & Jane, 2017b). In the physical world, hunting down a senior professional and yelling in her face that you are coming to her home to rape her with a military-grade weapon because you don't like her company's products is not regarded as acceptable conduct. From the target's perspective, too, guidelines for what qualifies as due process – how to proceed during or after such attacks – also exist. For instance, victims/survivors of physical and sexual assault can report to police, who follow procedures for interviewing victims and witnesses, for collecting evidence, and relatively clear and tested laws exist to guide the investigative process. These norms, standards and protections are not, however, extended to the online world, where women are routinely targets of noxious abuse and plausible threats, but yet struggle – for the most part in vain – to obtain assistance in staying safe and in bringing offenders to account.[9] Given that most offenders can attack with impunity, it is unfortunately not surprising that the number of girls and women subjected to cyberhate is rising rapidly.

For a glimpse of this growing and increasingly urgent problem, consider the following reports and trends. In 2015, the United Nations Broadband Commission warned that cyber violence had become "a global problem with serious implications for societies and economies around the world" (Cyber violence against women and girls: A world-wide wake-up call, 2015, 1). It noted that 73% of women and girls encountered some form of online violence; that women were 27 times more likely to be abused online than men; that

61% of online harassers were male; and that women aged between 18 and 24 were at particular risk (Cyber violence against women and girls: A world-wide wake-up call, 2015, 15). In November 2017, Amnesty International UK published research showing that one in five women in the UK, the US, New Zealand, Spain, Italy, Poland, Sweden and Denmark had experienced online abuse or harassment. Of these: more than a quarter (27%) had received direct or indirect threats of physical or sexual violence; almost half (47%) had experienced sexist or misogynistic abuse; and one-third (36%) felt their physical safety had been threatened (Amnesty International UK, 2017). In 2018, research by YouGov UK, found that four in ten female millennials had been sent unsolicited photographs of men's genitals – aka "dick pics" (Smith, 2018). More recently, the Australian Human Rights Commission has noted:

> that 76% of women under 30 years of age, have reported experiencing online harassment, and almost half (47%) of all women had been targets. Similarly, one in four lesbian, bisexual and transgender women reported targeted sexual orientation harassment. More recent research on the experiences of women in Australia found that, of those that had experienced online abuse and harassment, 42% of women said it was misogynistic or sexist in nature, and 20% said it had included threats of physical or sexual violence.
>
> *(2018, 20)*

Our own research also suggests that while gendered cyberhate was relatively rare and mild in the early decades of the internet, it has become far more prevalent, visible, noxious and directly threatening since at least 2010 (Jane, 2017a, 16–42).[10]

Contrary to portrayals of cyberhate as mostly innocuous, the widespread suffering it causes is very real and significant.[11] The coercive force of gendered cyberhate is causing many women significant social, psychological, reputational, economic and political harm.[12] Harassment, threats and abuse at the most extreme and sustained end of the spectrum are causing women debilitating fear and trauma, as well as profound life disruption. A number of women, for instance, have had to flee their homes and go into hiding out of warranted fear for their safety (Jane, 2017a, 31–32). Female cyberhate targets have also developed serious mental health problems or experienced psychological breakdowns in the aftermath of being attacked online (Jane, 2017a, 62–64).

Cyberhate is affecting women's ability to find – and keep – jobs, to market their personal brands and their businesses, to network socially and professionally, to participate politically (including to engage in activism in response to the problem of gendered cyberhate itself) and to freely enjoy key benefits of the Web 2.0[13] era in forms such as self-expression, self-representation, creativity, interactivity, collaborative enterprises and participation in civic life and democratic governance (Jane & Vincent, 2018, 39–40). Much of the cyberhate women receive while attempting to do their jobs involves abuse and harassment which would be in flagrant breach of multiple, international workplace-related regulations and guidelines if it occurred in physical spaces, offices and workplaces (Jane, 2018). Far from being "just words" or harmless "jokes," online abuse destroys women's reputations in ways that have significant and ongoing repercussions for their future employment and career-building prospects.[14] Further, as we have demonstrated gendered cyberhate infringes on at least ten Articles from the Universal Declaration of Human Rights (UDHR) (Jane & Vincent, 2018, 37–45). Finally, at the big picture level, the cumulative disadvantages of gendered cyberhate should be understood as constituting a new dimension of existing, gender-related digital divides[15] (Jane, 2017d; Jane & Vincent, 2018, 34–35, 40–41).

Despite the documented harms and damaging impacts of gendered cyberhate, however, a wide range of insidious structural, institutional and technological factors – as well as interactions between them – continue to make cyberhate into a problem that is difficult to notice let alone to understand and appreciate in its damaging entirety. Given that we have discussed these factors in detail elsewhere,[16] we will devote the remainder of this chapter to honing in on those issues of most direct pertinence to communications scholars, namely: (i) the need to accurately portray cyberhate, its harms and costs, and its victims, in finely-textured details, rather than using words and examples that will not cause offense to readers; and (ii) the need for caution around the use of narrow disciplinary lenses (which, as we are about to show, have the capacity to erase the phenomenon completely) and instead to consider a transdisciplinary approach to communications scholarship.

Inadequate Language and Examples

Our review of 30 years of scholarly literature shows that online hostility has historically been underplayed, overlooked, ignored, or otherwise marginalized by many researchers from across multiple disciplines.[17] A symptom as well as a contributing cause of these habits is that up until about 2014, this canon of scholarly work contained few vivid case studies as well as almost a complete absence of unexpurgated examples of cyberhate.[18] Instead, scholars used generic descriptors (such as "impolite," "uninhibited," "in bad taste," "profane" and so on), and relied heavily on vague, catch-all terms such as "flaming" (an anachronistic term for heated online exchanges) and "trolling" (a word associated with pranking in internet sub-cultures). Combined with just a few relatively tame and expurgated case studies – such as someone called "Dr Ski" being asked if their PhD was from a cereal box (Thompsen & Foulger, 1996, 243) and someone calling someone else a "jerk" (Kiesler et al., 1985, 89) – these often-playful, sometimes even euphemistic, and with almost no exception censored and linguistically pasteurized examples have added to the difficulties involved in having contemporary cyberhate recognized, let alone recognized as seriously inappropriate behavior, quite simply because they fail to convey its violent and threatening nature. We welcome a recent shift in particularly feminism-informed research in this respect (cf. Lewis et al., 2017; Mantilla, 2015; Vera-Gray, 2017), though we maintain that this practice needs to be taken up far more widely in academia.

To appreciate why we so emphatically stress the need to describe cyberhate using realistic words and unexpurgated examples, contrast our description of the attacks on Wu in the previous section – which we described as "a large number of graphic rape and death threats," to intentionally employ that same linguistically pasteurized and vague terminology we regard as so deeply problematic – to the following verbatim examples of messages that Wu received:

Your mutilated corpse will be on the front page of Jezebel[19] tomorrow and there isn't jack shit you can do about it.

I'm going to rape your filthy ass until you bleed, then choke you to death with your husband's tiny Asian penis.

I've got a K-bar and I'm coming to your house so I can shove it up your ugly feminist cunt.

If you have any kids, they're going to die too. I don't give a fuck. They'll grow up to be feminists anyway.

Women are the niggers of gender. If you killed yourself, I wouldn't even fuck the corpse.[20]

Unexpurgated examples of cyberhate are unsettling and unpleasant to read. Nevertheless, our case is that it is critical for scholars to quote them verbatim in their academic work – in peer-reviewed publications, talks, poster presentations and in coursework for their students. The *language* should be accurate, realistic, rich, nuanced and finely-textured – literally the same as the vulgar and confronting vitriol flung at cyberhate victims. Likewise, a wide range of detailed, realistic and evocative *examples* should be quoted verbatim – inclusive of grammatical and orthographic mistakes, slurs, and slang – to convey the diverse texture of (i) implicated behaviors, (ii) the harms caused, (iii) the range of victims and (iv) why these things are serious not innocuous.

As we noted early on in this chapter, part of the reason why the gross mischaracterization of cyberhate as a benign phenomenon is so difficult to dislodge is that it perpetuates itself. Against this backdrop, our insistence that scholars need to quote verbatim examples of cyberhate, could now be given the following critical/reactive rendition. Scholarly squeamishness about quoting unexpurgated examples of cyberhate has disturbing parallels with discussions that employ "enhanced interrogation" as a synonym for waterboarding – an irresponsibly opaque and misleading "euphemism that rationalizes and sanitizes torture" (Amnesty International, n.d.). Similarly, in the absence of accurate language and a diverse range of unexpurgated examples to draw upon, it is not just difficult but effectively impossible to convey what cyberhate is, why it is harmful, or to enable others to recognize it.

To supplement this critical/reactive rendition of our argument, we also offer the following distinctly constructive/proactive rendition of our argument. George Soros (2014) has observed that in systems which include thinking participants – what he calls "reflexive" systems, for reasons that shall readily become apparent – what people think and believe (i.e., epistemic features) has an impact on what occurs (i.e., ontological features). Reflexive systems, on Soros's account, are characterized by the presence of recursive causal loops in which people's beliefs perform a special function that connects the epistemic realm with the ontological realm. Because recursive phenomena are difficult to describe without picking an arbitrary point at which to enter into the description of the recursive system, Soros offers an illustration that compares non-reflexive to reflexive systems:

> Consider the statement "It is raining." That statement is true or false depending on whether it is, in fact, raining. And whether people believe it is raining or not cannot change the facts. … Now consider the statement "I love you." The statement is reflexive. It will have an effect on the object of the affections of the person making the statement and the recipient's response may then affect the feelings of the person making the statement, changing the truth value of his or her original statement.
>
> *(Soros, 2014, 312)*

We find Soros's concept and discussion of "reflexivity" very helpful for two reasons. First, it offers a neat way to explain why the view of cyberhate as "just words" and/or "just the internet" continues to have traction, though without having to spell out the concrete details of the five factors that we discussed around the start of this chapter. This is because what Soros's approach draws attention to are the structural relations among the interacting factors which create the recursive causal loops – loops that, in the case at hand, lead mischaracterizations of cyberhate to be self-perpetuating. Second, we also find Soros's concept of reflexivity (and the attendant theoretical machinery) helpful because it explains precisely how the scholarly work that we endorse can help. Initially, the self-perpetuating character of the mischaracterization of cyberhate can feel like a cause for pessimism – one reminiscent of the pointlessness of Sisyphus repeatedly rolling a huge stone up a hill, only to have it roll back down again. However, the optimistic corollary of Soros's insightful observation about how reflexive systems function is that ontological features are at least partly influenceable *via* epistemic features. Stated less abstractly, because people's beliefs and views can be influenced through evidence, reason and argumentation (i.e., what this chapter offers), the very same epistemic features in virtue of which this gross mischaracterization of cyberhate has remained entrenched in certain quarters, are also – yet again making reference to Greek mythology – this mischaracterization's own Achilles' heel. Such optimism underpins our efforts to set the record straight on cyberhate, and our emphasis on the need for scholars working in this field to abandon their squeamishness about quoting unexpurgated examples of cyberhate, since by doing this they will help to weaken the self-replicating character of the mischaracterization of cyberhate as a benign phenomenon.

Disciplinary Tunnel Vision

Up to this point, our discussion has focused on describing a specific feature of how cyberhate has been discussed,[21] on explaining what makes this feature problematic[22] and on suggesting a specific corrective measure.[23] However, to pre-empt the foreseeable objection that this particular problem only impacts on discussions of cyberhate and similar topics – and thus to forestall the critique that our observations and recommendations have only marginal utility, since they only benefit scholars who work on such boutique topics – in this section we shall address a broader issue of relevance to communications as a field. While the topic that we will use in our discussion will remain roughly the same, our discussion's focus will be the matter of how our own disciplinary lenses can unhelpfully stifle disciplinary progress, and what corrective measures we can take to avoid this.

Rather than speaking in the abstract, this section will discuss Patrick B. O'Sullivan's and Andrew J. Flanagin's article "Reconceptualizing 'flaming' and other problematic messages" (2003) in which these communications scholars aim to address what they see as a lack of "precise conceptual and operational definitions of 'flaming'" (69) in scholarly research into online hostility by proposing a rubric for identifying "true" flames. Our aim will be to demonstrate how O'Sullivan's and Flanagin's use of narrow disciplinary framings and theoretical lenses contributes to the problems discussed above, thus setting the scene for the following section's discussion of the benefits of employing a transdisciplinary approach in the field of communications.

Before continuing, though, two clarifications are in order. First, we have chosen this article not because we think its scholarship fails to meet the relevant disciplinary standards, but, on the contrary, because it meets and perhaps even exceeds them – it is a robustly cited, well-regarded and clearly influential scholarly work.[24] Precisely because it exemplifies a prominent and enduring ideal in communications scholarship – one that is observable in much other research into online hostility (Jane, 2015) – it offers a felicitous opportunity to tell our cautionary tale of the pitfalls of mono-disciplinary. Second, we also acknowledge that O'Sullivan's and Flanagin's article was published in 2003 –

a time when cyberhate was significantly milder and less prevalent than it is today. It is thus possible that they devised their rubric after considering far less graphic and explicitly threatening material than our unexpurgated examples. Critically, however, we simply don't know to what material they were responding because – as with the overwhelming majority of academic scholarship on this topic – they too do not quote a single verbatim example.

O'Sullivan's and Flanagin's "Interactional Norm Cube"

O'Sullivan and Flanagin express concern that definitions of flaming – generally understood to refer to "aggressive or hostile communication occurring via computer-mediated channels" – are "imprecise within, and inconsistent across research projects" (2003, 70). As such, their aim is to establish a meticulous, operational definition and taxonomy of flaming to reduce this imprecision and provide a solid foundation for future research into the prevalence, causes, social consequences, and remedies (2003, 70). To this end, they propose a rubric (which they call an "interactional norm cube") which defines "true" flames as only those communications that satisfy three criteria:

  i. the sender's intent is to violate norms, and
 ii. the message is perceived as a norm violation by the receiver, and
iii. a third-party observer (2003, 82).

O'Sullivan's and Flanagin's determination to avoid a hyper-focus on message content, and instead to focus on a relational (as opposed to a transmission) model of communication by accommodating factors such as interpretation, intentionality, context, situated and evolving norms, and channel choice, reflect the key disciplinary concern within communication studies to acknowledge the richness of communicative phenomena. Yet, despite their explicit aim to provide an operationalizable definition of "flaming," their interactional norm cube not only fails the test of operationalizability, but ethically it is also deeply troubling.

The most striking problem with O'Sullivan's and Flanagin's interactional norm cube is that its admirable scholarly even-handedness and precision inadvertently privileges the rights, experiences and claims-making of (potentially law-breaking) cyberhate senders at the expense of (potentially harmed and imperiled) cyberhate victims. This is because, by their definition, an utterance can only be classified as a "true" flame if researchers are able to obtain not just information directly from its author, but *veridical* information with regards to the sender's intention to violate norms. However, identifying and locating the (often anonymous) producers of hostile messages is notoriously difficult and is one of the reasons why only a small number of cyberhate senders anywhere in the world have ever been identified and tracked down by police. But even if a researcher managed to identify and locate, for example, the person who said they were coming to Wu's house to rape her with a hunting knife, what chance would they realistically have that this person would: (i) agree to a candid interview with an academic (given this may involve owning up to having committed a criminal offense); (ii) be willing to tell the truth about whether or not they intended their tweet to violate norms (presuming they know what this academic language even means); and (iii) that they would be cognizant of the full nature of all their intentions (given that people often have multiple and sometimes contradictory motivations, not all of which are even transparent to themselves)?

The few cyberhate producers who *have* been identified and asked to account for their actions have almost invariably claimed that their computer must have been hacked or used by someone else, or that their material was just a joke – a playful instance of edgy, insider humor misunderstood by oversensitives and hysterics (Jane, 2017a, 82). The fact that some scholars and other commentators concur with such self-serving self-assessments while others reject them, also raises questions about whether the warrant of a single, third-party assessor achieves anything close to the sort of scholarly rigor, impartiality and consistency the rubric is meant to achieve (Jane, 2015, 71).

O'Sullivan's and Flanagin's interactional norm cube sets the bar so high that few messages will ever even clear the first hurdle (owing to the difficulties of identifying and interviewing attackers and, to a lesser extent, targets), and the few remaining messages – if any – will be extremely unlikely to be classified as "true" flames (owing to the frequency with which cyberhate producers deny that their utterances were intended to cause offense). If an academic study of online hostility faithfully employed O'Sullivan's and Flanagin's rubric, its investigators might well conclude that there is close to *no* cyberhate on the internet.

The problem with O'Sullivan's and Flanagin's approach can be fruitfully recast as a specific instance of a general problem with a mono-disciplinary approach to academic research. A distinguishing feature of their article is their concerted effort to position online hostility first and foremost as a communication phenomenon. Accordingly, they conceptualize and investigate what they call "flames" (and what we call "cyberhate") through the disciplinary lenses of communication studies. However, while this is perfectly normal, justified, proper and perhaps even praiseworthy scholarly conduct when viewed from *within* the disciplinary norms and conventions of communication studies, their mono-disciplinary bias staggeringly under-plays what is at stake for victims in a way that is both scientifically and ethically problematic. After all, by O'Sullivan's and Flanagin's reckoning, if a sender maintains that they did not intend to violate norms, then the proper classification of a plausible threat to maim, rape or kill someone is as a mere miscommunication, rather than as morally and legally reprehensible misbehavior.

Admittedly, the "rather than" part of the previous sentence is our own. O'Sullivan and Flanagin do not maintain that such behavior is benign. However, when the vast majority of aggressive or hostile communications won't even register as instances of flaming, we find it difficult to draw another conclusion. Moreover, given the respected status of academic research – which, with increasing frequency, is encouraged to be translational (for instance by guiding government policy), and which is increasing often reported in the media especially when its findings are surprising or contentious – it would be naïve or plainly disingenuous for scholars to maintain that the results of such research are immune from being used as ammunition by cyberhate apologists. Finally, their narrow disciplinary framing sets no room aside to also classify the few utterances that might be diagnosed as "true" flames as something in addition to a communicative act (for instance, a human rights violation, or a criminal or civil offense).

Further, O'Sullivan and Flanagin also propose squeezing other high-stakes phenomena into a communication studies context in a similarly scholarly-unhelpful and ethically-questionable manner. For instance, they articulate the hope that their interactional–normative approach might be applied to "problematic interactions" in offline contexts such as sexual harassment and hate speech (2003, 88). Thus, for instance, if an office junior perceives her employer's lewd comments and breast touching as sexual harassment, but her boss maintains that he is just being friendly – or if a person of color perceives a dark-skinned effigy hanging from a tree as a form of hate speech, but the white supremacist who placed it there is adamant that it's just a bit of fun – then O'Sullivan's and Flanagin's analysis would have us categorize these acts as nothing more serious than miscommunications since neither perpetrator stated unequivocally that they intended to violate norms. We are reminded, here, of similarly dubious scholarship in anti-fan studies which positions hate speech in and about digital media as being on a par with a bad review of a television program (cf. Jane, 2014b; Jane, 2019). Consider, too, the political scientist Richard MacKinnon's suggestion – extrapolated from his consideration of a virtual rape – that victims of sexual assault could recode their sex organs as being on a par with other body parts, such that victims of sexual assault become victims of assault only, thereby experiencing "little if any social repercussions" (1997). We regard such suggestions as not just infeasible and quite frankly mind-boggling, but as highly insensitive and bordering on victim-blaming.

Analogous Examples From Other Scholarly Domains

To forestall potential misunderstandings, we by no means intend to imply – nor should what we have said above be taken to either explicitly or tacitly support the claim – that the problems we discussed above are endemic to the field of communications. To illustrate how analogous concerns imperil progress in other disciplines, we offer two brief examples.

For instance, we have argued elsewhere (Vincent & Jane, 2018) that because of the way the contemporary debate about "smart drugs" has been framed – namely, as a debate about the state regulation of pharmaceuticals and medical devices – it has become effectively impossible to point out (and to have the relevance and significance of this point noted by scholars working in the respective fields of neuroethics and bioethics) that *medically* safe smart drugs can still have *socially* adverse side-effects, and that the regulation of smart drugs would not *curtail* individual freedom but *enhance* diachronic self-control. However, because scholars in neuroethics and bioethics – disciplines which have an overtly medical focus – have claimed this topic as their own, *social* side-effects do not even register on their radar as effects. And because regulation has been framed as something that the state imposes on citizens in order to block prohibited choices – the idea of regulation as an instrument that citizens might use to enhance their diachronic freedom is not just heresy – it has become effectively unthinkable.

For another example, consider the work of Stephen J. Morse, who appeals to criminal law's incumbent definition of criminal guilt to rebuff attempts by scholars from the mind sciences to prompt a reconsideration of the criteria of criminal guilt (2013). In one way, Morse is clearly right that since the law makes no reference to neuroscientific criteria, unless neuroscientific facts can be translated into criteria that are relevant to criminal guilt – or unless their relevance to those criteria can be explained in some other way – then neuroscientific findings will make no difference to the criminal law. In another – and we think a more important – way, though, what Morse's retort completely seems to overlook is that his interlocutors are attempting to prompt him and other legal scholars to reconsider the adequacy of the criminal law's criteria for criminal guilt.

Just as the incumbent medical and individualistic framing of the contemporary debate about smart drugs blinds neuroethics and bioethics scholars from recognizing what are, in our view, clearly important considerations – and even prevents them from seeing regulation as a freedom-enhancing instrument – so, too, O'Sullivan's and Flanagin's narrow disciplinary approach treats the cases of aggressive and hostile communications that their rubric discards as irrelevant noise, and blinds them to the suffering that they should ideally at least treat as a potential reason to reconsider the usefulness of their rubric. And in exactly the same way as Morse's steadfast adherence to criminal law doctrine prevents him from even noticing the salience of evidence that should ideally prompt him to at least *consider* reappraising the criteria of criminal guilt, so, too, O'Sullivan's and Flanagin's rubric – and, more generally, the mono-disciplinary lenses of communications and other disciplines – also threatens to hinder progress within those disciplines.

Perils of Mono-Disciplinarity

Disciplinary framings are powerful and important. They enable scholars to dig down deep into detail that would otherwise overwhelm non-specialists. Without conceptual and terminology clarity, and expert tools that make it possible for the academy to achieve its distinctive and valuable brand of rigor, much would be lost due to vagueness and imprecision. At the same time, though, when disciplinary framings, conventions, and lenses prevent the respective disciplines' scholars from even recognizing certain things as potentially relevant to their field of inquiry – or when they filter out certain phenomena so that they can no longer even be seen, or seen *as* something that might warrant their attention and potentially even contemplate reshaping the boundaries of that discipline – then those disciplinary framings, conventions, and lenses become shackles that risk impeding rather than promoting scholarly progress.

We fully acknowledge that the norms and conventions of other disciplines have no less impact on how those disciplines' scholars frame their research, what questions they seek to answer, what methods they employ, and how much room they leave for the phenomena that they study to be understood in a different way, than the impact of the norms, conventions, and lenses in the field of communications. However, the fact that this practice is so commonplace and widespread in the academy is hardly a valid line of defense of this practice. We fully appreciate the importance and value of the autonomy of different disciplines, and that there *are* important reasons to leave some matters to the expertise of scholars in other disciplines who are better equipped to deal with those matters. At the same time, though, when disciplinary norms start preventing their own scholars from recognizing that some observations may furnish them with reasons to turn their sharp critical scholarly gaze to the fences we erect around our disciplines – to reconsider what falls within the purview of our disciplines – this risks hindering disciplinary progress.

As we foreshadowed earlier, we believe it is possible to address – or at least to mitigate – the problems with approaches such as O'Sullivan's and Flanagin's by adopting a transdisciplinary approach in communications scholarship, which is the topic to which we shall now turn our attention.

Attractions of Transdisciplinarity

Like communication studies, the (increasingly mature) field of internet studies is multi-disciplinary/interdisciplinary in nature (Dutton, 2013, 8). That said, the rigid approaches and orthodoxies associated with various disciplines have still tended to produce a fragmented and incomplete coverage of online hostility in the literature (Jane, 2015). This comports with the Chilean economist Manfred A. Max-Neef's observation that interdisciplinary practices such as "creating supposed teams conformed of specialists in different areas, around a given problem" risk producing "a series of reports pasted together" – an "accumulation of visions" instead of an "integrating synthesis" (2005, 6). Thus, while interdisciplinarity is often evoked as a selling point for research projects because it suggests "good relations between the disciplines among themselves," and/or is offered as a sort of "repair tool" for the perceived lack of utility and comprehensibility associated with the increasing particularization and complexity of subject matters and disciplines (Mittelstrass, 2011, 229–230), it is by no means a panacea.

Furthermore, even the most resolutely interdisciplinary research approaches are still likely to have pre-defined research agendas determined by and beholden to institutional imperatives, as well as framings or emphases on some aspects of the project at the expense of others, rather than being formulated as best-fit responses to objects of analysis. This has the potential to not only dramatically reduce the real-world usefulness of research projects because, as Garry D. Brewer wryly observes, "The world has problems, but universities have departments" (1999, 328),[25] but also to lead them to entirely miss the point of the phenomena under investigation as per our critique of O'Sullivan's and Flanagin's approach.

By contrast, transdisciplinarity is a "comprehensive, multiperspective, problem- and solution-oriented approach that transcends disciplinary boundaries and bridges science and practice" (Pohl, 2011). The term "transdisciplinarity" is associated with distinct yet interrelated contexts in the sciences (Osborne, 2015, 10–14), in which scholars have been motivated to overcome the limitations of specialization and institutionalized, disciplinary knowledge silos. "Transdisciplinarity" is not a mere synonym for some form of massive "inter-disciplinary" or "multi-disciplinary." Rather, the type of transdisciplinarity to which we refer has five distinct features that set it apart from the gathering-of-many-minds-and-disciplinary-approaches-around-a-single-problem approach sketched above. Specifically, it is concerned with:

i. tackling complex, real-world problems
ii. where the nature of the problem itself may be obscure or in dispute and that therefore
iii. demands that scholars transcend disciplinary paradigms and bridge science with practice
iv. partly by seeking out and integrating diverse views of multiple stakeholders (including stakeholders outside academia)
v. with the aim of devising analyses and interventions which promote flourishing.

We foresee that the motivation behind working closely with stakeholders and co-creating actionable problem-statements that address complex real-world challenges – core features of transdisciplinary research – could be easily misconstrued, criticized, and dismissed. For instance, as an instance of the increasing encroachment of inappropriate activist/political agendas that have no legitimate place in respectable academic scholarship. Personally, we take issue with the suggestion that entanglement with real-world problems tarnishes academic scholarship; quite to the contrary, our view is that it greatly enhances its value (Vincent, 2016). Nevertheless, setting aside our personal views on this matter, and anticipating this potential objection, we quote Florin Popa, Mathieu Guillermin, and Tom Dedeurwaerdere at length, to explain why this core feature of transdisciplinarity reflects not only an ethical commitment but also a deeply important *epistemic* consideration grounded in pragmatist thought – an epistemic consideration that, in our view, has much scientific merit:

> Pragmatism distances itself from both value neutrality and value relativism by conceiving of knowledge production as a social and reflexive process whereby criteria of scientific credibility and legitimacy are jointly defined within a community of inquiry… If the dominant discourse on interdisciplinarity… focused on articulating the contributions of different disciplines into a coherent framework, the more recent analyses of transdisciplinarity have shifted the focus towards the extended co-production of knowledge (by scientific and extra-scientific actors)… [T]ransdisciplinarity does not aim at establishing a common theoretical framework, but rather at fostering self-reflection; it calls for humility, openness to others, a contextualization of our own knowledge, and a willingness to engage with and be moved by others. Without an explicit reflexive dimension, transdisciplinarity [would be] confronted with the risk of either being reduced to formal social consultation… or evolving towards a politicized form of "democratic science" in which epistemic aspects are subordinated to procedures of social legitimation … [Thus, i]n the case of scientific research, the role of such reflexive processes is to encourage process of critical assessment and social learning on the background values and assumptions guiding research, and on the socio-institutional structures supporting particular norms and practices.
>
> *(2015, 47, internal quotation marks and citations omitted)*

In light of the above, from a transdisciplinary perspective, unless a problem-statement under investigation is itself left at least partly open to being co-shaped by those affected – i.e., the stakeholders – then even the most inter- and multi-disciplinary teams of investigators will run the risk of investigating problems and finding solutions that not only miss the bigger picture, but that stand on shaky epistemic foundations. In mono-, inter-, and multi-disciplinary projects, whether a problem-statement or solution finds resonance with stakeholders, and whether it improves their lives – rather than, for instance, mischaracterizing or obscuring the problem, as with the scholarly work on cyberhate that we criticized above – are not only things that get left up to chance, but they are viewed as optional extras. Even more cynically, they are viewed as the required after-thought add-ons that scholars must append to their grant applications to sell their research projects to grant funding bodies that are increasingly looking for evidence of the projects' potential to tangible benefits for society. By contrast, in transdisciplinary research, the inclusion of stakeholders reflects the pragmatist view that an analysis of a situation that either has no actionable outcomes – or, worse, that produces findings and/or interventions that stakeholders find aversive – would stand on shaky epistemic ground.

Rather provocatively, Richard Rorty – himself a pragmatist – summarizes pragmatist epistemology as follows:

> Pragmatists substitute the question "which descriptions of the human situation are most useful for which human purpose" for the question "which description tells us what the situation really is?" Pragmatism puts natural science on all fours with politics and art. It is one more source of suggestions about what to do with our lives.
>
> *(2007, 916–917)*

The image of science on "all fours with politics and art" – and, for all we know, perhaps also with alchemy, phrenology, astrology, tarot, religion, and poorly-considered lay views – will clearly sound abrasive to those with a background in science, and a commitment to the scientific method and scholarly rigor. However, looking beyond Rorty's (we suspect intentionally) abrasive provocation, to our mind, pragmatism's humility *vis à vis* our regrettably limited epistemic position, and its emphasis on the need to stay on the lookout for- and be prepared to re-evaluate our theories in light of observations – and, more generally, to achieve the best available fit among our different bodies and sources of knowledge – is an ideal to strive for, a paradigmatic commitment to eschew dogma in favor of scholarly rigor and integrity.

We concur with Popa, Guillermin, and Dedeurwaerdere's view that pragmatism offers a solid and principled foundation for transdisciplinary scholarship, and in what follows we discuss what a transdisciplinary approach to research on cyberhate might look like, and what we see as its attractions.

Transdisciplinarity and Cyberhate

For Max-Neef, complex contemporary issues such as forced migrations, poverty, environmental crises, violence, terrorism, and neo-imperialism represent some of the defining *problematiques* of the twenty-first century (2005, 5). They are challenges well-suited to transdisciplinary because they involve problems that cannot be adequately tackled from within any specific individual discipline or by combining them into multi-disciplinary units, or via interdisciplinary collaborations (Osborne, 2015, 10) – in part because of the aforementioned need to remain flexible about how a problem-statement should be defined, but in good measure also because complexity gives rise to what are sometimes called "wicked problems" in which due to the existence of multi-level causal feedback loops, neither an understanding of the problem nor adequate interventions can be designed if disciplinary silos remain the norm (Brown, Harris & Russell, 2010; Mazzucato, 2018). This coheres with the views of Gertrude Hirsch Hadorn et al who argue that transdisciplinary research is needed "when knowledge about a societally relevant problem field is uncertain, when the concrete nature of problems is disputed, and when there is a great deal at stake for those concerned by problems and involved in dealing with them" (2008, 34).

In light of the five criteria for transdisciplinarity we identified in the previous section, and Max-Neef's characterization of twenty-first-century *problematiques* (2005, 5), gendered cyberhate is an ideal candidate for transdisciplinary scholarship. It is an extremely socially relevant problem, yet there remain many knowledge gaps – as well as dispute – about its nature and impact. A transdisciplinary approach would be well-placed to recognize the pressing and high-stakes nature of the issue, both in terms of the serious harm caused to targets, as well as the impact on society at large (in terms of, for example, its undermining and erosion of ideals relating to equity, democratic inclusivity, protection of minority groups, and so on). A transdisciplinary approach can also recognize that cyberhate involves multiple stakeholders (such as victims, policy makers, police, ethicists, human rights activists, platform operators and corporations, scholars and educational institutions, technologists, concerned citizens, activists et cetera), and harness the expertise of these stakeholders to both better-understand the problem, and to devise novel interventions.

From O'Sullivan's and Flanagin's communication studies perspective, the main problems associated with hostility online are taxonomical and/or definitional – and from this narrow academic perspective they surely are. The problem, however, is that this runs the risk of making their work academic in a pejorative sense – i.e., irrelevant to the phenomenon under investigation. If they had engaged with literature from other fields as well as with stakeholders (including and especially cyberhate victims), they might well have reassessed their position and concluded that the most pressing problems associated with hostility online are actually ethical and/or political. Although we in no way deny that terms like "flaming" are elastic and vague, what O'Sullivan's and Flanagin's remedy – devising a rigorous rubric – effectively does is to privilege the *academic value* of precision and clarity over *the victims' needs*. And while we wholeheartedly agree that without clear definitions to guide empirical research, the results obtained will be ambiguous, what would have been a more helpful approach – from a perspective that gives adequate recognition to scholarly rigor and ethical matters – is if they had treated this vagueness as a reason to more closely investigate the phenomenon from the perspective of the many different stakeholders and using various disciplinary framings, instead of imposing criteria from outside and prior to such a finely-textured transdisciplinary investigation onto a clearly very complex phenomenon, which had the unfortunate effect of classifying most if not all flames as not "true" flames.

Input from feminist theory and gender studies, for instance, would have alerted these researchers to their failure to factor in the larger issues of power differentials, and structural inequity and oppression when analyzing individual messages. The critical importance of this big picture context would likely be affirmed by cyberhate victims who would probably also strenuously object to the reconceptualization moves O'Sullivan and Flanagin recommend *vis à vis* reclassifying what is arguably likely to be the majority of cyberhate, sexual harassment, and hate speech as nothing more than miscommunication. Having had first-hand experience of some of these phenomena ourselves as well as having interviewed a large number of other people who have had first-hand experience of these phenomena, we certainly find it very difficult to imagine these clinically detached and disembodied conceptual moves being endorsed as fair, or seeming even remotely feasible to targets/victims/survivors of these types of abuses. Given that the onus is placed on targets, victims, and survivors to do the requisite recoding of the various attacks against them, we also think these approaches have undertones of victim-blaming. If O'Sullivan and Flanagin had paid more careful attention to the perspectives of victims, they might have avoided this unfortunate situation, as well as discovering more productive and useful ways to frame the cyberhate problem, and to study it from a communications perspective but in a way that does not throw the baby out with the bathwater.

The researchers unquestioningly assume that flame senders will know whether or not they intended to violate norms and will tell the truth about this fact in an interview – thus making it extremely easy for them to establish intention. Yet there exists a multitude of insights into the etiology, nature and prevalence of self- and other-deceit that could be gained by considering theories and research findings from fields, disciplines, and contexts outside of communications studies (for example, social and applied psychology, jurisprudence, criminology, and sociology) which speak to the simple fact that intentionality is complex and people lie. Criminal law is particularly instructive in this regard. For instance, despite the difficulties involved in proving *mens rea* (that is, the degree of intention involved in the committing of a forbidden act) in criminal court cases, an accused person's insistence that they are innocent or did not mean to cause harm will not necessarily be accepted. Consider a recent example involving an Australian Army captain who anally penetrated a male colleague with a beer bottle in the toilets during a work function (Byrne, 2018). Given the captain's claim that this was just an instance of over-the-top tomfoolery (as opposed to, say, a norm violation), would it really be appropriate to classify this "problematic interaction" as merely a *miscommunication*? The court-martial panel hearing the case certainly did not think so. It rejected the defendant's account, found him guilty of rape and jailed him for three months (Back, 2018). Instructive, too, is the fact that – as with many court proceedings – the tribunal in this case considered a victim's impact statement from the man who was raped to assist in its deliberations and decision-making.

Another useful conceptual/theoretical perspective comes from literature on affect theory, such as Sara Ahmed's suggestive point that, "it is a common theme within so-called hate groups to declare themselves as organisations of love" in that they frame their activities not as hateful attacks but as a loving protection of their beliefs and ideals, a "defence against injury" (2004, 42). Thus, while it might appear that members of a white supremacist terrorist organization such as Aryan Nations *hate* mixed-race couples, Jews, immigrants and so on, they may claim – and even believe themselves – that they are motivated by *love* – of the average white man, white housewife, white farmer, white Christian and so on (Ahmed, 2004, 42). In the context of cyberhate, we could consider those members of anti-Islamic communities who routinely use social media to abuse and threaten the Australian lawyer Mariam Veiszadeh. These people might well claim they are *protecting* rather than *violating* norms. Yet, outside of the echo chambers of such groups, we suspect it would be difficult to find many people who would agree that messages calling Veiszadeh "a Satan worshiping [cunt]," a "fucking whiny baby cunt whore," and a "muslim loving cunt" who should have the "shit and fuck" kicked out her or be lynched if she doesn't "fuck off back to the sewers" (cited in Sargeant, 2016) are anything other than "true" cyberhate.

Finally, on pragmatist grounds there are serious epistemic problems with giving any one of three parties (sender, receiver, observer) the absolute power of veto over how an utterance should be categorized. Consider, for instance, Charles Sanders Peirce who critiques Cartesian methods of reasoning for relying on "a single thread of inference" that is no stronger than its weakest link (1992, 28–29). Peirce's recommendation for sound philosophical inquiry, in contrast, is to "trust rather to the multitude and variety of … arguments than to the conclusiveness of any one" (1992, 29). Reasoning, in Peirce's view, "should not form a chain which is no stronger than its weakest link, but a cable whose fibres may be ever so slender, provided they are sufficiently numerous and intimately connected" (1992, 29).

The transdisciplinary staples of stakeholder engagement and practical problem solving also appear – albeit in different guises – in the pragmatist literature. An example is Peirce's claim that the focus of epistemological inquiry should not be on showing how we can possess absolute certainty but instead ought to adopt methods of inquiry that contribute to our making fallible progress because inquiry is a "community activity" and the method of science has a self-correcting character (Hookway, 2016). James, meanwhile, advocates seeking knowledge that remains in contact with the exigencies of practice

(Hookway, 2016). These comments, in turn, are reminiscent of George Soros' writing on fallibility and reflexivity, in which he argues that people have imperfect knowledge about the world which – in virtue of thought's *manipulative* as well as *cognitive* function (roughly, that what we think, even if this departs from reality, alters reality) can influence the situation to which they relate through the actions of the participants (2014).

The transdisciplinary approaches we sketched above have the potential to expand the types of interventions or remedies that could be considered for gendered cyberhate. For instance, while victims and activists often call for increased regulation in response to cyberhate, many people have an in-principle opposition to increased regulation because it has the unsavory ring of "big government" restrictions on individual freedom. Apart from our earlier comments about why this view of regulations is heavily influenced by an unhelpful framing, we also wonder if at least some of those who view regulation as the arch-nemesis of liberty might reconsider their *a priori* opposition if they understood that the types of interventions that come under the umbrella of "regulation" do not need to involve a binary choice between prohibition and permission, but can be situated along an axis that includes at least the following: prohibit – discourage – permit – encourage – require (Human Rights and Technology Issues Paper: UTS Submission, 2018, 90). Finally, legislative responses need not focus solely criminal or civil sanctions targeting individual offenders but could, instead, mandate the teaching of and/or provide grant-based incentives for projects that employ Value-Sensitive Design (VSD) methods (Jane & Vincent, 2018, 58–65) to curb the frequency and severity of cyberhate.

Conclusion

This chapter has focused on the role academics have played – and risk continuing to play – with regard to misperceptions about gendered cyberhate, and the flow-on effect this is likely to have in terms of formulating interventions. It is not our intention to imply that scholars are the only people to blame for these solution-hindering misperceptions, or that they are more culpable than any other stakeholders. Setting aside the fact that this is a scholarly venue with a particular focus on conceptual and theoretical framings, we have focused on the academy because we maintain that if more researchers (i) choose to illustrate their work with more vivid case studies, and (ii) consider giving a rest to their tried-and-true but also potentially blinkered disciplinary lenses in favor of more transdisciplinary approaches, scholars could play a key role not only in tackling gendered cyberhate, and in responding to the multiple social challenges posed by emerging technologies in general, but also as an added bonus reimagine what could fall into the domain of phenomena studied (and the lenses used to study them) within the field of communications.

The impulse to position an object of analysis in our own fields – or only those fields about which we are knowledgeable and within which we are comfortable – is eminently understandable. It makes sense to attempt to make sense of things within frameworks with which we are familiar and within which we have expertise. Furthermore, the contours of contemporary scholarship – for example, the pressure to constantly justify our disciplinary approaches and to strenuously advocate for the reliability and validity of our research models and study results – encourages exaggerated claims relating to Archimedean vantage points and scientific or quasi-scientific objectivity (even in, possibly *especially* in, the Humanities).

However, once we become accustomed to doing things using certain approaches and to vigorously justifying these approaches, it is all-too-easy to forget our positionality and to assume our way is the *only* way. By this, we are not referring to the process of surveying a number of methodological, conceptual, and theoretical apparatus and making decisions about which we think are best-suited to our project and its goals (which, after all, is ultimately what *we* have done by deploying and advocating for transdisciplinarity in this chapter). Rather, what we mean is that we forget or fail to notice that we are even using lenses in the first place. But yet, as we noted early on in this chapter, our disciplinary lenses are just as much a way of seeing as they are a way of not seeing. This behooves every scholarly discipline to navel-gaze from time to time in order to do our best as academics to try to notice what our theoretical lenses, powerful as they may be, might be filtering out from our view. Our overall case, therefore, is that as scholars we must be mindful of how our various lenses might shape – both epistemically and potentially ontologically[26] – the sorts of things we see and (in cases such as the interactional norm cube) *not* see.

Notes

1 For a discussion of this phenomenon and the "gamification" of abuse online, please see: Jane (2012, 533–534, 2014a, 560–561, 2017a, 50).
2 Please see the section entitled *The Harms of Gendered Cyberhate.*
3 For a lengthy discussion of this topic and citations of examples of this discourse, please see: Jane (2017a, 76–87).
4 See, for instance: Jane (2012, 2014a, 2015, 2016, 2017a, 2017b, 2017c, 2017d), Jane and Vincent (2017a, 2017b), Vincent and Jane (2017a, 2017b) and most recently Jane and Vincent (2018).
5 For clarity's sake, our claim is not that this is the *only* reason, but that this is *an important* reason.
6 The government funding for this study was in the form of a Discovery Early Career Researcher Award (DE150100670) which was awarded to Dr Jane in 2015 and which funded a three-year project called "Cyberhate: the new digital divide?" The UNSW Sydney Research Ethics Committee reference for this project is HC15012.
7 For details about this Symposium, please see: Jane and Vincent (2018, 8–14).

8 Commutation tests are simple thought experiments in which one element of a text or concept is switched with another that is different but similar enough to expose those things which are "too obvious to see" (McKee, 2003, 107–106).
9 For a lengthy discussion of the failure of policy, policy makers, and corporate platform operations to respond to gendered cyberhate, please see: Jane (2017a, 88–111).
10 Also see: Duggan (2014), Powell and Henry (2015).
11 For coverage of the harms of gendered cyberhate, including the *embodied* harms of gendered cyberhate, see, for example: Citron (2014), Henry and Powell (2015), Mantilla (2015), Jane (2017c), Jane and Vincent (2018).
12 For a lengthy discussion of these issues, please see: Jane (2017a, 53–77, 2017c, 2017d), Jane and Vincent (2018).
13 "Web 1.0" refers to the early decades of the internet when content was mostly static and read-only. "Web 2.0," in contrast, refers to changes – from around 2006 – in the design and use of the web which has facilitated user-generated material, interactivity, collaboration, sharing, and so on.
14 For a lengthy discussion of these particular impacts of gendered cyberhate, cf.: Citron (2014), Jane (2017a), Jane and Vincent (2018).
15 "Digital divide" is a term used to discuss online equity and refers to differences between population groups in in terms of access to and use of information and communications technologies.
16 See, for instance: Jane (2016, 2017b), Jane and Vincent (2017a, 2018, 46–57), Vincent and Jane (2017a, 2017b), Vincent (2017).
17 See: Jane (2012, 2014a, 2015), Jane and Vincent (2018).
18 For discussion, see: Jane (2012, 2015, 2018).
19 "Jezebel" is a web site with a feminist approach to reporting pop culture.
20 The messages in this list are sourced from: Vingiano (2014) and Bianco (2014).
21 In particular, the use of vague, bland, and generic terminology, and the lack of unexpurgated examples.
22 That it fails to convey – and indeed conveys a misleading impression, as well as re-enforces that impression – that cyberhate is benign.
23 To use accurate terminology and quote verbatim examples.
24 According to Google Scholar, this article has been cited 313 times – see https://scholar.google.com.au/scholar?hl=en&as_sdt=0%2C5&q=%E2%80%9CReconceptualizing+%E2%80%98flaming%E2%80%99+and+other+problematic+messages%E2%80%9D+&btnG – and Flanagin's body of work has been cited 10,921 times – https://scholar.google.com.au/citations?user=nbvdjK0AAAAJ&hl=en&oi=sra
25 While Brewer says this in the context of highlighting the benefits of *inter*disciplinarity, our view is that the benefits are significantly more pronounced in *trans*disciplinarity.
26 For an example, we can consider one of the examples Soros provides to illustrate his ideas on reflexivity, namely, that if investors believe that markets are efficient, "then that belief will change the way they invest, which in turn will change the nature of the markets in which they are participating (though not necessarily making them more efficient)" (2014, 310).

References

Ahmed, S. (2004). *The Cultural Politics of Emotion*. New York: Routledge.
Amnesty International (n.d.) "Waterboarding Is Torture: 3 Things You Need to Know". Retrieved from www.amnestyusa.org/waterboarding-is-torture-3-things-you-need-to-know/on December 27, 2018.

Back, A. (2018, December 5). *The Canberra Times*. Retrieved from: https://www.canber ratimes.com.au/story/5998451/army-captain-jailed-for-raping-colleague-with-beer-bottle/

Bianco, M. (2014, August 30). Woman Turns the Tables on Misogynistic Troll in the Best Way Possible. *Mic*. Retrieved from: https://www.mic.com/articles/97640/woman-turns-the-tables-on-misogynistic-troll-in-the-best-way-possible

Brewer, G. D. (1999). The Challenges of Interdisciplinarity. *Policy Sciences*, 32: 327–337.

Brown, V. A., Harris, J. A. & Russell, J. Y. (2010). *Tackling Wicked Problems: Through the Transdisciplinary Imagination*. London and New York: Earthscan.

Burke, K. (1935). *Permanence and Change*. Berkeley and Los Angeles, CA: University of California Press.

Byrne, E. (2018, December 4). Army Captain Who Violated Colleague in "Tomfoolery Gone wrong" Found Guilty of Rape. *abc.net.au*. Retrieved from: www.abc.net.au/news/2018-12-04/army-officer-guilty-rape-beer-bottle-prank/10582462

Citron, D. K. (2014). *Hate Crimes in Cyberspace*. Cambridge, MA, and London, UK: Harvard University Press.

Cyber violence against women and girls: A world-wide wake-up call. (2015). *The United Nations Broadband Commission for Digital Development Working Group on Broadband and Gender*. Retrieved from: www.unwomen.org/~/media/headquarters/attachments/sec tions/library/publications/2015/cyber_violence_gender%20report.pdf

Duggan, M. (2014, October 22). Online Harassment. *Pew Research Center*. Accessed from www.pewinternet.org/2014/10/22/online-harassment/

Dutton, W. H. (2013). Internet Studies: The Foundations of a Transformative Field. In Dutton, W. H. (Ed.). *The Oxford Handbook of Internet Studies*. Oxford: Oxford University Press, pp. 1–26.

Henry, N. & Powell, A. (2015). Embodied Harms: Gender, shame, and Technology-facilitated Sexual Violence. *Violence Against Women*, 21(6): 758–799. DOI: 10.1177/1077801215576581

Hirsch Hadorn, G., Biber-Klemm, S., Grossenbacher-Mansuy, W., Hoffmann-Riem, H., Joye, D., Pohl, C., Wiesmann, U. & Zemp, E. (2008). The Emergence of Trans-disciplinarity as a Form of Research. In Hirsch Hadorn, G., Hoffmann-Riem, H., Biber-Klemm, S., Grossenbacher-Mansuy, W., Joye, D., Pohl, C., Wiesmann, U. & Zemp, E. (Eds.). *Handbook of Transdisciplinary Research*. Vienna: Springer, pp. 19–39.

Hookway, C. (2016, Summer Edition). In Zalta, E. N. (Ed). *The Stanford Encyclopedia of Philosophy*. Retrieved from: https://plato.stanford.edu/archives/sum2016/entries/pragmatism/

Human Rights and Technology Issues Paper: UTS Submission. (2018). Submission in Response to the Australian Human Rights Commission's *Human Rights and Technology Issues Paper*. Retrieved from: https://tech.humanrights.gov.au/submissions?mc_cid=3526cb82bd&mc_eid=8328e29bcc

iFactory. (2017, April 28). How old is the internet? Retrieved from: https://ifactory.com.au/news/how-old-internet

Jane, E. A. (2012). "Your a ugly, whorish, slut" – Understanding E-Bile. *Feminist Media Studies*, 14(4): 531–546. DOI: 10.1080/14680777.2012.741073

Jane, E. A. (2014a). "'back to the kitchen, cunt": Speaking the Unspeakable about Online Misogyny. *Continuum: Journal of Media & Cultural Studies*, 28(4): 558–570. DOI: 10.1080/10304312.2014.924479 (p. 563).

Jane, E. A. (2014b). Beyond Antifandom: Cheerleading, Textual Hate and New Media Ethics. *International Journal of Cultural Studies*, 17(2): 175–190. DOI: 10.1177/1367877913514330

Jane, E. A. (2015). Flaming? What Flaming?: The Pitfalls and Potentials of Researching Online Hostility. *Ethics and Information Technology*, 17(1): 65–87. DOI: 10.1007/s10676-015-9362-0

Jane, E. A. (2016, August 4). What Bit about the Wrongs of Sexual Threats against Women Do Courts and Men Not Get? *The Conversation*. Retrieved from: https://theconversation.com/what-bit-about-the-wrongs-of-sexual-threats-against-women-do-courts-and-men-not-get-63447

Jane, E. A. (2017a). *Misogyny Online: A Short (And Brutish) History*. Loss Angels, CA and London and New Delhi: SAGE.

Jane, E. A. (2017b). Gendered cyberhate, Victim-Blaming, and Why the Internet Is More like Driving a Car on a Road than Being Naked in the Snow. In Martellozzo, E. & Jane, E. A. (Eds.). *Cybercrime and Its Victims*. Oxon: Routledge, pp. 61–78.

Jane, E. A. (2017c). Feminist Fight and Flight Responses to Gendered Cyberhate. In Segrave, M. & Vitis, L. (Eds.). *gender, Technology and Violence*. London and New York: Routledge, pp. 45–61.

Jane, E. A. (2017d). Gendered Cyberhate: A New Digital Divide? In Ragnedda, M. & Muschert, G. W. (Eds.). *Theorizing Digital Divides*. Oxon: Routledge, pp. 158–198.

Jane, E. A. (2018). Gendered Cyberhate as Workplace Harassment and Economic Vandalism. *Feminist Media Studies*, special edition on Online Misogyny. DOI: 10.1080/14680777.2018.1447344

Jane, E. A. (2019). Hating 3.0 And the Question of whether Anti-fan Studies Should Be Renewed for Another Season. In Click, M. (Ed.). *Anti-Fandom, Dislike and Hate in the Digital Age*. New York: NYU Press, pp. 42–61.

Jane, E. A. & Vincent, N. A. (2017a, July 18). Women Online are Getting Used to Cyber Hate. They Need to Get Used to Reporting It. *The Sydney Morning Herald*. Retrieved from: www.smh.com.au/lifestyle/news-and-views/opinion/women-online-are-getting-used-to-cyber-hate-they-need-to-get-used-to-reporting-it-20170717-gxctr8.html

Jane, E. A. & Vincent, N. A. (2017b, July 7). *Gendered Violence Online – A Scholarly "Slam"*. Red Rattler Theatre), Sydney.

Jane, E. A. & Vincent, N. A. (2018). Cyberhate and Human Rights. Submission in response to the Australian Human Rights Commission's *Human Rights and Technology Issues Paper*. Retrieved from: https://tech.humanrights.gov.au/submissions?mc_cid=3526cb82bd&mc_eid=8328e29bcc

Kiesler, S., Zubrow, D., Moses, A. M. & Geller, V. (1985). Affect in Computer-mediated Communication: An Experiment in Synchronous Terminal-to-terminal Discussion. *Human Computer Interaction*, 1, 77–104. DOI: 10.1207/s15327051hci0101_3

Lewis, R., Rowe, M. & Wiper, C. (2017). Online Abuse of Feminists as an Emerging Form of Violence against Women and Girls. *The British Journal of Criminology*, 57: 1462–1481.

MacKinnon, R. (1997). Virtual Rape. *Journal of Computer-Mediated Communication*, 4(2). DOI: 10.1111/j.1083-6101.1997.tb00200.x

Mantilla, K. (2015). *Gendertrolling: How Misogyny Went Viral*. Santa Barbara, CA: Praeger.

Max-Neef, M. A. (2005). Foundations of Transdisciplinarity. *Ecological Economics*, 53: 5–16.

Mazzucato, M. (2018). Australia Is Uniquely up to the Task of Solving 'Wicked' Problems around Inequality. The Guardian, 14 December 2018. Retrieved from: www.the guardian.com/commentisfree/2018/dec/14/australia-is-uniquely-up-to-the-task-of-solving-wicked-problems-around-inequality

McKee, A. (2003). *Textual Analysis: A Beginner's Guide*. London: SAGE.

Mittelstrass, J. (2011). On Transdisciplinarity. *Trames*, 15(65/60), 4: 329–338.

More than a quarter of UK women experiencing online abuse and harassment receive threats of physical or sexual assault – new research. (2017, November 20). *Amnesty International UK*. Retrieved from: www.amnesty.org.uk/press-releases/more-quarter-uk-women-experiencing-online-abuse-and-harassment-receive-threats

Morse, S. J. (2013). Criminal Common Law Compatibilism. In Vincent, N. A. (Ed.). *Neuroscience and Legal Responsibility*. New York: Oxford University Press, pp. 27–52.

O'Sullivan, P. B. & Flanagin, A. J. (2003). Reconceptualizing 'Flaming' and Other Problematic Messages. *New Media & Society*, 5(1), 69–94.

Osborne, P. (2015). Problematizing Disciplinarity, Transdisciplinary Problematics. *Theory, Culture & Society*, 32(5-6), 3–35.

Peirce, C. S. (1992). Some Consequences of Four Incapacities (1968). In Houser, N. & Kloesel, C. (Eds.). *Selected Philosophical Writings – Volume I (1867-1893)*. Bloomington, IN: Indiana University Press, pp. 28–55.

Pohl, C. (2011). What Is Progress in Transdisciplinary Research? *Futures*, 43, 618–626. DOI: 10.1016/j.futures.2011.03.001

Powell, A. & Henry, N. (2015). Digital Harassment and Abuse of Adult Australians: A Summary Report. *Tech & Me Project*, RMIT University. Retrieved from: https://research.techandme.com.au/wp-content/uploads/REPORT_AustraliansExperience sofDigitalHarassmentandAbuse.pdf

Ramberg, B. (2009, Summer Edition). Richard Rorty. In Zalta, E. N. (Ed.). *The Stanford Encyclopedia of Philosophy*. Retrieved from: https://plato.stanford.edu/archives/spr2009/entries/rorty/

Rorty, R. (2007). Dewey and Posner on Pragmatism and Moral Progress. *University of Chicago Law Review*, 74(3), 915–927.

Sargeant, C. (2016, July 4). A Muslim Advocate Wrote About Pauline Hanson & Boy Did The Racists Go Wild. *Pedestrian TV*. Retrieved from: www.pedestrian.tv/news/a-muslim-advocate-wrote-about-pauline-hanson-boy-did-the-racists-go-wild/

Smith, M. (2018, September 8). Four in Ten Female Millennials Have Been Sent an Unsolicited Penis Photo. *YouGov UK*. Retrieved from: https://yougov.co.uk/news/2018/02/16/four-ten-female-millennials-been-sent-dick-pic/

Soros, G. (2014, January 13). Fallibility, Reflexivity, and the Human Uncertainty Principle. *Journal of Economic Methodology*, 20(34): 309–329.

Thompsen, P. A. & Foulger, D. A. (1996). Effects of Pictographs and Quoting on Flaming in Electronic Mail. *Computers in Human Behavior*, 12(2), 225–243.

Tuovinen, L. & Röning, J. (2007). Baits and Beatings: Vigilante Justice in Virtual Communities. *Published in the proceedings of the 7th International Conference of Computer Ethics*; Philosophical Enquiry, Trondheim, 397–405.

Vera-Gray, F. (2017). "Talk about a Cunt with Too Much Idle time": Trolling Feminist Research. *Feminist Review*, 115: 61–78.

Vincent, N. A. (2016). How Can Philosophy Contribute to Public Debates and Discourse? *Blog of the American Philosophical Association*, April 6, 2016. Retrieved from: https://blog.apaonline.org/2016/04/06/how-can-philosophy-contribute-to-public-debates-and-discourse-2/

Vincent, N. A. (2017). Victims of Cybercrime: Definitions and Challenges. In Martellozzo, E. & Jane, E. A. (Eds.). *Cybercrime and Its Victims*. Oxon: Routledge, pp. 27–42.

Vincent, N. A. & Jane, E. A. (2017a). Beyond Law: Protecting Victims through Engineering and Design. In Martellozzo, E. & Jane, E. A. (Eds.). *Cybercrime and Its Victims*. Oxon: Routledge, pp. 209–223.

Vincent, N. A. & Jane, E. A. (2017b, July 18). A Crime Is A Crime, Even if It's Online - Here are Six Ways to Stop Cyberhate. *ABC News*. Retrieved from: www.abc.net.au/news/2017-07-18/six-ways-to-stop-cyberhate/8721184

Vincent, N. A. & Jane, E. A. (2018). Cognitive Enhancement: A Social Experiment With Technology. In van de Poel, I., Asveld, L. & Mehos, D. C. (Eds.). *New Perspectives on Technology in Society: Experimentation beyond the Laboratory*. Oxon and New York: Routledge, pp. 125–148.

Vingiano, A. (2014, October 12). A Female Game Developer Was Driven From Her Home After Rape And Death Threats. *Buzzfeed*. Retrieved from: https://www.buzzfeednews.com/article/alisonvingiano/a-female-game-developer-had-to-flee-after-her-address-was-po

11

POLITICAL ECONOMY OF COMMUNICATION

The Critical Analysis of the Media's Economic Structures

Christophe Magis

A Critical Tradition of Media Analysis

There are many ways to engage in a theoretical critique of the media. A tradition that focuses on what media *say* is currently in vogue, due to the prominence of terms like "fake news" and "post-truth." This tradition analyses media texts and compares them to facts in order to identify lies and omissions. Another tradition focuses on media *effects* and the risks they pose to different categories of viewers. Some other traditions focus on technology and the way it shapes individuals' everyday experiences. These critiques usually consider the media for themselves and as an autonomous field of inquiry.

The critical approach that this chapter presents is of another kind. It starts by considering the media within the capitalist system, as a *function* of this system as well as a field for its investments. It is less interested in these media *messages* or these *effects* as such than in the moving strategies of global capital that the mutations of communication systems reveal, along with their overall influences on the production and circulation of meaning. This approach, called the Political Economy of Communication, is about half a century old. Particularly interested in examining how the economic organization of the communication[1] sector affects the production of symbols through which individuals make sense of society, it has historically incorporated the Marxian critique of political economy and its developments within the general framework of Western Marxism (e.g., the Critical Theory of the Frankfurt School ...).

Nonetheless, and although several works aiming at a general epistemology of the approach from a global or regional perspective have flourished since the mid-2000s (Bolaño et al., 2012; Calabrese & Sparks, 2004; Hardy, 2014; Mosco, 2009; Wasko et al., 2011; Winseck & Jin, 2012) uncertainty sometimes remains about its political scope. Of course, categorizations may differ from a national or regional perspective to another. But it may today seem quite hard to understand that some works labeled PEC are deeply rooted in the Marxian critique whereas some others are not at all. As James Curran (2014, xiv) points out, "media political economy has two wings," a left-wing as well as a right-wing, and the second, although more recent, is currently growing within the approach. We focus here on the "left-wing." Even if all the figures of this wing don't always strongly claim a Marxian heritage, it must be admitted that "[m]ost often, those working within a political economic approach in media and communication studies have adopted a Marxist/neo-Marxist theoretical framework and thus a critical perspective" (Wasko, 2014, 260). This chapter first presents a brief overview of how this Marxist framework has been adapted to the study of the media, focusing on the concept of *commodity*. It then introduces two core themes of the PEC: the issue of media ownership and the critique of the information society theories. In conclusion, it offers to discuss the issue of media texts as a possible current blind-spot of the approach.

The Commodity of Communications

Following the Marxian tradition, PEC theorists have tried to define the specific features of communications considered through the concept of *commodity*. It is well known that the commodity is analyzed by Marx as the "elementary form" of the capitalist economy (Marx, 1976, 125), comparable to the *cell* as the morphological unit in biology. Following the Aristotelian analysis of economy, "[E]very commodity has a twofold aspect, that of *use-value* and *exchange-value*" (Marx, 1904, 19). Commodities bear a qualitative dimension, "conditioned by the physical properties of the commodity" (Marx, 1976, 126) and a quantitative one, "the proportion, in which use-values of one kind exchange for use-values of another kind" (ibid.). Another aspect adds up from classical economic theory, that of *value* which is determined by the average socially necessary labor-time required to produce a particular commodity. Marx's analysis aims at deciphering the structure of the commodity's different forms of value. It thus points out how the principle of equivalence is the core dynamic of the commodity, transforming use-values into exchange-values. This principle rests upon a fundamental unit of quantification, that of the *labor-time*. In the capitalist mode of production, one can find this principle of equivalence at every level of the commodities' production and trading processes: through the exchange,

- The use-value of the commodities becomes "the form of appearance of its opposite, value" (Marx, 1976, 149).
- But then, *concrete* labor (which aims at producing use-values) "becomes the expression of abstract human labor [aiming at the production of exchange-value]" (ibid., 150).

– As a consequence, concrete labor itself becomes a commodity because "it presents itself to us in the shape of a product which is directly exchangeable with other commodities" (ibid.).

Indeed, the development of the social division of labor in the capitalist society led to considering every kind of labor as *abstract* labor, traded for a salary. Thus, the capitalist mode of production has integrated even the labor-power within the realm of the commodity. More than a concrete good, the latter actually is a *social relationship*.

Marx goes further in analyzing the origin of wealth in the circulation of commodities. Money is a specific commodity, which has appeared in history to represent the economic value: it is the universal equivalent. Capitalism consists in achieving permanent returns on investment by pouring money into the production of any kind of merchandise in order to finally exchange it back for (more) money, according to the process of circulation $M–C–M^2$, but with a quantitative change between the two extremes:

> More money is finally withdrawn from circulation than was thrown into it at the beginning. […] The complete form of this process is therefore M–C–M', where M' = M + ΔM, i.e. the original sum advanced plus an increment.
>
> *(Marx, 1976, 251)*

Marx calls "*surplus-value*" (*Mehrwert*) this difference between M' and M.

Another essential contribution of the German thinker resides in the explanation of the origin of this surplus-value, at the heart of the production processes. This origin, again, lies in the principle of equivalence when it applies to labor. Like every other commodity, labor-power has a use-value and an exchange-value. Through salary, the employer buys the productive implementation of this power, but for a time that isn't limited to the labor-time strictly necessary to reimburse the pay. Rather, there is a difference that Marx calls "*surplus-labor.*" Therein lies the difference between M' and M in the circulation process. Surplus-value is "nothing but objectified surplus labour" (Marx, 1976, 325). Again, one realizes that commodity is the fundamental social relation of capitalism (although ambiguous and concealed), even in the creation of surplus-value. And within the commodity, labor is the central concept upon which the whole Marxian critique of the political economy rests. The relationship between capital and labor in the production of commodities takes the form of a class struggle. PEC theorists aimed at analyzing how all these points apply in the particular forms that the commodities take in the communication sector.

Contents as Commodity

The idea that the commodity of the communication sector is constituted by the media contents and programs seems to stand to reason. Therefore, "[w]hen political economists think about the commodity form in communication, they have tended to start with media content" (Mosco, 2009, 133).

The most specific feature of the content-commodity is that a particular type of labor, "creative" or artistic labor, stands at the beginning of the production process with the consequence of a low possibility of industrialization within the processes of commodification (Lacroix & Tremblay, 1997, 69). This has further implications. First, the valorization of the media products is highly uncertain, mainly because they must retain some of an artistic imagination. Culture being "above all the sphere for the expression of difference," the use-values of the content-commodity are "difficult if not impossible to pin down in any precise terms, and demand for them appears to be similarly volatile" (Garnham, 1990, 161). This characteristic led to two main strategies within the sector, that of *formatting* – to make sure the programs at least respect the usual successful canons – and of *catalog building* – to "compensate failure with success (the flops with the hits)" (Miège, 1979, 305). Second, as Garnham (1990, 160) highlights, two further economic characteristics of the content-commodity turn out to be in contradiction: it is at the same time a *prototype* and also tends to enter the category of *public goods*. The need for artistic novelty limits the possibility of total abstract standardization. Therefore, each product requires an investment in the high costs of production compared to which the costs of reproduction are negligible: most of the costs are employed to produce the first copy, making the failures even more costly and the successes very profitable. This leads to "a powerful thrust towards audience maximization as the preferred profit maximization strategy" (Garnham, 1990, 160). At the same time, the cultural and media commodities don't meet the criterion of rivalry: they are not destroyed in the process of consumption. Thus, the consumption by an economic agent doesn't prevent simultaneous or deferred consumption by others, rendering the economic exploitation harder. Furthermore, the artistic and media contents can be easily reproduced. Therefore, several strategies have been developed to create artificial scarcity, whether technical (limited access to a resource) or legal (forbidding reproductions).

Following these three characteristics, the low level of possible industrialization of the content-commodity implies specific relations of production within the sector. Because of the necessary maintained artistic "aura" of the products, the creators "are granted considerable autonomy within the process of production – far more, in fact, than most workers in other forms of industry" (Hesmondhalgh, 2013, 24). As a consequence, the symbolic creation lies in pre-capitalist forms of organization: surplus-value cannot be derived from the exploitation of surplus-labor-time in a system of salaried workforce. But if this sometimes boosts romantic conceptions of a critique of capitalism naturally embodied in the sectors' structuration, PEC theorists have noted that this relative autonomy helps the cultural valorization of capital. In fact, rather than within the creation process the sources of profit are located at the downstream stages of the industry (production and circulation), and this is where corporate control is tight. The creators' autonomy, which can take the form of specific payment systems such as royalty fees through the copyright system, ensures the autonomous production of a reservoir of cultural artifacts "from which capital can choose without having to bear the risks and over-heads which are borne directly by labour" (Garnham, 1990, 37): creators remain unpaid so long as "they have not proved that their professional activity actually imparts use-value to their productions" (Huet et al., 1978, 97).

The social relations of production in the cultural sphere imply precarious-ness for most of the workforce in a cultural imagination that renders it acceptable: the glamour surrounding the media industries forbids an identifi-cation of creation (and its cultural ideas of freedom, emancipation and cre-ativity) with "real work." This is why, since the neoliberal turn in politics, the figure of the artist is at the *avant-garde* of every action toward more job "flexibility."

Audiences as Commodity

Defining the commodity of communication becomes problematic in the case of mass media such as television and radio broadcasting or the free newspapers. The contents are "offered" to audiences without any direct payment, render-ing their consideration as the source of surplus-value difficult. Does it mean that these mass media only produce use-value? Some Leftist idealist theorists may have answered "yes" to this question, theorizing the specific "sign-value" of the media commodities (e.g., Baudrillard, 1972; Enzensberger, 1974). But within the PEC tradition, Dallas Smythe produced a more materialist answer: the commodity form of mass-produced advertiser supported communications lies in their "audiences and readerships" (Smythe, 1977, 3), which are sold to advertisers along with specifications (such as socio-demographic data). For Smythe, although their quality may be of utmost importance in the competi-tion among media enterprises, contents must be considered as a mere "free lunch" offered "in order to optimize the 'flow' of particular types of audiences to one programme from its immediate predecessors and to its immediate suc-cessors" (ibid., 6). Furthermore, what is sold to advertisers is an active watch-ing-time (or reading-time) during which audiences are put to work. In the advanced capitalist economy that requires mass production, consumption is an essential output: goods must be sold. Through the watching of advertisements, audiences assume an important labor within the system, that of producing demand by convincing themselves of specific merchandises' utility. Smythe calls this their "*audience-power*" (comparable to the workers' "labor-power" in Marxian critique of political economy).

This argument has had many echoes within the PEC tradition. Some theor-ists such as Murdock (1978) or Miège (1987) criticized Smythe's tendency to neglect the importance of the media content – whose relevance is even more striking in the European media systems characterized by strong state intervention. But most of the discussion focused on this consideration of watching-time as labor-time. For Sut Jhally, if advertisers are indeed interested in buying an active "audience-power" in order to sell their goods, the media industries cannot ensure the production of such a commodity. They only control and guarantee the production of actual watching-time from which advertisers must extract the best possible use-value. And this commodity is co-produced by audiences themselves for if they wouldn't watch, then the media wouldn't produce any audience commodity. Thus, the production of surplus-value goes as follows:

> The networks buy (or licence) programmes from independent producers to entice the audience to watch. Networks then fill this empty time they control by buying the *watching-power* of the audience [...]. Having purchased this watching activity, this "raw material," they then process it and sell it to advertisers for more than they paid for it.
>
> *(Jhally, 1990, 75)*

Jhally considers that a part of the watching-time is employed to repay the cost of programming ("*necessary* watching-time"); the rest is watched over the cost of programming ("*surplus-time*"), meaning that the surplus-value is here produced through the exploitation of "surplus" watching-time.

For Richard Maxwell, this argument, just as Smythe's, tends to introduce a confusion concerning the actual labor that produces the value of the audience commodity. "Because the audience carries a price and can be bought and sold for profit, it appears as though money grows out of watching" (Maxwell, 1991, 31). But if audiences here *represent* the value, it doesn't necessarily mean that they *produce* it. In the continuation of the work of Eileen Meehan (1984, 1986, 1990), Maxwell considers that "the human labour that produces the value reflected in the audience commodity form" is located in the "ratings industry" and "marketing firms" (Maxwell, 1991, 32). Indeed, if contents as well as audiences can be considered as the commodity of communications, the importance of ratings in the production of the value of both shouldn't be overlooked. In the case of the audience commodity, these measurements are "the tangible 'proof' that the networks' intangible commodity – the audience – exists" (Meehan, 1986, 450).

All these debates showed how a second way to produce value in the communication sectors assumes a perfectly capitalist organization. It also showed the growing importance of surveillance in the production of the communication sector's value. Current works concerning the commodification of the Web and the *digital labor* in a Marxist perspective largely rely on these arguments (e.g., Fuchs, 2014).

A World of Commodities

After a review of these debates in his work of synthesis on the PEC, Vincent Mosco states that

> it is essential to move beyond the notion of finding *the* commodity in the media. It is more significant to foreground a process of commodification that connects a range of practices in a spiral of expanding exchange value that [...] draws all organizations into the orbit of the information business.
>
> *(Mosco, 2009, 143)*

He thus considers communications as producing "immanent commodities," i.e., helping the production of value for other commodities. This reminds how the industrialization of communication articulates several sectors and thus encompasses several forms of commodification and valorization.

For example, when Adorno provocatively affirms in 1938 that "Music [...] serves in America today as an advertisement for commodities which one must acquire in order to be able to hear music" (Adorno, 1991, 38) this is because, at the beginning of broadcasting, radio stations had been mainly created in order to sell receivers (Barnouw, 1978). Jhally (1982, 205) also points out that "in the earlier days of television, the commodity form of mass media was the technology." This argument is valid for much equipment. When Apple launched its *iTunes Store* in 2003, it was mainly to ensure some legal digital music offer compatible with its flagship product, the *iPod*, in order to sell it (Bouquillion, 2008). Several forms of the commodity of the media are here at play. Songs bought on *iTunes* are produced by record producers under the form of content-commodity (with relative autonomy granted to creators, strategies of building catalogs, etc.). Deals have been secured by Apple in order to sell each song for $0.99, where $0.70 is redistributed to the record companies. The other 30% is extorted by the Californian tech giant for the functioning of the service (that for long wasn't profitable). Highlighting some technical qualities of the *iPod* (that may also play some illegally downloaded music), Apple sells it around $400 with a high share (21.6%) of the portable music player market (figures of 2003).[3] Surplus-value on this technical product is created rather classically: through the exploitation of labor-power – especially low-paid labor in the developing countries.

Finally, one must observe that, since the 1970s and the more general crisis of industrial capitalism, communication is a structural phenomenon of central importance in the economic restructuring of the countries of the Global North (and especially North America and Western Europe) where it has "accompanied the redeployment of powers (and counter-powers) in the domestic space, school, factory, office, hospital, neighbourhood, region, nation …" (Mattelart & Mattelart, 1986, 8). This observation reveals a key feature of communication as a commodity: its capacity to generate surplus-value by extending the process of commodification to areas that had previously been kept out of its realm. Surplus-value is produced with content-commodities on creations that *have not yet been paid* (and only will be in case of effective valorization), because a previous romantic conception of the activity of creation makes it possible not to be totally considered as actual work producing an actual economic good. Also, the activity of watching for the audience commodity isn't yet experienced as alienating work, making it easier to exploit. The current development of information and communication technologies (ICTs) and networks relies on mobile applications that generate value by extending the realm of commodity to newer activities such as dating, offering hospitality, car sharing, etc. At the same time, the communications also are at the center of a broad-based movement of liberalization and commercialization of the public sphere with the commodification of public services (Bouquillion, 2008; Mosco, 2009).

Two Areas of Enquiry

As Jonathan Hardy (2014, 7) states, PEC "rests on a central claim: different ways of organizing and financing communications have implications for the range and nature of media content, and the ways in which this is consumed and used." Therefore, the approach has been particularly committed to analyzing the mutations of the communication sectors' organization and funding as well as the discourses that have accompanied these movements.

Ownership, and Media Concentration

Issues related to structures of ownership are a key feature of the general political economy analysis. Indeed, whereas the usual advocates of capitalism argue that a free market brings competition among many diverse suppliers resulting in greater innovation for higher quality and falling prices to the benefice of consumers, many sectors of the contemporary economy behave very differently. Paul Baran and Paul Sweezy have shown how twentieth-century capitalism is characterized by monopolies or oligopolies dominated by hierarchically organized "giant corporations," in order to maintain a high level of profitability with increasing industrial investments toward mass production (Baran & Sweezy, 1966). But these issues are also of utmost importance for the political-economic analysis of communication due to the socio-political significance of the media in contemporary societies: "the control of media corporations can provide a means to promote certain ideas and values" (Birkinbine et al., 2017, 4).

Concentration of ownership in the media is as old as industrial media themselves. The industrial press has developed at the turn of the twentieth century under the dominance of the press "barons" (such as the Harmsworth brothers). In 1937, four Lords "owned nearly one in every two national and local daily papers sold in Britain, as well as one in every three Sunday papers that were sold" (Curran & Seaton, 2010, 39). An explanation lies in the specific economic features of the media. The uncertain valorization and the resulting strategy of catalog building in an economy characterized by very high costs of production require a strong financial power in order to bear the risk and undertake the necessary investments. Plus, profitability being reached through audience maximization, companies need wide distribution networks as well as a capacity to keep funding promotion. Furthermore, on the demand-side the value of media products tends to depend on the number of other users or consumers. In short, the valorization principles, which in the sector are largely based on economies of scale, tend to favor the larger integrated structures whereas they discourage the smaller ones (Doyle, 2002). And the existence of larger firms finally creates "barriers" to market entry for newer competitors: in order to compete, they have to be already large and integrated enough.

But in the last 40 years, concentration of media ownership has been reinforced by an intense and continuous process of mergers and acquisitions. This is mainly due to two correlated factors. First, as we discussed above, in the wake of the Long Downturn beginning with the oil shocks of the 1970s, many industrialized countries began to reorient their economies toward sectors of higher added value that consumed less fossil energy – such as the cultural, telecommunications and media industries. Much of these sectors, which benefited from strong state regulation, underwent a period of liberalization, sometimes with a relaxing of antitrust laws. State-owned systems were privatized and the deregulation helped the development of transnational media conglomerates (see Winseck & Jin, 2012). A textbook case is the *Telecommunications Act* signed by US President William Clinton in 1996. Aiming "to let anyone enter any communications business – to let any communications business compete in any market against any other,"[4] the Act abolished separations between industries, relaxed ownership restrictions and allowed the creation of global media giants (McChesney, 2015). Second, the crisis of industrial capitalism also directed investors toward the finance, insurance and real estate sectors, resulting in the boom of financial market speculation (Magdoff & Sweezy, 1987). As an area of choice for financial strategies, the communication sector has been at the forefront of the financialization of the economy since the early 1980s and throughout the next two decades. Financialization also reinforced the processes of concentration on a global scale (Bouquillion, 2008; Winseck, 2012).

Concentration of ownership in any industry is a problem even for mainstream economics as it results in higher prices for poorer quality and hinders innovation. But for PEC theorists, the communication sector adds a further political dimension since the media "do not manufacture nuts and bolts: they manufacture a social and political world." (Bagdikian, 2004, 9) Media concentration thus raises concerns about the diversity and relevance of available opinions and, as a consequence, about the quality of democracy, since "the ideal media system in a democracy [...] would be accessible at a variety of levels, diverse, relevant and engaging" (McAllister, 1996, 44).

First, the concentration process has led the content industries to be mainly controlled by different kinds of interests, mainly outside the realm of culture and communication – such as heavy industry, luxury brands or technology (see Mattelart, 1976; McChesney, 2013). Plus, there are usually strong connections between the media "moguls" and other politically powerful actors (Freedman, 2014). This puts forward questions about the power of influence of such actors through the media contents, especially in the case of information networks. However, even if some studies have highlighted overt editorial interventions from big bosses such as Rupert Murdoch (Manning, 2001) or Vincent Bolloré (in France),[5] the argument of direct hierarchical control and the way it operates across the media organizations remains a matter of debate. Brian McNair's sociological analyses of journalism (2006) have for example challenged the presumption of direct owner control.

Second, even if *instrumental* control over media content is difficult to assess, some works have focused on more *structural* influences of the concentration of ownership such as the resulting corporate culture. Bigger media companies tend to promote producers and managers who have best integrated the values of commercialism and to consequently favor the most consensual media programs and those most likely to receive the assent of the larger share of audiences to the detriment of the least represented popular expressions (Gandy, 2000; Schiller, 1996). Media workers and managers usually share a similar educational background and are then likely to incorporate these values as marks of professionalism (Curran, 2002). Thus, control over senior appointment is a potential indirect way of maintaining editorial control that may not be perceived as such but rather as the rewarding of professional qualities (Murdock, 1982).

The integrated global media system is comprehended as "one that advances corporate and commercial interests and values, and denigrates or ignores that which cannot be incorporated into its mission" (McChesney, 2015, 103). Therefore, many PEC theorists have engaged in political or associative activism in order to promote different and more democratic forms of media organization.

Information Society Theories: A Critique

Not unlike the discourse of classical political economy was also a means of domination in that it served as an accompanying legitimation in the realm of ideas of the advent of a capitalist society – thereby justifying the Marxian *critique of political economy*[6] – the development of communications also came and still comes with many legitimating discourses that the PEC approach analyses and criticizes. Since the 1960s, many "slogans on communications," influenced by "philosophical, sociological and even economic schools of thought," regularly "play their part in the new narrative on communication and hence fuel the general myth" (Miège et al., 1986, 104, 107). Such theories have usually aimed at defining the type of society in which we currently live – when not prophesying the advent of a new one – considering and emphasizing socio-technical changes due to the development of the communication sector. Specific types of interplay between scientific or academic discourses, political discourses, international organizations or think tanks reports, and industrial comments have given these theories widespread diffusions as well as they helped them generate strong "reality effects," especially in times of crisis when recovery was to be seen through their paradigm.

If their histories can be traced back to the utopian project of a universal mathematical reasoning in the seventeenth and eighteenth-century humanism (Mattelart, 2003), it can be said that such theories mostly originate in the cybernetics' scientific program in the immediate post-war period (Mansell, 2009). The new centrality given to the treatment of "information" gave rise to the pervasive concept of "information society" (Porat, 1977). Thus, although other terms have flourished such as "knowledge" society (Machlup, 1962), "network" society (Castells, 1996), "post-industrial" society (Bell, 1973) or "post-modern society" (Lyotard, 1979) and so on, these theories are referred to as "information society theories." All of them generally rely on the claim that communication (considered in a very large sense) has *radically* transformed societies and economies, to the point that "existing power structures [...] are being rapidly eroded" (Bell, 1973, 37): communication and technologies have brought about a new organization of society where the centrality of immaterial and informational labor has supposedly reshaped economic and social relationships. PEC theorists have challenged these arguments on empirical grounds but also and mainly for their ideological character: defining a reshaping of the social structures, these theories decree the end of social struggles that were tied to previous structures (such as the struggle between labor and capital) and therefore urge social scientists to facilitate the social adaptation to new political realities (Webster, 2006). In a synthesis based on a precise review of Castells' *Rise of Network Society*, Garnham thus considers information society theories as "the favoured legitimating ideology for the dominant economic and political powerholders" (Garnham, 1998, 165).

Such critical investigations often raise counter-indictments, where PEC is portrayed as unable to understand the transformations of society: "if nothing is new under the sun, why bother to try to investigate, think, write, and read about it?" (Castells, 2000, 367). This portrait is unfair. Of course, it is not sufficient to limit the investigations to the saying that any new term is another repetition of the industrial society theories, that "nothing is new under the sun." It is necessary to conduct in-depth analyses of these theories and discourses – which, even as "mythical discourses," can be relevant for the critical theories as they "animate individuals" and sometimes lead them to engage in social struggle (Mosco, 2004, 3) – highlighting the realities they conceal or emphasize (with special regard to the evolution of labor), the real emancipatory possibilities they relate to the communication technologies and the individuals actually concerned.

PEC theorists of course assess that technologies and communication have brought about changes. However, they "want to warn that a reduction of the contemporary economy to the changes of the productive forces obscures the continued existence of capitalist class relations that are exploitative in character" (Fuchs, 2014, 144). Furthermore, obliterating the fact that capitalism is currently a globally organized mode of production with an international division of labor tends to ideologically hide that most of the physical and hard labor is still an important producer of surplus-value – exploiting the low-paid labor-power of individuals in the Third-World, sometimes under close-to-slavery working conditions.

Media Texts: A Current Blind-Spot?

In his seminal essay on the audience commodity, Smythe (1977) argued that communications had so far been the "blindspot of western Marxism." If this is no longer the case, some other key elements of the media have nonetheless been somewhat overlooked within the PEC. Media texts are a good example. For Hesmondhalgh (2013, 78), the issues concerning texts, their quality and the interests they promote represent "the most difficult and controversial aspects of the cultural industries." However, texts should be a key issue for an approach that analyses the media and cultural industries considering that they "play a pivotal role in organizing the images and discourses through which people make sense of the world" (Murdock & Golding, 2005, 70). How then not to work toward a concept of ideology that could study the mutation of texts within the evolutions of the media and alongside the wider movements of capitalism? Notwithstanding a few attempts (such as the famous example of Mattelart & Dorfman, 1975), it is not yet the direction that the PEC approach has taken.

This oversight can be explained. The PEC approach has emerged historically since the 1970s with the intention of bringing forward materialist studies of the economic structures of the media against the cultural idealism that had before been dominant in the field. The issue of texts has therefore been underexplored and disconnected from the analysis of ideology, following Garnham's statement whereby "[b]ecause capital controls the means of cultural production […] it does not follow that these cultural commodities will necessarily support, either in their explicit content or in their mode of cultural appropriation, the dominant ideology" (Garnham, 1990, 34). As a consequence, alongside the refinement of the economic analysis, PEC moved away from a possible apprehension of the media contents. Henceforth, as shown above, the approach only dares to evaluate them when considering the press (or the information society theories, which could be considered as media contents in many respects), namely when texts consist of *written* texts strictly speaking – apparently more easily relatable to authors' intentions or ideologies. Murdock and Golding's observation of the late 1970s is currently still valid:

> Most studies concerned with media output have not been concerned to discover its ideological underpinning, while those that have, have concentrated almost entirely on news, and have consequently neglected the main dramatic, fictional and entertainment forms which make up the bulk of most people's media fare.
>
> *(Murdock & Golding, 1977)*

It will be the task of future works within PEC to produce a concept of ideology that would be informed by textual analysis. Because, as Garnham states in a synthesis book, "no study of the media can bypass the complex and difficult questions posed by their content, by the symbolic forms they create and circulate" (Garnham, 2000, 138). Especially within an approach defined by "its focus on the interplay between the symbolic and economic dimensions of public communications" (Murdock & Golding, 2005, 70). This new consideration may take the form of an integration between PEC and other approaches such as Cultural Studies (Babe, 2009), structuralism (Buxton, 1990) or the Critical Theory (Magis, 2015). Of course, there is no question of abandoning the analysis of the economic structures of the media but rather of refining its linking with the evolution of social imaginations and symbols.

***

If it is time to "reimagine" the communication and media studies in a critical way, then the PEC approach could be extremely useful, especially in its way to consider the economic, social and political mutations of the media within the wider evolutions of contemporary capitalism. Analyzing how media and cultural production occurs within wider economic and political contexts, the latter possibly determining the former in various respects and the former justifying or co-constructing the latter, is a necessary starting point for any critical inquiry. This determination may though be refined. Further PEC explorations could work at this task, with regards to other critical approaches' arguments.

Notes

1 Like many other PEC theorists, along this chapter we will subsume every industry taking part in the overall mass production of cultural and media goods, services and technologies under the general term of "communication." We will thus choose to speak of Political Economy of *Communication*, instead of Political Economy "of Culture" (Calabrese & Sparks, 2004), "of Mass Communications" (Garnham, 1990) or

"of the Media" (Hardy, 2014). Within the tradition, as in the present text, all these terms though refer to comparable realities: that of the industrial production and circulation of symbols.
2 Where M = Money; C = Commodity.
3 See « The meaning of iPod », *The Economist*, June 10, 2004.
4 https://www.fcc.gov/general/telecommunications-act-1996 (accessed March 10, 2019).
5 https://www.lemonde.fr/actualite-medias/article/2015/09/12/les-medias-selon-vin cent-bollore_4754284_3236.html (accessed March 10, 2019).
6 "A Critique of Political Economy" is the subtitle of Marx's master work, *Capital*.

References

Adorno, T. W. (1991). *The Culture Industry. Selected Essays on Mass Culture*. New York, NY: Routledge.

Babe, R. E. (2009). *Cultural Studies and Political Economy: Towards a New Integration*. Lanham, MD: Lexington Books.

Bagdikian, B. H. (2004). *The New Media Monopoly: A Completely Revised and Updated Edition with Seven New Chapters*. Boston, MA: Beacon Press.

Baran, P. A. & Sweezy, P. M. (1966). *Monopoly Capital: An Essay on the American Economic and Social Order*. New York, NY: Monthly Review Press.

Barnouw, E. (1978). *The Sponsor : Notes on Modern Potentates*. New Brunswick, NJ: Transaction Pub.

Baudrillard, J. (1972). *Pour une critique de l'économie politique du signe*. Paris: Gallimard.

Bell, D. (1973). *The Coming of Post-Industrial Society*. New York, NY: Basic Books.

Birkinbine, B., Gomez, R. & Wasko, J. (Eds.). (2017). *Global Media Giants*. London and New York, NY: Routledge.

Bolaño, C., Mastrini, G. & Sierra, F. (Eds.). (2012). *Political Economy, Communication and Knowledge: A Latin American Perspective*. New York, NY: Hampton Press.

Bouquillion, P. (2008). *Les industries de la culture et de la communication : les stratégies du capitalisme*. Grenoble: PUG.

Buxton, D. (1990). *From "The Avengers" to "Miami Vice": Form and Ideology in Television Series*. Manchester, UK: Manchester University Press.

Calabrese, A. & Sparks, C. (Eds.). (2004). *Towards a Political Economy of Culture. Capitalism and Communication in the Twenty-First Century*. Boulder, CO: Rowman & Littlefield.

Castells, M. (1996). *The Information Age. Vol. 1: The Rise of the Network Society*. Malden, MA: Blackwell.

Castells, M. (2000). *The Information Age. Vol. 3: End of Millenium*. Malden, MA: Blackwell.

Curran, J. (2002). *Media and Power*. London and New York, NY: Routledge.

Curran, J. (2014). Foreword. In Hardy, J. *Critical Political Economy of the Media: An Introduction* (pp. x–xx). New York, NY: Routledge.

Curran, J. & Seaton, J. (2010). *Power without Responsibility: Press, Broadcasting and the Internet in Britain* (7th ed.). London & New York, NY: Routledge.

Doyle, G. (2002). *Understanding Media Economics*. London and New York, NY: SAGE.

Enzensberger, H. M. (1974). *The Consciousness Industry: On literature, Politics and the Media*. New York, NY: Seabury Press.

Freedman, D. (2014). *The Contradictions of Media Power*. London and New York, NY: Bloomsbury.

Fuchs, C. (2014). *Digital Labour and Karl Marx*. London: Routledge.

Gandy, O. H. (2000). Race, ethnicity and the segmentation of media markets. In Curran, J. & Gurevitch, M. (Eds.), *Mass Media and Society* (pp. 44–69). London: Arnold.

Garnham, N. (1990). *Capitalism and Communication*. London: Sage.

Garnham, N. (1998). Information society theory as ideology. In Webster, F. (Ed.), *The Information Society Reader* (pp. 165–183). New York, NY: Routledge.

Garnham, N. (2000). *Emancipation, the Media, and Modernity: Arguments about the Media and Social Theory: Arguments about the Media and Social Theory*. Oxford: Oxford University Press.

Hardy, J. (2014). *Critical Political Economy of the Media: An Introduction*. New York, NY: Routledge.

Hesmondhalgh, D. (2013). *The Cultural Industries*. London: Sage.

Huet, A., Ion, J., Lefèbvre, A., Miège, B. & Peron, R. (1978). *Capitalisme et industries culturelles*. Grenoble: PUG.

Jhally, S. (1982). Probing the blindspot: The audience commodity. *Canadian Journal of Political and Social Theory*, 6(1–2), 204–210.

Jhally, S. (1990). *The Codes of Advertising. Fetishism and the Political Economy of Meaning in the Consumer Society*. London & New York, NY: Routledge.

Lacroix, J. G. & Tremblay, G. (1997). The "information society" and the cultural industries theory. *Current Sociology*, 45(4), 1–154.

Lyotard, J.-F. (1979). *La condition postmoderne*. Paris: Minuit.

Machlup, F. (1962). *The Production and Distribution of Knowledge in the United States*. Princeton, NJ: Princeton University Press.

Magdoff, H. & Sweezy, P. M. (1987). *Stagnation and the Financial Explosion*. New York, NY: Monthly Review Press.

Magis, C. (2015). *La musique et la publicité: les logiques socio-économiques et musicales des mutations des industries culturelles*. Paris: Mare & Martin.

Manning, P. (2001). *News and News Sources*. London: Sage.

Mansell, R. (Ed.). (2009). *The Information Society. Critical Concepts in Sociology* (Vol. 1). New York, NY: Routledge.

Marx, K. (1904). *A Contribution to the Critique of Political Economy*. Chicago, IL: Charles H. Kerr & Company.

Marx, K. (1976). *Capital. A Critique of Political Economy* (B. Fowkes, Trans.) (Vol. 1). London: Penguin.

Mattelart, A. (1976). *Multinationales et systèmes de communication. Les appareils idéologiques de l'impérialisme*. Paris: Anthropos.

Mattelart, A. (2003). *The Information Society: An Introduction*. London: Sage.

Mattelart, A. & Dorfman, A. (1975). *How to Read Donald Duck*. New York, NY: International General.

Mattelart, A. & Mattelart, M. (1986). *Penser les médias*. Paris: La Découverte.

Maxwell, R. (1991). The image is gold: Value, the audience commodity and fetishism. *Journal of Film and Video*, 43(1/2), 29–45.

McAllister, M. (1996). *The Commercialization of American Culture*. London: Sage.

McChesney, R. W. (2013). *Digital Disconnect: How Capitalism Is Turning the Internet against Democracy*. New York, NY: The New Press.

McChesney, R. W. (2015). *Rich media, Poor Democracy. Communication Politics in Dubious Times*. New York, NY: The New Press.

McNair, B. (2006). *Cultural Chaos. Journalism, News and Power in a Globalised World.* New York, NY: Routledge.

Meehan, E. R. (1984). Ratings and the institutional approach: A third answer to the commodity question. *Critical Studies in Mass Communication*, 1(2), 216–225.

Meehan, E. R. (1986). Conceptualizing culture as commodity: The problem of television. *Critical Studies in Mass Communication*, 3(4), 448–457.

Meehan, E. R. (1990). Why we don't count: The commodity audience. In Mellencamp, P. (Ed.), *Logics of Television* (pp. 117–137). Bloomington, IN: Indiana University Press.

Miège, B. (1979). The cultural commodity. *Media, Culture and Society*, 1, 297–311.

Miège, B. (1987). The logics at work in the new cultural industries. *Media, Culture and Society*, 9(3), 273–289.

Miège, B., Pajon, P. & Salaun, J.-M. (1986). *L'industrialisation de l'audiovisuel : des programmes pour les nouveaux médias*. Paris: Aubier.

Mosco, V. (2004). *The Digital Sublime: Myth, Power, and Cyberspace*. Cambridge, MA: MIT Press.

Mosco, V. (2009). *The Political Economy of Communication*. Thousand Oaks, CA: Sage.

Murdock, G. (1978). Blindspots about western Marxism: A reply to Dallas Smythe. *Canadian Journal of Political and Social Theory*, 2(2), 109–115.

Murdock, G. (1982). Large corporations and the control of the communications industries. In Gurevitch, M., Bennett, T., Curran, J. & Woollacott, J. (Eds.), *Culture, Society and the Media* (pp. 114–147). New York, NY: Routledge.

Murdock, G. & Golding, P. (1977). Capitalism, communication and class relations. In Curran, J., Gurevitch, M. & Woollacott, J. (Eds.), *Mass Communication and Society* (pp. 12–43). London: Edward Arnold.

Murdock, G. & Golding, P. (2005). Culture, communications and political economy. In Curran, J. & Gurevitch, M. (Eds.), *Mass Media and Society* (pp. 60–83). London: Hodder Arnold.

Porat, M. U. (1977). *The Information Economy: Definition and Measurement*. Washington, DC: Office of Telecommunications.

Schiller, H. I. (1996). *Information Inequality: The Deepening Social Crisis in America*. New York, NY: Routledge.

Smythe, D. W. (1977). Communications: Blindspot of Western Marxism. *Canadian Journal of Political and Social Theory*, 1(3), 1–21.

Wasko, J. (2014). The study of the political economy of the media in the twenty-first century. *International Journal of Media & Cultural Politics*, 10(3), 259–271.

Wasko, J., Murdock, G. & Sousa, H. (Eds.). (2011). *The Handbook of Political Economy of Communications*. Chichester: Wiley Blackwell.

Webster, F. (2006). *Theories of the Information Society* (3rd ed.). New York, NY: Routledge.

Winseck, D. (2012). Financialization and the "crisis of the media": The rise and fall of (some) media conglomerates in Canada. In Winseck, D. & Jin, D. Y. (Eds.), *The Political Economies of Media. The Transformation of the Global Media Industries* (pp. 142–166). London: Bloomsbury.

Winseck, D. & Jin, D.-Y. (Eds.). (2012). *The Political Economies of Media. The Transformation of the Global Media Industries*. London: Bloomsbury.

12

THE PROPAGANDA MACHINE

Social Media Bias and the Future of Democracy

Sara Monaci

Introduction

The narratives concerning social media, in their first phase, focused– at the beginning of the twenty-first century – on the emancipatory value of the globalized means of expression which allegedly would introduce more democracy and freedom of speech. The first optimistic impetus gave way, however, to a more cautious and critical perspective on social media: the pervasive influence achieved by Facebook, Google, etc., recalled the public opinion on the private interests of those major technology corporations affecting such a broad dimension of the public debate. Moreover, the manipulative use of social media by multiple extremist and violent organizations questioned deeply the ingenuity of the net, highlighting the threats related to the newly available forms of propaganda. The chapter will debate the classical approach to propaganda through a set of theoretical questions: how could we reconceptualize propaganda in consideration of the social media phenomenon? What are the contemporary processes of "manufacturing consent" and how are they related to the social media bias?

The Propaganda Model in the Age of Social Media

Noam Chomsky and Edward Herman outlined the propaganda model for the first time in their 1988 book *Manufacturing Consent: The Political Economy of the Mass Media* (2010). The model was conceived from the perspective of the political economy of communication to explain the behavioral and performance patterns of the US mass media in relation to news production. The original version of the model focuses on the propaganda dimension of information by identifying five filters – ownership, advertising, information sourcing, flak and anti-communism – through which information must pass before seeing the light. The propaganda model was basically a mass media model, based on the intertwined power relations between US political and economic elites and mass media.

Despite the classic model having received several criticisms (Mullen, 2007, 2009; Pedro, 2011), it still represents a reference for the reflection on the propaganda phenomenon and it was recently revisited by Christian Fuchs in consideration of the spread of social media (2018). Fuchs discusses Chomsky and Herman's lesson trying to identify the elements of continuity and change with respect to the original model: regarding the issue of ownership and advertising, today as in the 1980s, the social media scene is characterized by an oligopoly articulated on a few technological players – Google, Facebook, Twitter and advertising has become the dominant social media business model. According to the advertising business, in fact, billions of users can enjoy free services – instant messaging, social networking, publication of personal audiovisual contents, etc. – while offering personal data and content (the same services that are accessed free of charge) that the platforms re-sell to advertisers:

> Google, Facebook and Twitter are not just sources of news and information. These websites are also among the world's largest advertising agencies. They are in the business of selling targeted ad space as a commodity and derive their revenues almost exclusively from targeted advertising.
>
> *(ibid., 75)*

Two other filters of the original model are discussed by Fuchs through a social media perspective: sourcing and flaking. In the first aspect, Fuchs insists that in spite of their greater openness and availability, social media are actually dominated by mainstream media as the main sources of information: subjects tend eventually to read on social media, the same sources of information that they use in traditional ways. The top visited social media pages turn out to polarize around the major newspapers such as *NY Times*, CNN, Fox, etc., as well as around a high number of movie stars, music performers, etc. Traditional news organizations remain powerful actors in online news dominating social media attention.

Chomsky defines flaking as a lobbying or containment activity on mainstream media:

> "Flak" refers to negative responses to a media statement or program. It may take the form of letters, telegrams, phone calls, petitions, lawsuits, speeches and bills before Congress, and other modes of complaint, threat, and punitive action. It may be organized centrally or locally, or it may consist of the entirely independent actions of individuals.
>
> *(Chomsky & Herman, 2010, 26)*

According to Fuchs, in the digital age, lobbying for certain interests has been extended to social media and is no longer simply aimed at centralized media organizations, but now aims to directly transmit political messages to as many internet users as possible:

> The basic difference between computer networks and broadcasting is that the network is a universal machine, at once a technology of production, distribution and consumption. Combined with its global reach and significant bandwidth rates, this allows the phenomenon of user-generated content. User-generated content does however not automatically imply political plurality and diversity.
>
> *(Fuchs, 2018, 78).*

In spite of the outdated rhetoric concerning the "user-generated contents" related to the first development of the web 2.0 (O' Reilly, 2009), Fuchs highlights a central element of propaganda 2.0: that is the appearance on the social media scene of non-institutional actors who are not part of the elite talked about by Chomsky in his classic model. These new actors, thanks to social media, can express their political ideas, their opinions on climate change or on their favorite football teams. Nevertheless, both these new subjects and their counter-narratives present an ambivalent nature: on the one hand they express the possibility to be issuers of their own messages and to be precisely subjects of the mass self-communication (Castells, 2013), on the other hand those new voices can only illusively give rise to new different points of view. Fuchs emphasizes how the social mediascape is in fact dominated by an homogeneous aesthetics of entertainment inspired by a dominant neo-liberal ideology that prevents the formation of a mature and conscious public debate. This superficial and entertainment-oriented public sphere enhances in fact emotional rather than rational debate and it may even facilitate visceral forms of social communication that can also assume the drifts of extremism, hate speech, etc. As a matter of fact, Herman and Chomsky also argued in a recent interview that right-wing media, including Fox News, right-wing talk radio and blogs, form "a right-wing attack machine and echo-chamber." In the current political climate of nationalism, racism, xenophobia and elements of fascism, social media is certainly a right-wing attack machine. It must, however, also be seen that the political left is skilled at using social media, which maintains online politics as a contradictory space (Mullen, 2009).

The reading of Fuchs therefore denotes, in the transition from a propaganda mass media model to that of social media, very marginal changes that do not significantly influence the dynamics and relations of political and economic power affecting the manufacturing of consent. According to Fuchs, social media as well as the mass media are still dominated by an oligopoly of actors – only partly different from the traditional media networks of the 1980s – that operate equally powerful and binding filters on the conditions for developing a real open public debate.

The Social Media Bias

As well as the Chomsky's inspired elements referred to in the classic propaganda model, other dimensions should be included as the proper bias of social media affecting new forms of propaganda in the twenty-first century.

As Harold Innis maintained in the 1950s, the bias of communication – the characterizing traits of a certain medium – are fundamental in marking the character of an era: its culture, its arts and above all, its power and control over knowledge (Innis, 2008).

More recently, Neil Postman (2000), referring to the tradition of the Toronto School, wrote:

> A medium is a technology within which a culture grows; it is to say, it gives form to a culture of politics, social organization, and habitual ways of thinking. Beginning with that idea, we invoked another biological metaphor, that of ecology.

In Postman's reflection, a conception of the media is developed as an "environment" within which a culture grows and influences different aspects of society: politics, social relations, the way of thinking in general. Thinking of contemporary forms of propaganda in these terms means therefore going beyond an analysis focused on the forms of economic and political relations and economic power, but considering the contemporary propaganda as the result of a media environment – that of social media – which has its own specific bias influencing public opinion, attitudes and voting behaviors.

This chapter will highlight the social media bias affecting the contemporary processes of dissemination and information sharing within the public debate, and it also aims to discuss how such biases can produce new forms of propaganda: new content that becomes more and more a characteristic manifestation of the social media debate. But before outlining those bias, a preliminary topic should be addressed:

What do we talk about when we talk about propaganda today?

What Do We Talk about When We Talk about Propaganda Today?

The concept of propaganda may appear obsolete in contemporary Western democracies where journalism practices adhere to ethical standards and rules of conduct that guarantee in most cases transparency, reliability and confidentiality of the sources; nevertheless, the term propaganda has persistently returned in recent years as an analytical category to describe diverse social and political phenomena which seem to have one common denominator: the role of social media – e.g., Facebook, Twitter, Instagram, etc. – as an enabling technological environment capable of significantly conditioning not only the processes of dissemination of news and information online, but also the actions of subjects influenced by propaganda. This is the case, for example, of the recent Cambridge Analytica scandal related to US presidential elections in 2016 which involved the exploitation and the deliberative manipulation of millions of social media Facebook profiles as the target of ad-hoc propaganda, or the case of systematic propaganda that several authoritarian regimes (China, Russia, Egypt, Saudi Arabia) implement to strengthen the consensus among their citizens, or even the increasingly decisive role that social media seem to have in the forms of socialization of extremist organizations that use networks to recruit new affiliates, to spread hate messages online, and to reinforce their echo chambers. In these phenomena that I will describe below, social media turn out to be at the center of the propaganda dynamics while the mainstream media, once at the center of debate, appear in the background in the role of authoritative counterpart with respect to the fragmented and chaotic voices arising from social media.

The Facebook Targeted Propaganda

At the beginning of 2018, the facts related to Cambridge Analytica scandal, the articulated and intertwined relations network among prominent politicians and spin-doctors – US President Donald Trump and his former communication consultant Steve Bannon – the social media giant Facebook and a group of academics from Cambridge University, highlighted the complexity of the propaganda phenomenon at the beginning of the twenty-first century.

In March 2018, the testimony of Christopher Wylie – previously Cambridge Analytica head of research – along with the comments of the former Cambridge Analytica CEO Alexander Nix collected by Channel4 News, revealed that the data-mining company, harvested data of up to 87 million Facebook users thanks to a quiz application realized by a Cambridge University researcher, Aleksandr Kogan. The mentioned application – myPersonality – now banned by Facebook, collected data not only from its direct users but also from the users' network of friends exploiting the network connections structure of the social media platform. As well as exploiting personal data from users, accusations toward Cambridge Analytica stated that personal data were illegally used for actively influencing users in the 2016 US elections and in the context of the Brexit referendum with targeted contents supporting pro-Trump votes and Leave votes.[1] As a result of the scandal, Cambridge Analytica closed its London headquarters in May 2018, also due to an investigative report which revealed the company's involvement in influencing the elections of at least eight different countries such as India, Malta, Kenya, Mexico, etc.[2] Facebook CEO Mark Zuckerberg had been repeatedly summoned to clarify to the US Senate, his company's regulations in the subject of protecting users privacy.[3] Very recently Facebook suspended other 400 applications on the allegation that they misused users' personal data.[4]

The Cambridge Analytica case is nowadays an ongoing trial and its consequences have still to be defined both in a legal and economic sense. Moreover, the case revealed to the outlines of a propaganda machine which radically differs from that used to in the twentieth century: cultural and socio-technical consequences of the Cambridge Analytica case will be more impactful on propaganda conceptualization rather than on its business implications.

The Authoritarian Regimes Propaganda and the Exploitation of Social Media

In 2009, 2010 and 2011, social media were hailed as new fundamental tools for different revolutionary movements that had shaken several countries in the Middle East (Iran, Tunisia, Egypt, Syria): the so-called Arab Springs movement. In particular, the facts of Tahir Square in Egypt, or the video of the young Tunisian who burnt himself to protest against the grievance shared by a whole generation of unemployed youngsters, have recalled through a cascade of tweets, posts and YouTube videos, recalled the Western opinion to give attention to those dramatic uprisings. Even if the attention of the major Western media indeed focused in those weeks on the streets in Egypt as in Tunisia, the final outcome of the uprisings has not been – except in Tunisia where the premier Ali was forced to resign and to go into exile – revolutionary in a proper sense (Abouzeid, 2011; Eltantawy & Wiest, 2011; Khondker, 2011; Kirkpatrick, 2011). In some cases, as in Iran in 2099 authoritarian regimes have exploited the attention of Western media, accusing the West of interference in their internal politics and have even solicited the population to re-publish on social media, counter-narratives supporting the government against the individuals supporting the protests (Morozov, 2012). In short, the disruptive use of social media in the so-called revolutionary phase has led – in Iran and in Egypt for example – to a subsequent tightening of control measures and repression by the authoritarian regimes who then became promoters of their targeted counter-propaganda. This counter-propaganda effort has also been accomplished thanks to the automated social bots: automated software able to create and to spread online propaganda contents as if they were real social media users. As Morozov claims (ibid.), these facts show that a first wave of optimism and technological utopianism linked to the democratic properties of the use of social media has given way to a more cautious consideration of social media as tools, whose contribution to the democratic life depends on a set of contextual factors and not on the alleged "participatory" qualities of the media themselves.

What seems to be a scandal in Western democracies – that is, the manipulation of social media for political purposes – is nevertheless a widespread practice in authoritarian regimes such as China or Russia where the instrumental use of social media as a means of strengthening consensus among citizens, is a practice usually carried out by civil servants. These authoritative regimes systematically use social media to disseminate, also through the use of automated chat-bots, messages and opinions favorable to a precise political position trying to orient the opinions and attitudes of citizens in a unilateral way (Woolley & Howard, 2018).

Various authoritarian regimes such as Russia and China have adopted ambivalent policies in the social media management: on the one hand they practice a censorship aimed at the Western platforms (Google, Facebook, YouTube) but on the other hand, they are aware of the enormous potential of social media as a business opportunity and also as a monitoring tool for public opinion. They designed and implemented their own social media platforms – e.g., Vkontakte or Weibo – which resemble Facebook, Google, etc. in many senses – but are controlled and managed by the government. This allows the authoritarian states to exploit the business potentials related to the use of social media – e-commerce or simply the flywheel effect that social applications have on the purchase of smart-phones and internet traffic thus benefiting the big ICT companies – and at the same time it allows the authoritarian regimes to protect themselves from the commercial colonization of Western players such as Facebook, Google and their possible interference in their status quo.

Violent and Extremist Propaganda

Social media are described by many as enabling factors for the spreading of extremist propaganda often characterized by extreme right inspired xenophobic and racial discourses or, on the opposite side, by a violent opposition to Western imperialism, as in the recent Jihadist online phenomenon dominated by ISIS (Islamic State of Iraq and Syria) (Stern & Berger, 2015). Alongside the political forms of expression of discontent, new forms of extremism and radicalization emerge online assuming the form of terrorist propaganda. A typical trait of this phenomenon is the use of violence – verbal and visual – as a propaganda tool used to engage emotionally and cognitively its alleged target (Meleagrou-Hitchens & Kaderbhai, 2016, Hoskins et al., 2011, Atwan, 2015). I am referring to the recent phenomenon of ISIS propaganda and in particular to the appeal aimed mainly at young Europeans to become Foreign Fighters and to join the construction of the Islamic State in the Levant. The young propaganda targets, born and raised in Europe – as various studies have shown – decide to adhere to such an extreme cause because they do not have a job, because they feel disappointed with expectations of personal growth and social affirmation or because, especially in the case of the second and third-generation immigrants, they feel socially and economically marginalized (Carter et al., 2014; Neumann, 2015; Roy, 2016).

Many studies analyzed the role of social media as powerful propaganda tools able to conquer the hearts and minds of such young people by offering them basically two elements: a set of radical narratives, finely elaborated and distributed online in the aesthetic canons of Western communication (Maggioni & Magri, 2015; Monaci et al., 2017); online echo-chambers within which the identities of individuals are radicalized, polarized on opinions and attitudes that do not include doubts or comparisons with different opinions, re-strengthening the bonds within the group and with the charismatic leaders.

These groups, even if virtual, offer many young people the illusory opportunity to build a new identity in the radical rejection of what they had been before: that's why many Foreign Fighters consider themselves as "born again" (Awan, 2017; Farwell, 2014; Mahood & Rane, 2017; Monaci et al., 2018). Moreover, echo chambers facilitate the emergence of de-individualized collective identities: identities which lose their biographic and personal traits and are increasingly polarized and concentrated on ideology, a shared narrative which represents the main bond and the common ground of the extremist group (Sunstein, 2018, 236–342)

To what extent are these phenomena referable to social media? What are the social media bias enabling the targeted forms of propaganda, as in the Cambridge Analytica case, or the manipulation of propaganda implemented by authoritarian regimes, or what makes social media so effective tools in spreading the violent propaganda of terrorist groups?

Social Media User Commodification

The concept refers to "audience commodification" formulated originally by Dallas Smythe's (1981) and discussed more recently by Fuchs (2012). For social media sites like Facebook and Twitter, consumption is subsumed by commodity production. That is, as the service is consumed, new commodities (user-generated data) are produced that in turn can be sold to advertisers, therefore generating a twofold commodification of the user that Fuchs labels the "digital labor prosumer commodity" (ibid.).

What bias is the phenomenon linked to?

Since their emergence at the end of the 1990s, social media and then web 2.0 has been characterized by persistence and traceability (Boyd, DM, & Ellison, NB, 2007): data left by users once "written" were no longer owned by the users but become an integral part of the platform (even if at that time there was no main platform at all and Facebook was still a promising PhD research).

What was not clear to early observers was how these data would be used, by whom – which players would have taken advantage of that huge stream of user-generated content – and especially why and under what conditions users would have continued to insert their own data in the form of comments, blogs, posts, etc. Until 2006–2007 there was not a single subject able to optimize the enormous availability of user data; in 2007 Facebook appeared online with an innovative business model: it offered to the subscribers the free opportunity to connect to other subjects (at the time only students of qualified American private universities) offering them a free entertainment information service and the opportunity to sift into other student social profiles. Based on the subscriber's data – photos, messages, personal opinions – Facebook developed an advertising business model that exploited the users' social profiles as advertising targets: Social media users became "the target" for buying advice, friends suggestions, brands offers, etc. Rapidly this model became a *de facto* standard: the commodification of social network profile data became a standard model in the functioning of the various emerging social media platforms: Twitter, Pinterest, Google+, etc. In other words, social media have become the only media in which the subjects of online messages are both issuers and content at the same time: online profiles and identities are in fact reduced to traceable, manipulable, commodified data exploited by platforms for different purposes: targeted propaganda, online marketing, social dating, etc.

Moreover, even if users registered on social networks are aware of this mechanism, they seem to accept a profitable compromise between privacy and the advantages freely made available by the social networks: the enormous expansion of the main social platforms in recent years has not stopped but has rather been re-articulated among old and new social networks (Fuchs, 2017).

The commodification of data therefore results, despite the many critical voices, in a growing trend: a constant bias in the current social media landscape.

The Cambridge Analytica phenomenon – of which the substantial results in relation to the election trend is still to be assessed – is nevertheless symptomatic of the commodity phenomenon: the exploitation of online identities translated into data has made possible the manipulative and distortive propaganda whose outcomes will have long-term consequences on the democratic political life of the United States, UK and also in the rest of the world.

Social Media Computational Data

If the subjects of social media communication can be commodified as data, can they still be referred to as an actual individual? Couldn't they be replicated and or created fictitiously? This is the case of computational propaganda BOTs, where fake profiles are invented to multiply messages to support this or that organization or to force voters to opt for a specific party.

Computational propaganda is a term that refers to the recent phenomenon of digital misinformation and manipulation. As a communicative practice, computational propaganda describes the use of algorithms, automation and human curation to purposefully manage and distribute misleading information over social media networks (Woolley & Howard, 2016). "As part of the process, coders and their automated software products (BOTs) will learn from and imitate legitimate social media users in order to manipulate public opinion across a diverse range of platforms and device networks" (Woolley & Howard, 2018). These bots are built to behave like real people – for example, automatically generating and responding to conversations online – and then let loose over social media sites in order to amplify or suppress particular political messages. These automated social actors can be used to bolster particular politicians and policy positions supporting them actively and enthusiastically, while simultaneously drowning out any dissenting voices (Abokhodair et al., 2015). They can be managed in conjunction with human troll armies to "manufacture consensus" or to otherwise give the illusion of general support for a (perhaps controversial) political idea or policy, with the goal of creating a bandwagon effect (Woolley & Guilbeault, 2017). This practice interests both authoritarian countries where social media platforms are a primary means of social control but it even affects the democratic life of Western countries especially during political and security crises. The consequences and the impact of the use of computational propaganda are the objects of study and analysis: Woolley and Howard (ibid.) observed that, for example, during the 2016 US presidential campaign, the computational propaganda practice had been exploited by both parties – Republican and Democratic – and that there's also little control and still approximate legislation on this aspect.

Automated Selectivity and Echo Chambers

Which social media bias facilitates the creation of online echo chambers ?

Some elements have always characterized the participation of individuals in online environments: e.g., the ease and affordability of access to online environments. Since the 1990s, in the first virtual communities and then later in blogs or in collective projects such as Wikipedia, anyone could access and contribute to a collective dialogue without any economic or cultural threshold. Another element characterizing "online participation" is anonymity: the possibility of assuming a fictitious identity or even multiple different identities based on the different types of online environments that one attends, which may be the multi-player gaming platform or the political blog or the community of interest related to their hobbies. Another element is the opportunity to reach millions of users thanks to the growing diffusion of online social platforms and the practices oriented to "prosumerism": the creative re-working of the content generated by users.

Starting from the second half of the 2000s, social media added to these prerogatives, some new and important dimensions. First of all, social media favors what anthropologists define as selective sociability, or the possibility of connecting only with selected subjects that the individual can decide whether to re-enter or exclude from his or her friends' network or followers (Miller et al., 2016). This dynamic favors the definition of relatively closed groups characterized by a certain level of homophilia: individuals who share the same tastes, opinions, lifestyle and consumption styles. Moreover, homophilia is enhanced by the logic of the "social network algorithm" that profiles our networks of friendships and systematically suggests new pages or profiles similar to ours, reinforcing the similarities instead of the differences among our social networks. Does this mean that anyone active on social media lives in an echo chamber? Not so, everyone is free, regardless of the recommendations of the platforms, to expand and diversify his own network thanks to the serendipity in following contacts and relationships very different from those that reflect their tastes or their political orientations.

However, it is evident how these characteristics might greatly help the possibility of spreading messages within selected networks homogeneous for propensities and political views, expanding these networks thanks to the social algorithm encouraging homophilia and popularity, and channeling to these networks personalized messages, visually and emotionally adapted to the type of target and subjects that are to be intercepted. Through the mechanism of selective sociability these networks can also remain relatively closed to outsiders or those who want to enter uninvited, thus reinforcing the cohesion and identification of the subjects within the group, essential to the good impact of propaganda.

Many recent studies on the phenomenon of recruitment and indoctrination of Foreign Fighters by ISIS have highlighted evidence that supports this thesis. First, the process of online radicalization is most of the time a bottom-up process: that is the young people search the web for information and news about a particular organization to get closer to this, and not vice versa (Marone, 2016; Vidino & Marone, 2017). After an initial approach the subjects can be co-opted within virtual groups (Facebook or Telegram groups) in which propaganda content may circulate: extremist magazines, ad-hoc texts of the Koran, leaders accounts. Moreover the subjects are introduced gradually inside to a closed social network – an echo chamber precisely – aimed at strengthening the individuals' ideological positions and their commitment to the cause.

Moreover, social media make available to the extremist propaganda, formats, languages and audiovisual styles widely appreciated at the emotional level: graphic memes, popular hashtags, short videos integrated by infographics. In recent years, ISIS has, for example, taken up these propaganda formats with great freedom and creative capacity, perfectly emulating the styles and aesthetics of Hollywood trailers and commercial marketing. Through the use of

Tweets on the daily life under the Caliphate, or the motivational videos in the Dabiq propaganda magazines recalling the aesthetics of highly successful video games such as Call of Duty (a particularly popular title especially in Saudi Arabia) (Maggioni & Magri, 2015), ISIS has shown to know how to exploit the most popular social media trends to achieve its own goals (Weiss & Hassan, 2016; Winkler & Pieslak, 2018).

Conclusions

The emergence of social media as a central platform in modern public life raises new questions and problems for the advancement of contemporary democracies. If, in some totalitarian states, social media are routinely used as new forms of social control, even in the most evolved democracies of the West, the social media bias may significantly produce ambiguous and pervasive forms of propaganda, often at the service of extremist, anti-democratic organizations that represent a serious threat to the public life for millions of citizens.

Recently the governments of many Western countries (Europe, US, Canada) established actions and various legislative protocols of agreement with the major online players (Facebook, Twitter, YouTube) to deal with the threat of global terrorism online propaganda in which ISIS, between 2014 and 2016, proved to be one of the most active subjects. The online platforms have therefore developed various tools to contain the flow of messages of indoctrination or aimed at recruiting young people in the West (Gillespie, 2018). In this case the collaboration between governments and the major private players has shown proof of effectiveness and positive cooperation.

Other cases – the most striking is Cambridge Analytica – revealed the ambiguous relations between technological players such as Facebook, a number of lobbying subjects related to the presidential campaign and multiple private software companies.

Computational propaganda is now one of the most powerful tools against democracy. Social media firms may not be creating this nasty content, but they offer a safe harbor for them. Social media have the power to control what information or news people see; that gives them the responsibility for making sure this information is not harmful, harassing or false. This is especially true during pivotal political events like elections, but also true in general.

Social media platforms must play a central and a new role in the mediation, moderation and also in the redesign of the bias which significantly affects the democratic debate online.

As Woolley, S. C., & Howard observed:

> they cannot rely upon tired defenses about being technology not media companies. Trending features, algorithmic curation, and personalized news feeds mean that companies do, to use their language, arbitrate

truth. Because they control information flow, they are media companies. To solve these problems, social media companies must confront their role as media platform. They must design for democracy.

(ibid., 3025 (kindle pos.)

Notes

1 For a detailed report of the scandal cf. The Cambridge Analytica Files, *The Guardian* www.theguardian.com/news/series/cambridge-analytica-files;
2 https://en.wikipedia.org/wiki/Cambridge_Analytica
3 For a full transcript of Marc Zuckerberg hearing on April 11, 2018 cfr. www.washingtonpost.com/news/the-switch/wp/2018/04/10/transcript-of-mark-zuckerbergs-senate-hearing/?utm_term=.a49c7bb5cd7c
4 https://gizmodo.com/facebook-bans-app-that-inspired-cambridge-analytica-and-1,828,548,982

References

Abokhodair, N., Yoo, D. and McDonald, D. W. (2015, February). Dissecting a social botnet: Growth, content and influence in Twitter. In *Proceedings of the 18th ACM Conference on Computer Supported Cooperative Work & Social Computing* (pp. 839–851). ACM.

Abouzeid, R. (2011). Bouazizi: The man who set himself and Tunisia on fire. *Time Magazine*, 21.

Atwan, A. B. (2015). *Islamic state: The digital caliphate.* Los Angeles, CA: University of California Press.

Awan, I. (2017). Cyber-extremism: Isis and the power of social media. *Society*, 54(2), 138–149.

Boyd, D. M. and Ellison, N. B. (2007). Social network sites: Definition, history, and scholarship. *Journal of Computer-Mediated Communication*, 13(1), 210–230.

Carter, J. A., Maher, S. and Neumann, P. R. (2014). # Greenbirds: Measuring importance and influence in Syrian Foreign Fighter Networks.

Castells, M. (2013). *Communication power.* Oxford: OUP.

Eltantawy, N. and Wiest, J. B. (2011). The Arab spring social media in the Egyptian revolution: Reconsidering resource mobilization theory. *International Journal of Communication*, 5, 18.

Farwell, J. P. (2014). The media strategy of ISIS. *Survival*, 56(6), 49–55.

Fuchs, C. (2012). Dallas Smythe today-the audience commodity, the digital labour debate, Marxist political economy and critical theory. *tripleC: Communication, Capitalism & Critique. Open Access Journal for a Global Sustainable Information Society*, 10(2), 692–740.

Fuchs, C. (2017). *Social media: A critical introduction.* London: Sage.

Fuchs, C. (2018). Propaganda 2.0: Herman and Chomsky's propaganda model in the age of the internet, big data and social media. In Pedro-Carañana, J., Broudy, D. and Klaehn, J. (Eds.) *The propaganda model today: Filtering perception and awareness.* London: University of Westminster Press. Available at: 10.16997/book27. f.

Gillespie, T. (2018). *Custodians of the Internet: Platforms, content moderation, and the hidden decisions that shape social media.* New Haven, CT: Yale University Press.

Herman, E. S. and Chomsky, N. (2010). *Manufacturing consent: The political economy of the mass media*. London: Random House.

Hoskins, A., Awan, A. and O'Loughlin, B. (2011). *Radicalisation and the media: Connectivity and terrorism in the new media ecology*. London: Routledge.

Innis, H. A. (2008). *The bias of communication*. Toronto: University of Toronto Press.

Khondker, H. H. (2011). Role of the new media in the Arab Spring. *Globalizations*, 8(5), 675–679.

Kirkpatrick, D. D. (2011). Tunisia leader flees and prime minister claims power. *New York Times Online*, 14.

Maggioni, M. and Magri, P. (Eds). (2015). *Twitter and Jihad: The Communication Strategy of ISIS*. Available at: www.ispionline.it/sites/default/files/pubblicazioni/twitter_and_jihad_en.pdf. Accessed January 10, 2019.

Mahood, S. and Rane, H. (2017). Islamist narratives in ISIS recruitment propaganda. *The Journal of International Communication*, 23(1), 15–35.

Marone, F. (2016). Italian Jihadists in Syria and Iraq. *Journal of Terrorism Research*, 7, 1.

Meleagrou-Hitchens, A. and Kaderbhai, N. (2016). *Research perspectives on online radicalisation*. London, UK: A Literature Review.

Miller, D., Costa, E., Haynes, N., McDonald, T., Nicolescu, R., Sinanan, J. … Wang, X. (2016). *How the world changed social media*. London: UCL Press.

Monaci, S., Mazza, C. and Taddeo, G. (2017). Designing a social media strategy against violent extremism propaganda: The# heartofdarkness campaign. *Digitale Medien und politisch-weltanschaulicher Extremismus im Jugendalter*, 213.

Monaci, S., Mazza, C. and Taddeo, G. (2018). Mapping the impact of ISIS propaganda narratives. Evidences from a qualitative analysis in Italy and France. *Comunicazioni Sociali* 3/18.

Morozov, E. (2011). *The net delusion: The dark side of Internet freedom*. Phildaelphia,PA: Public Affairs.

Mullen, A. (2009). Editorial: The Herman-Chomsky propaganda model twenty years on. *Westminster Papers in Communication and Culture*, 6(2). Retrieved from: www.wmin.ac.uk/mad/pdf/WPCC-Vol6-No2-Andrew__Mullen.pdf.

Neumann, P. R. 2015. Victims, perpetrators, assets: The narratives of Islamic state defectors, *ICSR*, 9.

O' Reilly, T. (2009). What is web 2.0. O'Reilly Media, Inc.

Pedro, J. (2011). The propaganda model in the early 21st Century (Part II). *International Journal of Communication*, 5, 21.

Postman, N. (2000). The humanism of media ecology. *Proceedings of the Media Ecology Association*, 1(1), 10–16.

Roy, O. (2016). *Le Djihad et la mort*. Paris: Seuil.

Smythe, D. W. (1981). On the audience commodity and its work. *Media and cultural studies: Keyworks*, 230–256.

Stern, J. and Berger, J. M. (2015). *ISIS: The state of terror*. London: Harper Collins.

Sunstein, C. R. (2018). # *Republic: Divided Democracy in the age of social media*. Elizabeth, NJ: Princeton University Press.

Vidino, L. and Marone, F. (2017). The Jihadist threat in Italy: A primer analysis, *Italian Institute for International Political Studies (ISPI)*, November, 48–63.

Weiss, M. and Hassan, H. (2016). *ISIS: inside the army of terror*. New York: Simon and Schuster.

Winkler, C. and Pieslak, J. (2018). Multimodal visual/sound redundancy in ISIS videos: A close analysis of martyrdom and training segments. *Journal of policing, Intelligence and Counter Terrorism*, 13(3), 345–360.

Woolley, S. C. and Guilbeault, D. R. (2017). Computational propaganda in the United States of America: Manufacturing consensus online. *Computational Propaganda Research Project*, 22.

Woolley, S. C. and Howard, P. N. (2016). Automation, algorithms, and politics. Political communication, computational propaganda, and autonomous agents - Introduction. *International Journal of Communication*, 10, 9.

Woolley, S. C. and Howard, P. N. (Eds.) (2018). *Computational propaganda: Political parties, politicians, and political manipulation on social media*. Oxford, UK: Oxford University Press.

13

FROM FANS TO FOLLOWERS TO ANTI-FANS

Young Online Audiences of Microcelebrities

Maria Murumaa-Mengel and Andra Siibak

Introduction: Established Theories, Emerging Phenomena

Traditional media, dominated by public figures and mainstream celebrities, is increasingly overshadowed by young social media microcelebrities, i.e., popular online personae who produce and create online content to gain followers or fans (Abidin, 2018; Marwick & boyd, 2011). These microcelebrities have built up solid fan bases (and simultaneously, masses of "haters") and have attracted millions of followers. Much of the fame of microcelebrities is due to the fact that microcelebrities not only engage in textual and visual storytelling about their personal lives and lifestyles but also "engage with their following in 'digital' and 'physical' spaces so as to sustain and amplify their shared sense of camaraderie" (Abidin, 2016a, 3). However, celebrity studies scholars (e.g., Mendick, Allen & Harvey, 2015) argue that currently too few researchers focus on analyzing "real, empirical audiences" (Holmes, 2004, 169), such as the everyday audiences of (micro)celebrities.

Different authors have created various labels in order to refer to such online personae: social media influencers (Abidin, 2015, 2018), blogebrities (Hopkins, 2009), online celebrities (Strafella & Berg, 2015), market mavens (Goodey & East, 2008) and social media opinion-makers (Fuchs, 2017), to name a few. We will use the term "microcelebrity" throughout this chapter to talk about online content creators who have gained the attention of relatively large audiences, but some clarification is necessary. Although the term has been widely used by many to describe these influential online content creators (e.g., Abidin, 2015; Marwick & boyd, 2011; Senft, 2013), it can be somewhat misleading to the reader. Take, for instance, the most popular YouTubers, who have tens of millions of subscribers and whose videos have been watched billions of times, raking in thousands of comments and reactions (e.g., the most successful YouTuber, PewDiePie). At this moment in digital culture, there is nothing "micro" about the celebrity of these new stars. The term microcelebrity can also denote specific practices and techniques used to gain and maintain status and attention online (Khamis, Ang & Welling, 2017; Senft, 2008), so these new media influencers practice microcelebrity, even if they are full-blown celebrities in terms of numbers. In fact, Abidin has described the four main characteristics that make a microcelebrity different from a traditional celebrity (Abidin, 2018, 11). First, microcelebrity is based on a sense of belonging and active interaction with audiences. Second, microcelebrities are not primarily known for their extraordinary skills or performances; rather, audiences are drawn to them because microcelebrities are often perceived to present themselves as they are: they are authentic. The third characteristic, according to Abidin (2018), is the fact that microcelebrities need not appeal to large global mass audiences, as the depth of the audience's relationship with the celebrity is more important. And last, microcelebrities are almost "obligated" to foster strong interactional relationships with their audiences, whose labor is then "exploited in the accumulation of communicative capital" (Nixon, 2014, 732).

Young people are spending more and more time online, continuously creating, seeking, modifying and interpreting what is perceived as attention-worthy content and, in effect, providing the attention that "makes" a microcelebrity. In that sense, the specific audiences at the heart of this chapter form a good example of "active audiences," who are actively "intellectually engaged *with* the text, rather than an intervention *in* a text" (Tincknell & Raghuram, 2002, 200). When systematic scientific attention was first focused on audiences, the receivers of messages, publics were seen as rather passive entities for a long time (for an overview of the history and shifts in thinking about the concept of audiences, see McQuail, 2013 or Livingstone & Das, 2013). One-directional mass media audiences were often thought of as unidentifiable more or less homogeneous masses who could not respond to the broadcaster (Marwick & boyd, 2011); one message could reach many, producing a uniform effect (McQuail, 2013). Then, step by step, audience studies and reception theory began to include various aspects of communication, e.g., the social structure of audiences, mediators, self-determination, technological innovations and diversity in many senses (McQuail, 2013), and we have steadily shifted to an active audience paradigm.

A turn to an active audience paradigm does not necessarily mean we need radically new approaches and theories to think about these new formations of audiences; well-established theories and frameworks are often valuable as theoretical tools for researchers. One such seminal work that we believe is still helpful to consider is Stuart Hall's en/decoding framework, where the (re)production and sustaining of social structures is kept up through the articulation of linked but distinctive moments: production, circulation, distribution, consumption and reproduction (Hall, 1980/2008). Although Hall's ideas were originally focused on television as a medium, we still find them relevant today – perhaps quite unoriginally, as Hall's thoughts have found a fruitful foundation for audience scholars focusing on very different media for decades (cf. Jenkins, 2018). In the case of microcelebrities, we need to consider how the messages are encoded, how their content is strategically and/or authentically produced and how their audiences decode it. To some extent, the audiences of microcelebrities are unable to decode the production and circulation means in novel and unexpected ways, e.g., the channel, the genre, visual language or the "form of appearance" (Hall, 1980), limit and exclude some possibilities of decoding. The findings of our previous empirical studies (e.g., Murumaa-Mengel & Siibak, 2017, 2018) suggest that audience members of contemporary microcelebrities are used to following certain rules and predetermined ways of "reading texts," but they also actively resist, negotiate and modify these messages. In that sense, we align well with the approach of another well-established theory, structuration by Anthony Giddens (1984), where agency is essentially people's free will and the capability of making independent choices that are continuously mirrored in and by their unconscious motives, and discursive and practical consciousness. So, aspects of social structure can be perceived and reproduced completely differently, including deliberately starting a counter-practice and thus allowing for a change in the structure.

Stuart Hall (1980, 509) was mainly talking about television when he stated that "circulation and reception are, indeed, 'moments' of the production process … and are reincorporated, via a number of skewed and structured 'feedbacks,' into the production process itself." In the context of online media and social media, we clearly see that the feedback loop has become more important than ever. Furthermore, in structuration theory "a system is maintained through the use or application of structures … agency is characterized by an innate ability to imagine different outcomes" (Wiggins & Bowers, 2015). According to Giddens, agency also includes unintentional acts of a person, it "refers not to the intentions people have in doing things but to their capability of doing those things in the first place" (Giddens, 1984, 9). So, for example, if oppositional reading is widely practiced, it can gradually lead to a larger shift in the structure, the accepted norms of communication in society or the culture of communication.

As demonstrated above, Stuart Hall (1980) has invited us to think about the various articulations in which encoding and decoding can be combined. And, so far, many researchers (e.g., Abidin, 2018; Jenkins, 2006; Khamis et al., 2017; Marwick, 2015; Marwick & boyd, 2011; Westenberg, 2016; Senft, 2008) have already given us valuable information from the encoding phase of microcelebrity-produced content: how social media content creators try to engage their audiences and use textual, visual and audio messages to gain attention. In the current chapter, however, we aim to take a closer look at the decoding phase by exploring the viewing practices of young online audiences of microcelebrities.

When we turn our attention to fan studies and see how different authors have mapped the multifaceted landscape of audiences where fans are located, it is clear that defining what fandom is has proved to be a hard task for scholars. Hills (2003) has pointed out that this could be due to the fact that "fan" and "fandom" are a part of everyday language, thus signifying a broad spectrum of people and behaviors, and are terms laden with connotations. These are often negative connotations, as the root of this word – Latin word *fanaticus* for "insane, mad, possessed by the gods" (Cochran, 2008, 239–240) – has a strong "otherness" (Linden & Linden, 2016) coded into even the denotative layers. Additionally, specific online environments have used the term "fan" rather loosely for a long time (e.g., Facebook, and the now-defunct Orkut), thus once again modifying the connotations associated with it. In these social media networks, one could become "a fan" of anything: banks, smoothies, types of make-up or sportswear. Linden and Linden (2016, 16–17) argue that the "utilisation of the word 'fan' has arguably contributed to the broader and more holistic use of the term, and thus promoted – for better or worse – a more casual approach towards fans, fan cultures and fandom."

Various typologies have been suggested, for example, Abercrombie and Longhurst's (1998) taxonomy of audiences, which includes consumers, fans, cultists, enthusiasts and petty producers, or Wasko's (2001) audience archetypes: fanatics, fans, consumers (enthusiastic, admiring or reluctant), cynics, uninterested audiences, resisters and antagonists. We will follow the example of Jonathan Gray (2003), who has called for a simpler categorization that will allow us to think beyond fan identities, as fans are only a small fraction of all audience positions. Gray suggested three categories – fans, non-fans and anti-fans – which in general fit our analysis. We have decided to rename non-fans "followers" and thus utilize Tulloch and Jenkins' term (1995) in this overview, but in a new context. For Tulloch and Jenkins (1995), fans claim social identity and followers do not. Online audiences do bring something new to the field and to the meaning of a follower: one can follow a microcelebrity rather passively, as information is inserted into our social media feeds. Thus, interaction and connection to a microcelebrity is there, but the whole relationship can be a mundane part of life, maintained by the fear of missing out. Such audiences provide constant attention but do not claim a connected social identity.

Although fandom and fan culture have been thoroughly studied and spectra of identities and experiences have been mapped by many (Duffett, 2013; Geraghty, 2015; Linden & Linden, 2016; Reijnders, Zwaan & Duits, 2014), the knowledge of the nature and extent of the different types of audiences of online celebrities has just begun to grow. Most of the existing research has focused generally on microcelebrities or social media influencers (Abidin, 2018, 2015; Fägersten, 2017; Jenkins, 2006; Khamis et al., 2017; Marwick, 2015; Senft, 2008). Research on specific microcelebrities' audiences and their perspective on the phenomenon has received less academic attention so far (e.g., Erhart, 2014; Westenberg, 2016) and this is where our chapter strives to contribute. Furthermore, as argued by some (e.g., Gray, 2003; Mendick et al., 2015), intentionally or not, audience research often equals fan research, as anti-fans and non-fans are often ignored or assumed. In fact, scholars (e.g., Gray, Harrington & Sandvoss, 2007, 10) argue that anti-fan studies could actually help to broaden the scholarship related to audiences and thereby "tell us something about the way in which we relate to those around us, as well as the way we read the mediated texts that constitute an ever-larger part of our horizon of experience."

Relying on Stuart Hall's (1980, see also Gray, 2003) en/decoding framework, we will discuss three hypothetical decoding positions (as this chapter is focused more on the "decoders" of initial texts, i.e., audiences) in the context of young people's "reading" of microcelebrities: the dominant-hegemonic position (here, mostly "fans"), the negotiated position ("followers") and the oppositional position ("anti-fans"). The empirical examples in this chapter come from numerous smaller studies: surveys, content analysis of (audio-)visual and written texts, focus-groups and interviews with microcelebrities, as well as members of their audiences, carried out over the past five years by our research team at Tartu University, consisting of different faculty members and students at various levels. By combining ideas and notions from well-established theories and novel research, we aim to offer a multifaceted look at the topic and to spark some ideas on how to approach, theorize and re-imagine dynamic contemporary online audiences.

Fans

The dominant-hegemonic position (Hall, 1980/2008) of audiences is represented by fans, often the "ideal readers" of multi-modal texts (Marwick & boyd, 2011; Murumaa-Mengel, 2017), in which audience members decode the messages and connotative meanings more or less as they have been encoded. Ideal audiences are often modeled in our minds based on ourselves and those close to us. When information is shared online, it is often posted for members of the audience who are the "mirror image of the user" (Marwick & boyd, 2011, 120). Microcelebrities that we have interviewed have repeatedly said that they tend to make content for people like themselves, but at the same time have to consider the reality of having many fans who do not fit the profile of that ideal:

> YouTuber, male, 16, mostly humorous and lifestyle content: "*yeah, they* [very young audiences] *are f****** cool! I don't have anything against them. Just that it would be awesome if there were more people who were my age.*"

Many interviewees in our audience studies have described themselves as fitting the profile of an "ideal reader," acknowledging microcelebrities as important others in their lives and creating a "strange familiarity," as Senft (2013) has called it. By sharing their own personal stories with the audience, microcelebrities create a sense of intimacy and often surround themselves with people who have experienced similar life events, ordeals or troubles. Thus, followers feel supported, understood and cared about by the microcelebrity. which strengthens the bond between them and reinforces the fan identity. Furthermore, it has been argued that "fans' investment in certain practices and texts provides them with strategies which enable them to gain a certain amount of control over their affective lives" (Grossberg, 1992, 65):

> Boy, 15: *my all time favorite post is a YouTube video by Conan Gray in which he talks about femininity, masculinity and him coming to terms with being a "girly boy." I understood him because I myself am not a masculine guy. I felt connected to him. I felt like he was talking to me.*

In other words, the degree of understanding in the communicative exchange is high and the positions between encoder-producer and decoder-receiver are symmetrical (Hall, 1980/2008). So if microcelebrities are producing content and encoding their texts, they expect and hope the majority of their audiences will decode these texts in these given dominant positions. A large part of their audiences, very often fans, do. As an example, we present an excerpt from one of our focus group interviews:

> Girl, 16: *when you see that they* [young microcelebrities] *do not have to study for school and … Not do anything and you have to … then it sometimes like not actually annoys me but I am a bit cross about it. I don't want to do it at all and they don't have to do the same thing every day … it's like you know how some people are naturally skinny; they can eat whatever they want and you can't really do that […] maybe yeah, they don't have to study and all …*

We see that the lifestyle presented by the microcelebrity is taken "as it is presented," not questioning the strategic presentation of edited life. The performed nature of authenticity might be lost on some members of the audience, perhaps partly due to deficiencies in digital literacies or just the uncritical acceptance of offered dominant-hegemonic discourses. In our empirical work, we have found evidence of some young audience members not recognizing sponsored posts, product placement and all the creative and aspirational labor (Duffy, 2015; Duffy & Wissinger, 2017) that is hidden at first glance.

Paying attention to fans and followers, "communicating reciprocal intimacies" (Abidin, 2015), is a crucial element of practicing microcelebrity. The audience expects the microcelebrity to address them, ask for their comments and likes, answer fan mail and interact with them in other ways. Thus, in order to study this interaction between fans and celebrities, media scholars (e.g., Abidin, 2016b; Schroath, 2016) have often relied on Horton and Wohl's (1956) para-social framework, which suggests that through media audiences are able to receive only the information which has been carefully considered and constructed, information which fits the fictional roles the personas have created for themselves. However, audiences must perceive this information to be credible and realistic in order to enable them to feel sympathy, sociability, intimacy and empathy toward the persona. As argued by Duffy and Hund (2015, 9), all these "socially mediated representations of affective pleasure and compulsory sociality" should also be interpreted in the context of emotional labor and viewed as elements of "deep acting" (Hochschild, 1983).

These audience members are more likely to include their attachment to their idols as a strong part of their identity and they can sometimes be considered to be super-fans (Abidin, 2018), fanatics (Wasko, 2001) or even "#teamXers," as the hashtag publics (Rambukkana, 2015) have structural affordances that allow collective fan identities to concentrate. By that we mean sub-audiences and sub-cultures that are based on heavy involvement with the celebrities they follow, as we have witnessed forming tight-knit communities in the case of Trekkies, Twihards, Directioners, Lady Gaga's Little Monsters or Beyoncé's BeyHive. Microcelebrities use and enforce specific labels in a similar vein to foster the emergence of a community and address them. These engaged audiences are happy to internalize these labels as well, for example, the YouTuber Logan Paul's fans make-up the "Logang," Tessa Brooks has the "Brookters," KSI's followers form the "KSI army" and, an Estonian example from our studies, the popular YouTuber Istoprocent's fans have taken on the collective label of "#teamisto."

Oftentimes screen-mediated communication can still be perceived as a "personal thing," very individual and focused on the microcelebrity–fan relationship, rather than a fan community building on the creation of a shared fan identity. Thus, one's social identity as a fan (Tulloch & Jenkins, 1995) may also be more individual and hidden. However, some fans are "activated" by the dynamics and opportunities of social media, are eager to reach out to their idols for reciprocal intimacies (Abidin, 2015) and thereby have the opportunity to become visible and known via the overall communicative mode of "conversation," where the strict separation of sender and receiver is blurred (Schmidt, 2014).

It is extremely important to emphasize that the dominant-hegemonic position does not describe all fans, far from it. Fan positions and media texts are rarely stable or static (Gray et al., 2007, 314), and in the case of youth (Jorge, 2011) and dynamic online spaces and digital culture, even less so. Fans can be and often are producers of cultural meanings, contesting the cultural value of media texts (Jenkins, 1992), interpreting content and communicating meanings with others, thus producing meanings that rework and refashion the original text (Fiske, 1989). Fandom is most often seen as a participatory culture (Linden & Linden, 2016), consisting of active audiences whose reading aligns well with what Hall (1980/2008) referred to as the negotiated position, which enables fans to explicitly re-articulate "culture in unique and empowering ways" (Pullen, 2000, 53), and in a mixture of preferred and resistant readings (Shaw, 2017).

For example, a significant number of 9–13-year-olds whose viewing practices and favorite microcelebrities we set out to explore actively follow You-Tubers whose content can, on some occasions, be described as sexist, racist or simply problematic. Our analysis suggests that although the members of these young audiences might not always agree with all the viewpoints and beliefs presented, they still consider themselves fans of these content producers. As argued by Hall (1980, 516),

> decoding within the negotiated version contains a mixture of adaptive and oppositional elements: it acknowledges the legitimacy of the hegemonic definitions to make the grand significations (abstract), while, at a more restricted, situational (situated) level, it makes its own ground rules – it operates with exceptions to the rule.

In the case of microcelebrities and their audiences, even very young members of these audiences (e.g., interviewed 9-year-olds) can express the decoding logics that are inherent to the negotiated position: "I really like him/her, even though I do not agree with everything." Also, using the logic of Gerbner (1970, cited in Hall, 1980/2008), if a microcelebrity expresses racist views, such content can be decoded as a representation, "a message about racism," not racism itself.

In other situations, fans may also take some messages with a grain of salt, for example, in the case of commercial and sponsored content. When a microcelebrity raves about some product, some young audience members recognize the underlying economic motivation and contextualize such content in the framework of their own "ground rules," e.g., skipping the part where the microcelebrity is endorsing something, while acknowledging that such information can be useful and that is just how their idols earn a living.

When thinking about great examples of fandoms from older media – books, films, TV-shows, etc. – "fans often constitute an elite fraction of the larger audience of passive consumers" (Grossberg, 1992, 52), putting serious effort into the cultural work. They seem to follow "slow principles," such as reflexivity and mindfulness (Poirier & Robinson, 2014, 700), within the fan community. In our studies, though, we have noticed a sort of "new passivity" among fan audiences. Even when the young audience members of microcelebrities describe very engaged and dedicated fan practices, they do not engage in the negotiation of these texts as actively as one would assume. Paradoxically, the media that are hailed for endless opportunities for participation and the realization of truly (inter)active audiences have actually resulted in greater passivity by the audience, at least on the level of visible engagement. At least the findings of our empirical studies seem to hint that these new audiences in the new media have truly taken on a fast-paced lifestyle and, hence, are experiencing the social acceleration of time (Rosa, 2013), which has normalized imperatives of "efficiency" (Hassan & Purser, 2007) and taken-for-granted instantaneity (Poirier & Robinson, 2014), all of which also applies in the context of producing and consuming information. Microcelebrities themselves have also made strategic use of these imperatives by obtaining easy access to highly visible and dynamic metrics (followers, likes, comments, etc.) of their audiences (Marwick, 2015). However, this always-on lifestyle, as well as regular and frequent updates, make it hard for audience members to concentrate on specific texts for longer periods of time and provide less time for negotiated meaning-making. In fact, much of the negotiated decoding takes place directly in communication with the original sender of the message (thus strengthening the dominant decoding frameworks), as audience engagement through interaction is central to practices of microcelebrity.

Followers

In our empirical studies, we have also noticed that even when young members of microcelebrities' audiences describe what could be perceived as fan practices as part of their daily routines, they take extra care to convey that they are not fans, but rather merely casual followers, as if considering oneself to be a fan would be perceived as too eager and not "cool." For a long time, the dominant discourse has defined fans as different from "people like us" (Jenson, 1992). For example, Kozinets (2001) has pointed out that sometimes fans fear revealing their obsession to others, feeling embarrassed and stigmatized when they do so, except when socializing with like-minded fans.

Another reason for refusing fan identity could lie in the very fact that the many-to-many communication model has broadened and fragmented audiences. Young social media users tend to follow numerous microcelebrities and influencers, without "committing" to one, binge-watching one for hours and then jumping on to the next one:

> Girl, 13: *on weekends, I can be on YouTube for eight hours straight, because I start watching the videos and before I know it, it's two o'clock in the morning.*

Within these less or more uninterrupted following practices, audience members move swiftly and even unconsciously between all three decoding positions that Hall (1980/2008) spoke of. Gray (2003) has pointed out about television –

but it is fitting for social media audiences as well – that being a fan requires discipline, whereas being a non-fan involves considerable flow in and out of different viewing positions. Dedicating undivided attention to a single, idolized microcelebrity is rare, and thus the fan identity can dissolve in an endless stream of content from various microcelebrities. Gray (2003) wrote about non-fans, by whom he meant those audiences who do not have intense involvement with texts, and this information can also help us when describing followers. Parallel and (often) superficial activities common to multitasking make people switch continuously from one role to another: "audiences and publics, producers and produsers citizens and consumers are converging and diverging in parallel with the media texts and technologies with which they engage" (Livingstone & Das, 2013, 3), so a distinct self-identification as a "fan" can be a status that is hard to achieve in the current social acceleration of time (Rosa, 2013). This does not mean, of course, that these invisible, often "lurker" audiences do not interact with the content and are not involved in the meaning-making process. Negotiation, opposition or reinforcement of discourses may take place and may be internalized unconsciously; they may take place in other realms, other contexts, and still be part of the larger interpretative community (McQuail, 2013).

Sometimes a fan can shift to being a follower, and even to being an anti-fan. McRae (2017, 14) has argued that dedicated followers or fans have become "genre experts" and some kinds of posts become so familiar that they appear unoriginal or inauthentic. This, in fact, can slowly change the perception of the microcelebrity who fails to bring new, exciting and entertaining information to the relationship. Thus, some audience members start to move away from their content, to the point where following some microcelebrities is seen as "so 2017," "lame" and "SSDD" (same s*it, different day), i.e., unoriginal and out of fashion.

The previous example was about audience expectations that require change and evolution. On the other hand, shifts can occur precisely because things have changed and reciprocal intimacies (Abidin, 2015) have been lost. For example, although most of the young audience members we have interviewed understand that it is not humanly possible to answer every comment after the microcelebrity has "made it big," some still harshly criticize the failure to respond.

Anti-Fans

And finally, there are anti-fans. Some members of microcelebrities' audiences can and have taken an oppositional decoding position (Hall, 1980/2008), where they have understood the literal and the connotative inflection of a discourse but decode the message in a contrary way. Oppositional decoding can also occur when audience members are unable to decode the text completely: "the viewer does not know the terms employed, cannot follow the complex logic of argument or exposition, is unfamiliar with the language, finds the concepts too alien or difficult" (Hall, 1980/2008, 514). They miss jokes and do not get the cultural references (Gray, 2003). Of, course, such decoding positions can be taken by fans and followers as well, but we see it most often with anti-fans.

Matthew Collins and Danielle Stern (2015) have argued that some fans can be classified as narcissistic, as they want to "see themselves" within the text. Similarly, anti-fans can find gratification in texts that they strongly oppose, as they can decode the texts as "beneath them," frivolous, dumb or completely "other" in some way. Or, as posed by Gray (2003: 70): "this is the realm not necessarily of those who are against fandom per se, but of those who strongly dislike a given text or genre, considering it insane, stupid, morally bankrupt and/or aesthetic drivel." In fact, in such decodings of texts, when engaged in hate-watching, i.e., watching content that one does not actually enjoy, anti-fans are also "effectively organizing themselves, other viewers, and fans into the hierarchical construct of the viewing audience" (Gilbert, 2019, 74). In fact, as argued by Jane (2019, 58), by studying the culture of anti-fans we also "bring critical attention to the fact that much of the vitriol and hateful speech currently proliferating in digital cultures is indissolubly interwoven with popular entertainment."

From the microcelebrities' points of view, these members of the audience could be referred to as "nightmare readers" (Marwick & boyd, 2011) or "nightmare audiences" (Murumaa-Mengel, 2017), i.e., the opposite of ideal audiences. Such members of the social media audience usually represent different spheres of life, have some control over or, in most cases, are just very different from the person who originally posts the messages. Nightmare audiences might have different values (e.g., materialistic vs spiritual), senses of humor (black gallows humor vs lighthearted slapstick), standpoints on potentially polarizing issues (e.g., gender or race issues), or perceived norms of netiquette (e.g., private vs public info), and they can de-totalize the message in the preferred code in order to re-totalize the message within some alternative framework of reference (Hall, 1980/2008).

In public communication, this means the emergence of "haters" and "trolls," as microcelebrities from our studies often call them. Such members of the audience, those who are active and interact with the microcelebrity, usually do so in the form of negative comments and responses, whether they are communicated in a direct and private manner to the microcelebrities or publicly in social media.

> Blogger, female, 22, fitness and food: *at the beginning, I took everything to heart, like … how can they say these things, why are they saying this. But then I grew a thicker skin and I think that the best antidote to such people is just ignoring them because their greatest aim is to hurt someone through mean words.*

Even very young members of microcelebrities' audiences have accepted contemporary "mean" internet and communication culture where the jokes are often borderline, vicious comments and extreme trolling is seen as entertainment or accepted form of self-expression (Laineste, 2013; Murumaa-Mengel, 2017):

> Boy, 10: *you will get hate* [used the word "heit" in Esto-English hybrid language that is widely used in younger age groups], *you have to be prepared for that, they will hate on you and you just have to learn from it, you shouldn't quit right away, it's just that you should make better content, maybe then you will not get hate?*

Although some scholars have separated "haters" and anti-fans (Giuffre, 2014) into different audience types, we will try to keep it simple, as suggested by Gray (2003), and see them both as part of the oppositional anti-fan audience type. Giuffre (2014, 53) argues that "unlike hate, which is arguably a destructive process, anti-fandom can be a constructive form of engagement," but we feel it is important to note that hate can also be the basis of a shared identity.

Often, anti-fans are especially attentive to evidence which points to the constructedness of a microcelebrity persona (McRae, 2017), and spend a remarkable amount of time and energy picking apart the content microcelebrities are updating, with the aim of finding "glitches": contradictions, exaggerations, signs of non-authenticity and unethical behavior. Anti-fans may even launch or join a massive online public shaming campaign, as computer-mediated communication can make the somewhat anonymous other "feel less human." (Hall, 1980/2008, 511) has pointed out that the visual discourse in television translates a three-dimensional world into two-dimensional planes ("the dog in the film can bark but it cannot bite!") and this holds true in social media, too. The comment of an anti-fan may be hateful, but to the encoder, and sometimes the decoder, too, it "barks" on the representational level, but it is perceived not to "bite" the flesh-and-blood person behind the screen. Young people we have interviewed over the years have, however, taken a surprisingly passive stance toward produsers (Bruns, 2008) when encountering such malicious practices. Even if they consider themselves to be fans of the microcelebrity who is being publicly attacked, they usually just scroll on and do not report disturbing content to the platform manager or step in to defend their idol:

> Girl, 13: *if I see something disturbing, I just scroll on because YouTube will know. They control everything.*

"Technology from a strictly Giddensian viewpoint cannot be an agent, and can only exhibit 'structural properties' when utilized as a resource in social practice by human agents," noted Rose and Jones (2005, 22). Although it deviates from the Giddensian concept that we find theoretical support from, it can be argued that technology takes the form of mediator-agency as an enabler, the structural capability to make a difference. When people act in very large groups, feeling separated by screens and thus somewhat anonymous, their agency is diluted, and even attributed to the technology.

Nearly all of the young participants in our microcelebrity-audience studies have said that two of the main qualities they are looking for in these multimodal texts is entertainment value and humor. The entertainment may come from laughing at someone, rather than laughing with them, as "YouTube drama" is enjoyed by most of the young followers and sometimes even instigated by them:

> Boy, 13: *if I dislike a video I usually just scroll past it, but there have been cases where I go to another YouTuber that uploads "drama alerts." I just have to leave a comment with a link to the sketchy video and he will do his own video and alert everyone else about it.*

Microcelebrities themselves are not blind to these expectations and, although they might have an "ideal reader" profile in mind when creating content, "it is not enough to pick one type of reader and tailor the text exclusively to this reader's measurements" (Gray, 2003, 78). Conflict brings a sharp increase in attention (Sanderson, 2008; Fullwood, Melrose, Morris & Floyd, 2013) and can bring "eyes" to their "digital estates" (Abidin, 2016b), which in turn are exchangeable for different forms of capital. So in that sense, microcelebrities have strategically double-encoded these texts, as a technique of social steganography (boyd, 2010; Murumaa-Mengel, 2017), offering anti-fans a predetermined and hidden dominant discourse that anti-fans themselves perceive as oppositional and appropriate to their identity as anti-fans.

Conclusion

As digital culture often follows a participatory logic this also means that fandom and fan culture in a larger sense have gone through a transformation. Additionally, in the fragmented mediascape, audiences are fragmented as well. We argue that, although there are noticeable shifts and transformations, established theories are also still applicable and enable us to understand contemporary dynamic audience formations.

In this chapter, we have relied on the encoding/decoding framework posed by Hall (1980/2008) and various studies with empirical audiences of microcelebrities and have proposed three different "readings" of microcelebrity practices online: the dominant-hegemonic position (mostly associated with fans), the negotiated position (often taken by followers) and the oppositional position (mainly by anti-fans). Although all of these different readings could lead to the formation of collective alignment – a loyal fan base or an anti-fan base – the findings of our empirical studies indicate that microcelebrity-audience practices are mainly individual rather than collective. In fact, studies indicate that the pleasures of viewing microcelebrity content is built on an affective relationship between the microcelebrity and each fan, follower or anti-fan. Maintaining the affection and thus attention of one's audiences, however, is a tricky task, especially in the era of attention economy, when everybody is fighting for their audience share. Microcelebrities constantly make use of different strategies and invest their invisible labor in holding the interest of their audiences. However, as changes in the digital realm are both constant and fast, new empirical studies on microcelebrity audiences are needed to provide a more detailed and nuanced look at this fascinating phenomenon.

Acknowledgments

We are very thankful to our inspiring and hard-working students who have joined us in exploring this fascinating research field over the recent years. Our young colleagues have dedicated their time and effort to further the knowledge on this topic and have provided us with fresh ideas, insiders' look and countless hours on "the field." Thank you, Kristel Kaljuvee, Liisa Johanna Lukk, Liisi Maria Muuli, Marget Miil, Kaari Perm, Kätriin Viru and Piia Õunpuu!

References

Abercrombie, N. & Longhurst, B. J. (1998). *Audiences: A sociological theory of performance and imagination*. London: Sage Publications.

Abidin, C. (2015). Communicative intimacies: Influencers and perceived interconnectedness. *Ada: A Journal of gender, New media, and Technology*, 8(1). Retrieved June 25, 2019, from: https://adanewmedia.org/2015/11/issue8-abidin/

Abidin, C. (2016a). "Aren't these just young, rich women doing vain things online?": Influencer selfies as subversive frivolity. *Social Media + Society*, 2(2), 1–17. doi:10.1177/2056305116641342

Abidin, C. (2016b). Visibility labour: Engaging with influencers' fashion brands and# OOTD advertorial campaigns on Instagram. *Media International Australia*, 161(1), 86–100. doi:10.1177/1329878X16665177

Abidin, C. (2018). *Internet celebrity. Understanding fame online*. Bingley: Emerald Publishing Limited.

boyd, D. M. (2010). Social steganography: Learning to hide in plain sight. *Danah boyd's webpage*. Retrieved June 25, 2019, from: www.zephoria.org/thoughts/archives/2010/08/23/social-steganography-learning-to-hide-in-plain-sight.html

Bruns, A. (2008). *Blogs, Wikipedia, second life and beyond: From production to produsage*. New York: Peter Lang.

Cochran, T. R. (2008). The browncoats are coming! Firefly, serenity, and fan activism. In R. V. Wilcox & T. Cochran (Eds.), *Investigating firefly and serenity: Science fiction on the frontier* (pp. 239–249). London: IB Tauris.

Collins, M. R. & Stern, D. M. (2015). Fan response of community's unlikely fifth season. In A. F. Slade, A. J. Narro & D. Givens-Carroll (Eds.), *Television, social media, and fan culture* (pp. 109–128). Lanham, Maryland: Lexington Books.

Duffett, M. (2013). *Understanding Fandom: An introduction to the study of media fan culture.* London: Bloomsbury.

Duffy, B. E. (2015). The romance of work: Gender and aspirational labour in the digital culture industries. *International Journal of Cultural Studies*, 19(4), 441–457. doi:10.1177/1367877915572186

Duffy, B. E. & Hund, E. (2015). "Having it all!" on social media: Entrepreneurial femininity and self-branding among fashion bloggers. *Social Media + Society*, 1(2), 1–11. doi:10.1177/2056305115604337

Duffy, B. E. & Wissinger, E. (2017). Mythologies of creative work in the social media age: Fun, free, and "just being me". *Internal Journal of Communication*, 11, 4652–4671. Retrieved June 26, 2019, from: https://ijoc.org/index.php/ijoc/article/view/7322/2185

Erhart, J. (2014). 'Mr G is deffinately bringin'Sexy back': Characterizing Chris Lilley's YouTube audience. *Continuum*, 28(2), 176–187. doi:10.1080/10304312.2014.888039

Fägersten, K. B. (2017). The role of swearing in creating an online persona: The case of YouTuber PewDiePie. *Discourse, Context & Media*, 18, 1–10. doi:10.1016/j.dcm.2017.04.002

Fiske, J. (1989). *Understanding popular culture.* Boston: Unwin Hyman.

Fuchs, C. (2017). *Social media: A critical introduction.* London: Sage Publications.

Fullwood, C., Melrose, K., Morris, N. & Floyd, S. (2013). Sex, blogs, and baring your soul: Factors influencing UK blogging strategies. *Journal of the American Society for Information Science and Technology*, 64(2), 345–355. doi:10.1002/asi.22736

Geraghty, L. (Ed.). (2015). *Popular media cultures: Fans, audiences and paratexts.* London: Palgrave McMillan.

Gerbner, G. (1970). Cultural indicators: The case of violence in television drama. *The Annals of the American Academy of Political and Social Science*, 388(1), 69–81.

Giddens, A. (1984). *The constitution of society: Outline of the theory of structuration.* Cambridge: Polity Press.

Gilbert, A. (2019). Hatewatch with me: Anti-fandom as social performance. In M. A. Click (Ed.), *Anti-fandom: Dislike and hate in the digital age* (pp. 62–80). New York: New York University Press.

Giuffre, L. (2014). Music for (something other than) pleasure: Anti-fans and the other side of popular music appeal. In L. Duits, S. Reijnders, and K. Zwaan (Eds.) *Ashgate research companion to fan cultures* (pp. 49–62). Ashgate: Farnham.

Goodey, C. & East, R. (2008). Testing the market maven concept. *Journal of Marketing Management*, 24(3–4), 265–282. doi:10.1362/026725708X306095

Gray, J. (2003). New audiences, new textualities: Anti-fans and non-fans. *International Journal of Cultural Studies*, 6(1), 64–81. doi:10.1177/1367877903006001004

Gray, J., Harrington, L. & Sandvoss, C. (2007). *Fandom: Identities and communities in the mediated world.* New York: New York University Press.

Grossberg, L. (1992). Is there a fan in the house? The affective sensibility of fandom. In L. A. Lewis (Ed.), *The adoring audience: Fan culture and popular media* (pp. 50–65). London: Routledge.

Hall, S. (1980/2008). Encoding/decoding. In N. Badmington & S. Thomas (Eds.), *The Routledge critical and cultural theory reader* (pp. 234–244). London, New York: Routledge.

Hassan, R. & Purser, R. E. (2007). *24/7: Time and temporality in the network society.* Stanford: Stanford University Press.

Hills, M. (2003). *Fan cultures.* London: Routledge.

Hochschild, A. R. (1983). *The managed heart: Commercialization of human feeling.* Berkeley: University of California Press.

Holmes, S. (2004). "All you've got to worry about is the task, having a cup of tea, and what you're going to eat for dinner": Approaching celebrity in *Big Brother.* In S. Holmes & J. Jermyn (Eds.), *Understanding reality television* (pp. 111–135). London: Routledge.

Hopkins, J. (2009, July). Blogging field notes: Participatory innovation or methodological dead end. Paper presented at *Australian and New Zealand Communication Association Conference*, QUT, Brisbane.

Horton, D. & Richard Wohl, R. (1956). Mass communication and para-social interaction: Observations on intimacy at a distance. *Psychiatry*, 19(3), 215–229. doi:10.1080/00332747.1956.11023049

Jane, E. A. (2019). Hating 3.0: Should Anti-fan studies be renewed for another season. In M. A. Click (Ed.), *Anti-fandom: Dislike and hate in the digital age* (pp. 42–61). New York: New York University Press.

Jenkins, H. (1992). *Textual poachers: Television fans and participatory culture.* New York: Routledge.

Jenkins, H. (2006). *Fans, bloggers, and gamers: Exploring participatory culture.* New York: NYU Press.

Jenkins, H. (2018). Fandom, negotiation, and participatory culture. In P. Booth (Ed.), *A companion to media and fan studies* (pp. 13–26). Oxford: Wiley.

Jenson, J. (1992). Fandom as pathology: The consequences of characterization. In L. A. Lewis (Ed.), *The adoring audience: Fan culture and popular media* (pp. 9–29). London: Routledge.

Jorge, A. (2011). Young audiences and fans of celebrities in Portugal. *Comunicação & Cultura*, 12, 47–60. Retrieved June 26, 2019, from: http://comunicacaoecultura.com.pt/wp-content/uploads/03.-Ana-Jorge.pdf

Khamis, S., Ang, L. & Welling, R. (2017). Self-branding,'micro-celebrity'and the rise of Social Media Influencers. *Celebrity Studies*, 8(2), 191–208. doi:10.1080/19392397.2016.1218292

Kozinets, R. V. (2001). Utopian enterprise: Articulating the meanings of Star Trek's culture of consumption. *Journal of Consumer Research*, 28(1), 67–88. doi:10.1086/321948

Laineste, L. (2013). Funny or aggressive? Failed humour in Internet comments. *Folklore: Electronic Journal of Folklore* (53), 29–46. Retrieved June 26, from: www.folklore.ee/folklore/vol53/laineste.pdf

Linden, H. & Linden, S. (2016). *Fans and fan cultures: Tourism, consumerism and social media.* London: Palgrave Macmillan.

Livingstone, S. & Das, R. (2013). The end of audiences? Theoretical echoes of reception amidst the uncertainties of use. In J. Hartley, J. Burgess & A. Bruns (Eds.), *Blackwell companion to new media dynamics* (pp. 104–121). Oxford: Blackwell.

Marwick, A. & boyd, D. (2011). To see and be seen: Celebrity practice on Twitter. *Convergence*, 17(2), 139–158. doi:10.1177/1354856510394539

Marwick, A. E. (2015). Instafame: Luxury selfies in the attention economy. *Public Culture*, 27(75), 137–160. doi:10.1215/08992363-2798379

McQuail, D. (2013). The media audience: A brief biography—Stages of growth or paradigm change?. *The Communication Review*, 16(1–2), 9–20. doi:10.1080/10714421.2013.757170

McRae, S. (2017). "Get off my internets": How anti-fans deconstruct lifestyle bloggers' authenticity work. *Persona Studies*, 3(1), 13–27. Retrieved June 26, 2019, from: https://ojs.deakin.edu.au/index.php/ps/article/view/640/611

Mendick, H., Allen, K. & Harvey, L. (2015). Turning to the empirical audience: The desired but denied object of celebrity studies?. *Celebrity Studies*, 6(3), 1–4. doi:10.1080/19392397.2015.1063757

Murumaa-Mengel, M. (2017). *Managing imagined audiences online: audience awareness as a part of social media literacies*. Doctoral dissertation. University of Tartu, Institute of Social Sciences. Retrieved 26 June, 2019, from: https://dspace.ut.ee/bitstream/handle/10062/56324/murumaa_mengel_maria.pdf?sequence=1&isAllowed=y

Murumaa-Mengel, M. & Siibak, A. (2017). Performance of authenticity: A case-study of microcelebrities and their followers. Paper presented at *#AoIR2017 "Networked Publics"Annual International and Interdisciplinary Conference of the Association of Internet Researchers*, Estonia, Tartu.

Murumaa-Mengel, M. & Siibak, A. (2018). Engaged and ashamed – Audiencing practices of Estonian YouTubers' young followers. Paper presented at *ECREA 7th European Communication Conference*, Switzerland, Lugano.

Nixon, B. (2014). Toward political economy of "audience labour" in the digital era. *Communication, Capitalism & Critique*, 12(2), 713–734. doi:10.31269/triplec.v12i2.535

Poirier, L. & Robinson, L. (2014). Informational balance: Slow principles in the theory and practice of information behaviour. *Journal of Documentation*, 70(4), 687–707. doi:10.1108/JD-08-2013-0111

Pullen, K. (2000). I-love-Xena.com: Creating online fan communities. In D. Gauntlett (Ed.) *Web.Studies: Rewiring media studies for the digital age* (pp. 52–61). London: Arnold.

Rambukkana, N. (2015). *Hashtag publics: The power and politics of discursive networks*. New York: Peter Lang Publishing.

Reijnders, S., Zwaan, K. & Duits, L. (2014). *The Ashgate research companion to fan cultures*. London: Ashgate Publishing, Ltd.

Rosa, H. (2013). *Social acceleration: A new theory of modernity*. New York: Columbia University Press.

Rose, J. & Jones, M. (2005). The double dance of agency: A socio-theoretic account of how machines and humans interact. *Systems, Signs & Actions*, 1(1), 19–37.

Sanderson, J. (2008). The blog is serving its purpose: Self-presentation strategies on 38pitches.com. *Journal of Computer-Mediated Communication*, 13(4), 912–936. doi:10.1111/j.1083-6101.2008.00424.x

Schmidt, J. -H. (2014). Twitter and the rise of personal publics. In K. Weller, A. Bruns, J. Burgess, M. Mahrt & C. Puschmann (Eds.), *Twitter and society* (pp. 3–14). New York: Peter Lang.

Schroath, K. (2016). *#Parasocial Interaction: Celebrity Endorsements*. Doctoral Dissertation, Kent State University, College of Communication and Information/School of Communication Studies. Retrieved 26 June, 2019, from: https://etd.ohiolink.edu/!etd.send_file?accession=kent1480336010417442&disposition=inline

Senft, T. (2008). *Camgirls: Celebrity and community in the age of social networks*. New York: Peter Lang Publishing.

Senft, T. (2013). Microcelebrity and the branded self. In J. Hartley, J. Burgess & A. Bruns (Eds.), *Blackwell companion to new media dynamics* (pp. 346–354). Oxford: Wiley Blackwell.

Shaw, A. (2017). Encoding and decoding affordances: Stuart Hall and interactive media technologies. *Media, Culture & Society*, 39(4), 592–602. doi:10.1177/0163443717692741

Strafella, G. & Berg, D. (2015). The making of an online celebrity: A critical analysis of Han Han's blog. *China Information*, 29(3), 352–376. doi:10.1177/0920203X15608130

Tincknell, E. & Raghuram, P. (2002). Big brother. Reconfiguring the "active" audience of cultural studies?. *European Journal of Cultural Studies*, 5(2), 199–215. doi:10.1177/1364942002005002159

Tulloch, J. & Jenkins, H. (1995). *Science fiction audiences: Doctor who, star trek, and their fans*. London, New York: Routledge.

Wasko, J. (2001). *Understanding disney: The manufacture of fantasy*. Cambridge: Polity.

Westenberg, W. M. (2016). T*he influence of YouTubers on teenagers: A descriptive research about the role YouTubers play in the life of their teenage viewers*. Master's thesis, University of Twente. Retrieved 26 June, 2019, from: https://essay.utwente.nl/71094/1/Westenberg_MA_BMS.pdf

Wiggins, B. E. & Bowers, G. B. (2015). Memes as genre: A structurational analysis of the memescape. *New Media & Society*, 17(11), 1886–1906. doi:10.1177/1461444814535194

14

REIMAGINING MEDIA EDUCATION

Technology Education as a Key Component of Critical Media Education in the Digital Era

Anne-Sophie Letellier and Normand Landry

Introduction[1]

While the association between education and the media dates back to the 1920s during which organizations such as "regional offices for educational cinema" devoted to "education and propaganda through cinema" were established in France (Laborderie, 2012 in Landry, 2017, 12), the field was formalized as an academic discipline from the second half of the twentieth century through pedagogical practices and public policy.

In Canada, the work of Marshall McLuhan, which established the field of communication as an academic discipline (Landry & Letellier, 2016), paved the way for the first media educators in the country between the 1960s and 1980s (Anderson, Duncan & Pungente, 1999; Landry, Basque & Agbobli, 2016). From the 1980s onwards, media education was substantially developed and institutionalized in many countries both as a research field and as a set of educational practices and policies. Organizations such as UNESCO and the Council of Europe were the first to simultaneously work on the topic on an international scale and played a foundational role in research transmission and dissemination of knowledge. The following decades, for their part, were mostly characterized by the consolidation of public policies, establishment of a research field associated with media education themes and structuring of organizations tasked with its implementation both on the educational front as well as public policy deployment (Hoeschmann & Poyntz, 2012).

Since the early 2000s, however, the field has undergone profound mutations due to transformations of the media and technology landscape. This has led to an increased interest in media education and resulted in important growth in research and training programs. In fact, in an environment where mediatization of social relations evolves alongside "fake news," cyberbullying, massive collection of personal data and the filtering and organization of online content, citizens are confronted with an increasingly complex media and digital environment on which a growing number of daily tasks depend (Shariff, 2008; Livingstone & Haddon, 2009). The necessity to offer tools to younger as well as older people to deal with issues that structure these complex and multiform environments becomes the pillar of the social and democratic preoccupation linked to the usage of digital devices as well as the production and consumption of media content. Thus, it goes without saying that the growing interest of educators, policy makers, researchers and activists toward media education is in line with the convergence of interests and preoccupations surrounding the development of skills considered essential for our hypermediatized societies (UNESCO et al., 2013).

Following this reflection, this chapter focuses on three objectives. First, it seeks to delineate media education as an academic research field. Second, it identifies the main debates and issues structuring this field. Among these issues, the technological mutations that necessitate the development of technical knowledge and skills is a prominent one. Finally, we argue that the contemporary technological and digital landscape initiates the need for the reevaluation and reimagination of media education in order to systematically integrate the development of knowledge and skills related to the use, functioning and governance of digital technologies. We consider that those skills directly relate to the integration of learners to the job market and, more importantly, to the exercise of citizenship and protection of human rights in a democratic context. In other words, by reviewing the field's structuring dimensions and main debates, we are proposing a reimagination of media education, its theoretical foundations and its pedagogical operationalization through technology education.

Defining Media Education

We must first delineate and define the term "media education." The details of the debates on its theoretical, conceptual and practical foundations (Von Feilitzen & Carlsson, 2003; Hobbs, 2011; Martens, 2010; Potter, 2013) are explained later in this chapter. Here, it should be noted that the term is especially polysemic and subject to controversy since its theorization and implementation is oftentimes included in a social and political process. Therefore, this requires one to take a stand on debates that take place in the media, within the education field and, more broadly, in our societies (Gonnet, 2001; Yousman, 2008). Hence, most of the debates on the subject focus not only on the skill sets and forms of knowledge that structure the field, but also on the biases, preferences and priorities of those who put it into practice.

Consequently, we approach media education as a praxis since this conception highlights the complex, iterative relationship it maintains between theory and practice. In this approach, theoretical tools are in constant relationship with reality and develop the learner's knowledge and skills that are in line with a social and political project without necessarily assuming their nature. This method of education is thus modulated in time according to environments, cultures, needs and contexts. In other words:

> "Media education" refers to the set of pedagogical and didactic processes for the development of skills and forms of knowledge specific to the media as it relates to social, political, economic and cultural issues.
>
> *[our translation] (Landry & Roussel, 2018, 6)*

Therefore, the main objective of media education is to develop the learner's cognitive, metacognitive, technical and behavioral skills to address media content as "text" critically. It is also concerned with the infrastructure as well as the political economies underpinning the production of media messages. It seeks to provide individuals and communities with the skills they need in order to distance themselves from these messages, to question and deconstruct them, and to interrogate the forms and shapes that they take. To this end, it uses active pedagogical processes in which learners are called upon to use various media devices in order to produce and broadcast content of various formats and on various platforms.

This understanding of media education in its reflexive and critical dimension simultaneously states what it is not. As outlined by Landry and Basque (2015):

> Media education is neither *training on the media* aimed at the development of skills for using technical tools nor an *education through the media* that positions it as a device that promotes the teaching of knowledge outside of the media itself (Buckingham, 2003). Additionally, it is not limited to *teaching about the media*, meaning the study of the characteristics of different media.
>
> *[our translation] (p. 49)*

Therefore, understood as a praxis and a "pedagogy of interrogation" (Macdonald, 2008), media education articulates knowledge with public policy and pedagogical practices. As we will demonstrate in the following pages, the development of theoretical and conceptual knowledge related to media is anchored both in educational sciences and communication studies. They also are directly related to the deployment of public policies and specific pedagogical practices. Its goal is to be operationalized in a framework prominently calling for social and political action or, at the very least, the exercise of responsible citizenship and job market integration.

The Theoretical Foundations of Media Education

The numerous themes and forms of knowledge carried by media education require a multidisciplinary field. While it finds its conceptual and thematic support in several fields including sociology, political science, cognitive science and psychology (Buckingham & Sefton-Green, 1994; Kellner & Share, 2005; Hammer & Kellner, 2009; Potter, 2013), media education remains mostly theorized within the fields of communication studies – more specifically cultural studies – and educational sciences. Since it is in line with an active citizenship approach, it favors both the development of knowledge and skills that allow for distanced and critical relations and reflexivity on media content through a pedagogical approach that fosters critical autonomy (Council of Europe, 1989; Thoman & Jolls, 2004; Share, 2009a; Kellner & Share, 2009a). It thus becomes inseparable from the notions of citizenship and democracy (Gonnet, 2001). As such,

> It requires exchange, experimentation, dialogue and problem resolution with the ultimate purpose of "liberation of human intelligence." Its essential function is related to the development of the learner's autonomy by empowering their agency and their capacity to take their lives into their own hands in the outside world (Landry, 2013). The Brazilian pedagogue Paulo Freire (2000) proposes a "pedagogy of liberation" for individual and collective emancipation. Educators and learners both position themselves as partners enrolled in a relation towards mutual discovery and learning. Educators and learners both position themselves as partners enrolled in a relation towards mutual discovery and learning. In this context, education must participate in the emergence of a conscience of the oppression vector that the learner submits to while providing them the necessary tools to act on their living conditions.
>
> *[our translation] (Landry & Basque, 2015, 52)*

Based on the work of Paolo Freire (2000), media education positions the individual as a free *subject*, conscious of their interests and capable of taking charge of oneself and acting on their environment. Thus, the purpose of pedagogical activities is to equip the learner with necessary tools that helps them identify the "hegemonic myths that are produced, disseminated and reified" by economic actors participating in the dynamics of reproduction of the dominant ideologies (Landry & Basque, 2015, 52). To this end, it mobilizes methods centered on the development of the learner's autonomous capacities (Masterman, 1985; Share, 2009a) and distances itself from the common lecture-based pedagogical methods that are considered inefficient and inadequate. It concentrates on approaches that require the learner to undertake an introspective, creative and collaborative process oriented toward the analysis and deconstruction of media "texts" and production (Share, 2009b; Bragg, 2002; Hobbs, 2016). In other words, the autonomy-based approaches employed frequently concur in a playful and creative manner with the development of critical thinking and unveiling of dynamics and structures favoring social and cultural reproduction and domination (Kellner & Share, 2005, 2007).

Media education is concerned with the study and demystification of key concepts from communication studies such as culture, ideology, identity and subjectivity. It uses theoretical tools from cultural studies since the field is dedicated to culture. Roland Barthes' writing emphasizes the point that the field is aimed at "demystifying what in culture comes to seem natural by showing that it is based on contingent and historical constructions. In analyzing cultural practices, he identifies the underlying conventions and their social implications" (Culler, 2006, 43).

Cultural studies is extremely useful to deconstruct "normalized" images and representations by using numerous theories and perspectives such as Marxist, feminist and postcolonial (Buckingham & Sefton-Green, 1994; Kellner & Hammer, 2009) to name a few. Hence, the mobilization of these concepts and their juxtapositions lay the foundations that are necessary to question the political and social issues related to media representations while addressing and deconstructing them critically. By aligning itself with critical pedagogy framework, media education provides the learner with the agency to understand and act when experiencing complex relationships with media, culture and ideological systems that are in place.

Media Education Approaches

It goes without saying that the large array of themes and issues included under the umbrella of media education reveals both a theoretical and conceptual complexity and a multitude of approaches that can be mobilized in pedagogical practices. In his work, Piette (1996) identifies various approaches that address the issues at the center of the debates surrounding media education and the pedagogical practices that accompany them.

Media education is concerned with media effects (Potter, 2012). The first approach, at times labeled as "protectionist," seeks to mitigate the "effects" characterized by parental or school authorities as detrimental. As such, it interrogates the influence of texts and media actors on the values, attitudes, behaviors and perceptions of learners. It positions the pedagogue in a situation of power from the outset by allowing them to intervene between the learner and the detrimental effects of media.

The critical approach interrogates the interests, actions and impacts of media actors on society. It positions mass media productions within a capitalist political economy of production and circulation of culture and information (McChesney, 1995, 2000). It consequently seeks to make visible the relationships between media productions, economic and political interests, ideology and power (Knight Abowitz, 2000; Kellner & Share, 2005).

The semiotic approach is concerned with the meaning and symbolic power of language as well as the visual content of media products. Borrowing from cultural studies, it is mainly anchored in a cultural and language-based perspective due to its manner of demonstrating an interest in culture through language, discourses and visual symbols that accompany them. As Hall (1999) emphasizes:

> There's something always decentred about the medium of culture, about language, textuality, and signification, which always escapes and evades the attempt to link it, directly and immediately, with other structures. And yet, at the same time, the shadow, the imprint, the trace, of those other formations, of the intertextuality of texts in their institutional positions, of texts as sources of power of textuality as a site of representation and resistance, all of those questions can never be erased from cultural studies.
>
> *(Hall in During, 1999, 106)*

This approach lays the foundations for analyzing and questioning the semiotic elements accompanying the formal text. It also converges with the critical and synthetic postures. While the first approaches media as sites of representation, the second is interested in the relationships of power and domination that it underpins. Therefore, it seeks to offer resources to the learner for the development of their critical judgment on the texts, images and videos that they consume.

The ethical approach stands among the most well-spread ones in media education. Emerging from the axiological branch of philosophy, it is concerned with the actions that follow certain representations. In addition, this approach enables the learners to question their relationships with advertising messages while enabling them to "examine the values transmitted by the media and their repercussion on society" [our translation] (Corriveau in Landry & Letellier, 2016, 135). The goal for learners is to reflect on sensitive topics (such as violence, stereotypes, etc.) and position themselves as reflective and active agents.

The artistic and production approach highlights that any production can also be the object of an esthetic appreciation and critique (Piette, 1996). It also situates the perspective of content production as an integral part of media education. By doing so, it enables the development of know-how and creativity relating to the production of multimodal texts integrating multiple media forms and formats (Lebrun et al., 2012). This approach is sought by public agencies through the capacity to write and understand computer code. It is the topic of renewed interest in the context of the development of so-called digital competencies.

Finally, the approach concentrating on uses focuses on the processes of appropriation of digital devices. It positions the learner as an active receiver, producer and broadcaster of media texts. The processes through which media devices are used by individuals are related to socialization as well as political and cultural integration activities. This "appropriation" of media technologies calls for responsible consumption and the learner's reflexivity on their own practices (Corriveau in Landry & Letellier, 2016).

In sum, although these approaches mobilize different processes and pursue distinct pedagogical objectives, the perspective presented in each case favors the integration of themes specific to media education for a large variety of learning activities (mathematics, French, art, history, etc.). They also converge and influence one another in a dynamic manner through the pedagogical activities mobilized in media education.

What are the Public Policies Related to Media Education?

The development and analysis of public policies in media education originated in the 1980s when the first programs benefiting from a formal institutional and political recognition were structured (UNESCO, 1982; Masterman, 1988; Piette, 1996). The consolidation of the field and the strong interest developed during the past decades are namely illustrated through the multiplication of international declarations, charters and recommendations (Schwarz, 2005; Frau-Meigs & Torrent, 2009).

UNESCO plays a federative role among active international organizations in the field of media education. Organizations that are currently invested in the field of public policy related to media education nowadays include international organizations (the European Commission,[2] UNESCO[3]), national agencies (Belgium's *Conseil supérieur de l'éducation aux médias,*[4] the UK's Office of Communications [Ofcom[5]]), expertise and broadcast centers (CLEMI[6]), non-governmental organizations (National Association for Media Literacy Education,[7] HabiloMédias,[8] Association for Media Literacy[9]) and professional associations (Canadian Teachers' Federation[10]). Experts confirm that civil society now occupies a "leading role" for the development and implementation of media education activities (European Audiovisual Observatory, 2016).

The inclusion of media education within school curriculums aims at ensuring that the practice is mainstreamed in learning environments. This, in turn, contributes to the development of prioritized forms of knowledge and skills with regard to media and responds to issues and preoccupations raised by the mediatization of societies. Viewed in this manner, school curriculums are at the heart of educational policies since they establish educational priorities, define disciplinary orientations, attribute responsibilities to different school actors and define assessment measures (Livingstone et al., 2012; Hartai, 2014). More literature on the comparative analysis of public policies in media education is emerging (Buckingham & Domaille, 2002; Frau-Meigs & Torrent, 2009; UNESCO et al., 2013; Frau-Meigs et al., 2017). Most European and North American countries have developed such policies (European Audiovisual Observatory, 2016).

Actors Involved in Media Education

Scientific literature and the vast majority of public policies identify elementary and high schools as the preferred locations for teaching media education skills. Nevertheless, the situation differs greatly from one country to another and even from one region or locality to another. Some programs are the result of community initiatives, some are positioned as voluntary approaches developed by committed teachers, and others are formally integrated (to various degrees) in school curriculums (Hartai, 2014; EAO, 2016).

By its very nature, media education calls upon the development of multidisciplinary skills. As will be demonstrated later, the development of skills associated with the understanding and mastery of technical devices – such as programming and coding, for instance – is an eloquent example by which knowledge can be introduced into a school setting. However, it is equally beneficial to address them in extracurricular contexts that are taken on by community organizations and civil society. In line with the large number of public policy sectors directly or indirectly involving media education, it is definitely relevant to identify other locations that can include media education practices, namely through the action of civil society members and organizations, interest groups or even public libraries. In fact, giving value to a diversity of locations where pedagogical practices are developed would surely fall within its definition as praxis. In other words, in order to address how this field is in constant communication between theory and practice, it is necessary that a variety of groups can appropriate its tools and participate in the sharing of skills and knowledge.

Reimagining Media Education

The notion of "media education" evokes complex and varied processes that are constantly redefined. As a field of research, its renewal requires a critical analysis of the current state of knowledge. In the context of this chapter, we consider three main challenges for the development of media education knowledge. The first is related to the articulation between the knowledge and skills that are constitutive of "media literacy" and the pedagogical practices that are taking place in school environments. The second is related to the fragmentation of the forms of "literacy." This tends to create conceptual noise both in the definitions mobilized to speak of the different skills developed through media education activities, and in the effort to identify priorities and evaluation methods for the skills within the field itself. The third challenge refers to the current digital transformations that call for a reconsideration of the skills and objectives constitutive of the field.

Difficulties in Operationalization and Articulation

The first challenge highlights the lack of research aimed at analyzing the articulation between knowledge, practices and public policy, namely in relation to the school environment (Hartai, 2014). The development of knowledge in media education requires an interest in the conditions, environments and actors that structure its practice. However, the research accomplished by Hartai and his collaborators demonstrate that the knowledge of curricular contents in media education remains fragmented and incomplete. This results in proscribing a critical analysis of the quality of contents and an evaluation of their binding with scientific knowledge and the practices deployed by professionals on the field.

In the same vein, research conducted by Landry and Roussel (2018) leads to two conclusions. First, they find that very few research projects were conducted on the practice of media education taking place in a school framework and the factors that influence its practice in this environment. This is indeed surprising when considering the fact that the consolidation and entry of media education in educational policies in a context where it is conceived as praxis dates back to over four decades. Second, they observe that reinforcing the articulation of media education requires the development of knowledge on school curriculums and their teaching practices.

For these reasons, the disjunctions observed among the themes of media education – the partial autonomy of each of these elements, their approximate mutual binding, and the recurrence of gray areas in the knowledge to which they are related – remains a limit that the field still has to overcome. In order to overcome this, the field of media education must consolidate through the reinforcement of the articulation between knowledge, pedagogical practices, skills and objectives (hereby integrated within curriculums).

Conceptual Confusion and Heterogeneity of Literacies

The second challenge highlighted in the literature is related to the fragmentation of the literacies and to the inconsistent manner of defining, categorizing and creating hierarchies for the knowledge and skills required for media education. If the presence of multiple literacies (Addison & Meyers, 2013; Stordy, 2015) can easily reflect the dynamism of the field coupled with the diversity of research projects, many see a risk of complexifying their application and evaluation process through pedagogical practices.

As such, Landry and Basque (2015) identify a tendency related to this *conceptual noise*. For the past 15 years, the types of literacies have multiplied, mainly in an effort to reflect technological changes. To this end, authors have conceptualized notions such as ICT Literacy (Markauskaite, 2006), Internet/Web Literacy (Bawden, 2001; Fahser-Herro & Steinkuehler, 2010), Cyberliteracy (Gurak, 2008), Computer Literacy (Horton, 1983). They have also developed more inclusive "meta-literacies," such as the multiliteracies (Provenzo et al., 2011), Metaliteracies (Mackey & Jacobson, 2011), new literacies (Greenhow & Robelia, 2009) and media and information literacy (UNESCO et al., 2013). The literature consulted is unanimous regarding the necessity to organize the multiplicity and heterogeneity of concepts and approaches used by different authors (Bawden, 2001; Aharony, 2010; Addison & Meyers, 2013; Erstad & Amdam, 2013; Lee et al., 2013; Stordy, 2015). This work has yet to be accomplished.

The majority of these literacies tend to report the technological disruptions that impact the knowledge about, and practice of, media education in order to integrate new skills and challenges. As such, some point out that "media literacy relates to 'all media,' including television and film, radio and recorded music, print media, the Internet and other new digital communication technologies" (Silver, 2009, 12). Others argue that the expansion of the required knowledge and skills calls upon the development of "new" literacies, and consequently of more specific concepts that invoke the advances resulting from media education (Carrington & Robinson, 2009). The media transformations of the past decades nonetheless require an expansion of the field and, more specifically, the integration of technology education. This includes the development of new knowledge and skills regarding initiation to the computer code, understandings of the materiality (software, protocols, networks, etc.) of digital networks as well as their modes of governance.

Four Themes That Call for the Renewal of Media Education

The historical development of knowledge related to media education was developed using theoretical and conceptual tools – namely critical pedagogy and cultural studies – aimed at accompanying the learners in a media environment dominated by mass media. The advent of digital media and the networked societies call for the utilization of a conceptual framework that allows for the adaptation of the issues specific to media education and the knowledge and skills that can respond to issues specific to the digital context. This adaptation, or need for renewal, is operationalized based on four themes.

The first theme is oriented around the political economy of media and showcases the collapse of 1) traditional economic models for the production and broadcasting of culture and information, 2) the emergence of new economic models based on accumulation, capitalization and the selling of personal data and 3) an industrial convergence around high-tech companies positioning themselves as both device producers and content, platform and services diffusers. Hence, concepts of concentration, convergence, property, cultural industry, ideology and hegemony – inevitable to critical approaches to media education – are updated in the context of a radically transformed political economy of media production and diffusion (Kellner & Share, 2005; Potter, 2013).

The second theme relates to media technology itself. It is oriented toward interactive and personalized communications, a capacity for the user to reach vast audiences or targeted groups, a marked democratization of the production and broadcasting of contents, and an openness toward the programming (and reprogramming) of mobile, networked and connected devices. The skills necessary to their mastering evolve concurrently and gain increased importance as they are deployed and generalized within the different sectors of social life (Lessig, 2008; Jenkins, 2009). To this end, the mastering of information and digital devices takes on considerable importance for media education (Lee et al., 2013; UNESCO et al., 2013).

These skills are directly related to digital inequality. They are assumed to reduce the existing gaps between the uses and practices of mobilizing digital devices (Van Deursen & Van Dijk, 2011). In turn, these technological innovations nourish the usage and practices of new media. This is the subject of our third theme. It is oriented toward the capacity for appropriation (authorized or illegal), production and diffusion of content as well as personalization and reconfiguration of the devices themselves (Hands, 2011; Landry et al., 2017). These usages and practices are constitutive of an integral cultural and political modes of expression within highly mediatized societies.

Finally, the fourth theme refers to the combination of industrial, technological and social factors mentioned above. This blending contributes to the affirmation of new issues and revitalization of secular preoccupations associated with the media. It follows the idea that the revitalization of skills required for rapidly evolving media environments becomes imperative (Buckingham, 2008). This is in addition to the development of skills and knowledge specifically oriented toward the issues of the digital world such as online surveillance and cyberbullying (Shariff, 2008; Livingstone & Haddon, 2009).

Indeed, changes associated with the digital world result in a necessary renewal of media education. The following section argues that technology education must be established in continuity with the pedagogical preoccupations of media education.

Technology Education

What Is Technology Education and Technology Literacy?

It is important to conceptually locate technology education in relation to its parent concepts such as critical media education and technology literacy. In this section, we first address technology education as a dimension of critical media education. As a reminder, we position media education as a field that borrows theoretical tools both from cultural studies and critical pedagogy in order to develop the learners' knowledge and skills as part of a social, cultural or political integration project. By providing citizens with the tools to be critical about the produced and consumed information, media education positions itself as a pedagogical process that favors the development of skills allowing for the analysis of relations between media contents, information and power (Kellner & Share, 2007).

We argue that the themes specific to cultural studies – ideology, culture, power relations – combined with critical pedagogy practices can, and must, be mobilized in relation to technological mutations that are taking place within the contemporary media environment. In this sense, we consider that it is necessary to connect technological issues to the educational objectives of media education. For this matter, some scholars immediately position digital technologies as media (Silver, 2009). For our analysis, we would like to step further and consider that the contemporary technological environment involves issues and themes requiring special attention. It is necessary to connect technological issues with pedagogical objectives that are put forward by media education. To demonstrate this, we address technology education as a dimension of media education.

The sum of knowledge and skills developed through the processes of technology education converges with the notion of "technology literacy." We refer to this term in order to evoke the "gains" resulting from the specific pedagogical teaching and learning activities devoted to this purpose. The notion of technology literacy is addressed in a complementary manner to the concept of digital literacy. The latter is often associated with the ability of the learner to use computers, the internet, social media and other digital technologies (Hobbs, 2010) by developing the "personal, emotional, technological and intellectual skills necessary to evolve in a digital world" [our translation] (Plante in Landry & Letellier, 2016, 117). This helps us understand and use technologies in a context that is undergoing a constant mutation. However, it sometimes tends to relegate to second place critical thinking "when programs unproblematically teach students the technical skills to merely reproduce hegemonic representations or express their voice without the awareness of ideological implications or any type of social critique" (Kellner & Share, 2007, 7).

Therefore, including these learnings "within a field that applies critical research to media," proves to be an essential component of this research since it "produces concrete knowledge aimed at developing not only the praxis of critical thinking among learners, but also a plethora of cognitive, technical, ethical and behavioral skills anchored in critical approaches to communication and education" [our translation] (Landry & Basque, 2015, 48). In this respect, technology literacy not only focuses on the areas of competence related to the use of digital tools targeted by digital literacy, but also integrates itself into a pedagogical project that raises the necessity of problematizing issues inherent to technological infrastructure and their modes of governance.

In line with the objectives originating from the development of media education in the second half of the twentieth century – which included the development of critical thinking among citizens regarding issues and problems related to the development of mass media (McLuhan, 1994; McChesney, 1995) – we argue that technology literacy must deepen the knowledge related to digital infrastructure and issues related to their concentration, convergence and modes of governance in an increasingly connected world. This is in addition to focusing on the development of skills related to usage that is ethical and coherent with the personal values of the learner.

In sum, it must favor the development of skills for the use and creation (namely, through coding) of digital technologies (Eshet-Alkalai, 2004; Buckingham, 2008). However, it must simultaneously initiate the learner into understanding the infrastructural and programmatic materiality (databases, routers, protocols, software, etc.), and the power relations that shape digital environments in order to evoke a comprehension of how they function, how they are governed, and what issues are related to them.

As such, the following paragraphs seek to identify the themes, issues and perspectives structuring the pedagogical objectives related to the technological dimension of critical media education that we name "technology education."

Digital Infrastructures: The End of Neutrality and of Abstraction Layers

Considering digital infrastructure as a constitutive element of technology education rests on the premise that technologies are not neutral. This means that the economic, social, cultural and political context in which technologies are deployed, as well as their technical characteristics, are not fully malleable to the desires and skills of their users. This premise wittingly falls within the scope of critical communication studies (Feenberg, 2012) as well as the literature from science and technology studies (Winner, 1980; DeNardis & Musiani, 2016; Eipstein et al., 2016; Musiani, 2015). The latter is concerned with the co-construction of technical devices and their underlying power relations (Jasanoff, 2004). This literature is in opposition to the discourses articulated by corporations that own the dominant digital platforms which often present themselves as neutral, non-ideological or, at the very least, deprived of political interests.

This technology "neutrality" stems from a discursive and ideological construction driven by economic and political biases and interests (Galloway, 2004; Galloway & Thacker, 2013). Indeed, a number of militants, jurists and academics argue that digital technologies carry values and are part of power relations inherent to the political, legal, economic and cultural systems within which they reside (Lessig, 2008; Nissenbaum, 2011). Be it through the computer code directly controlling the world of possibilities for a program, or through the choice of physical locations through which fiber-optic cables can pass through for the distribution of high-speed internet, digital networks possess a materiality that is often dissimulated and obscured by discourses of dematerialization. An example of this is the virtual or cloud computing which is a topic of constant negotiations relating to their design choices (by engineers and programmers), network governance (by corporations, governments or different legal decisions) and usage and appropriations (e.g., by interest groups, citizens or hackers).

Indeed, the material and programmatic components of digital networks constitute an architecture through which a number of interests, at times conflictual, are exercised. They are themselves defended by economic, cultural and political discourses and power relations (Hogan, 2015). For instance, computer coding is conceptualized by Alexander Galloway as an expressive form, an operational language, with political implications both through semantics and action. Similarly, the jurist Lawrence Lessig (2009) has famously argued that coding is a form of law in cyberspace due to its selection and organizational role for the functionalities of a program or an application. The interface for the popular social network Facebook is a great example. Here, programming limits the possibilities available to users who use the spaces where they can publish their comments, posts and status. Coding frames the possibilities available to react to publications and allows the development of algorithms that are responsible for the direct organization of information presented to users. This example is explored in more depth below.

Power relations inherent to the programming and algorithmic functions are increasingly theorized and problematized. However, little attention has been given to the material components that these networks depend on – including the data centers, the antennas, the servers, the underwater cables and other "stuff you can kick" (Parks, 2015, 335). Moreover, their absence in the literature contributes to the creation of layers of abstraction masking physical infrastructure related to data storage (Hu, 2015). As such, they contribute to the reinforcing of a perspective of the Internet as a cloud, that is, a deterritorialized and distributed flow of information. Paradoxically, in order to allow for the data to be fluid and transnational, it must travel, and be centralized and reunited in a physical space. This particular relation between the location and the non-location, between the material and the immaterial, characterizes the internet and constantly affects the contemporary organization of power structures (Hu, 2015) underlying the governance of digital networks.

Addressing power relations underlying these infrastructures is even more essential when considering the central role they play in our contemporary societies. It is also essential when facing the fact that this broadly privatized infrastructural environment represents a form of materialization of the forms of knowledge and governance (Hogan & Shepherd, 2015). To illustrate this statement, Hu (2015) advances an analogy between digital networks and aqueduct systems:

> [W]ater is still largely regarded as a public resource, while the cloud is almost entirely owned by private companies. When there is a drought, it does not seem unreasonable to invoke collective solutions, such as water conservation and water rationing; yet when digital culture runs into a snag – for example, over privacy – the default response in the United States is to appeal to Silicon Valley start-ups to build better apps. This is what happens when today's dominant metaphor for digital space, "the cloud," is actually a metaphor for private ownership.
>
> *(p. 147)*

Beyond the metaphor presented by Hu, this excerpt highlights a crucial issue related to technology education: the ways in which digital networks are apprehended and understood not only affect crisis management, but also, more broadly, involve rules and regulation modes that structure information and digital services within a capitalist and neoliberal ideological paradigm.

Technology education allows for the framing and understanding of the internet as a socially constructed apparatus for which the material and programmatic form that we know of today are largely influenced not only by usage, but also by a heterogeneous assembly of economic, political and cultural interests (Fish, 2017). The goal of critical media education is to guide learners to acknowledge that "the constructed nature of media texts, and thereby media representations [reinforce] the ideologies of dominant groups in society" (Buckingham in Hobbs, 2011, 427). Technology education is part of this pedagogical project and is oriented toward praxis. It develops aptitudes for the ethical and coherent use of digital technologies among its learners. It also seeks to show and problematize the different interests and power relations structuring the material form as well as modes of governance (Musiani et al., 2016) of the technologies that we use daily. Focusing on infrastructure leads to understanding the way devices and networks linking them together function. Additionally, acknowledging the distributed structure of the Internet facilitates the user "to imagine the numerous roads that can be taken by an information request, and thus, conceiving how this data can be intercepted or tracked at different stages of its itinerary" [our translation] (Letellier, in Landry & Letellier, 2016, 182).

To summarize, considering the various layers of abstraction of digital technologies, "technology education" is concerned with the issues related to usage, while demystifying the issues related to the governance and design of information and communication technologies.

Discrimination and Socio-Economic Issues of Technology Education

Numerous socio-economic issues structure the contemporary digital environment. Digital technologies represent both the promise of the convenience of networked societies, the inherent efficiency of a rapid treatment of information in the fields of health or environmental sciences, the expression and personal emancipation of individuals and the diffusion of, access to, and sharing of, content and information on a transnational scale. At the same time, these digital networks and technologies allow for the erosion of citizens' privacy (Shade & Shepherd in Landry & Letellier, 2016; Solove, 2008; Nissenbaum, 2011) through data collection on digital media and through the "Internet of Things" (Schneier, 2018). In doing so, they facilitate infringements on freedom of expression (Wu, 2018), censorship, discrimination practices (Ferguson, 2017; Eubanks, 2018) or political manipulation. Furthermore, the development of networks is exercised in a context of media concentration and convergence that results in the concentration of informational and economic power (Mayers-West, 2017; Wu, 2018) in the hands of the digital giants – Google, Amazon, Facebook, Apple and Microsoft.

From this perspective, technology education does not seek to be either *techno-optimist* or *technophobe*. Rather, it seeks to uncover the opportunities and challenges inherent to the contemporary media and digital environment. This critical understanding of platforms constitutes a way of empowering citizens who interact with digital technologies (Taylor, 2014). Virginia Eubanks (2018), whose research is concerned with the use of databases for the "management" of economically disadvantaged American populations, highlights that an optimistic perspective on technological promises can, at times, reveal the absence – or the opacity – of power relations created by data collection:

> High-tech economic development was increasing economic inequality in my hometown, intensive electronic surveillance was being integrated into public housing and benefit programs [...]. Nevertheless, my collaborators articulated a hopeful vision that information technology could help them tell their stories, connect with others and strengthen their embattled communities.
>
> *(p. 12)*

For Eubanks, this situation is directly related to social justice issues. Her volume, *Automated Inequalities* (2018), demonstrates how innovations from technological revolutions usually remain tools that reinforce social disparities. Additionally, she argues that data collection often reinforces the vulnerability of marginalized groups through the color of their skin, ethnic origin, sexual orientation or low income. These groups are then confronted with "higher levels of data collection when they access public benefit, walk through highly policed neighborhoods, enter the health-care system, or cross national borders" (Eubanks, 2018, 7). In addition, the growing popularity of "predictive policing" – thus the use of "powerful computers crunching large data sets almost instantaneously" by police forces to operate "the decisional work of identifying criminal actors, networks, and patterns" (Ferguson, 2017, 3) – risks exacerbating and making some forms of discrimination operated by algorithmic biases more opaque and neutral.

Technology is only one aspect of the problem, and its apparent neutrality contributes to operationalizing some forms of discrimination and rendering them opaque, mechanical and non-problematic. Technology education, then, becomes a tool for the contestation and de-ontologizing of these relations and a more efficient transfer of the ethical discussions related to the technological field.

Digital Infrastructures, Human Rights and the Exercise of Citizenship

The mechanisms through which material and programmatic infrastructure exercise forms of control operate at once through the institutional, governmental and corporative spheres (Galloway, 2004) and directly affect the exercise and protection of human rights. By addressing the right to privacy and freedom of expression as an example, and by clarifying the consequences related to human rights, we argue that technology education is closely associated with citizenship education.

The protection of the right to privacy is one of the main issues related to a ubiquitous digital environment. Be it through web browsing, interactions on social media, use of smart devices and home automation or personal sensors, the user creates a large quantity of data and metadata every day.[11] The data is then stored on servers, sometimes sold to third parties or shared with business partners. Its collection is operated by corporations (Angwin, 2014), government actors (Greenwald, 2014) and peers (interpersonal surveillance) (Cardon, 2010). This echoes two conceptions of privacy. First, informational privacy refers to the control exercised by an individual on corporate or governmental actors for the confidentiality or the protection of data and personal information (Shade & Shepherd in Landry & Letellier, 2016). Second, privacy, in its social (boyd & Marwick, 2011) and interpersonal (Cardon, 2010) dimensions, focuses on social norms that allow for the "choice of which individuals can access information, photographs and hyperlinks" [our translation] (Shade & Shepherd in Landry & Letellier, 2016) published by a user on social media. As such, Shepherd and Shade argue that

> The combination of critical media education – a discipline that emphasizes teaching that the media consist of texts and technologies that shape our cultural understandings – with citizenship education leads to

positioning privacy at the center of pedagogical objectives aimed at a responsible use of digital technologies.

[our translation] (p. 197)

Therefore, technology education is aimed at the development of knowledge and skills on what constitutes data and metadata, their capture methods, their storage and, the consequences of and issues related to their potential use. This knowledge not only enables actors to instantly act in a more ethical manner, but also, more importantly, better apprehends the risks related to massive data collection. Examples are the leakage of data for fraud or identity theft purposes (Crypto. Quebec, 2018) and the mobilization of personal data for purposes of political micro-targeting in the context of an electoral campaign.

Freedom of expression and access to information are also essential points of interest. The literature on this topic notes the existence of a consensus on the potential for the democratization of societies through digital networks. Indeed,

considering that information is the starting point of the democratic process, the idea that large quantities of it are made available to a large number of people (at least, to those who have access to the networks) implies an increased political participation potential [our translation].

(Gingras, 2009, 216)

Nonetheless, here as well, the knowledge related to digital infrastructure leads to a better understanding not only of the censorship mechanisms operated by state actors (Deibert, 2013), but also in a more subtle manner, of the processes for the organization of information through the search engines and platforms used daily by millions of individuals.

Search engines and social networks will henceforth hold power equivalent to that of the media due to their capacity to determine the information that is broadcast to citizens. The algorithms that define Facebook's newsfeed content for each user are eloquent examples of this. Mark Zuckerberg declared in 2010 that the algorithm constructing newsfeeds could reveal that

"a squirrel dying in your front yard may be more relevant to your interests right now than people dying in Africa." The question remains as to what extent such platforms, through their prescriptive effects, affect the plurality of information, social conflict and the constitution of the public space. This is even more true as a platform such as Facebook is becoming one of the main mediums used by the press to defuse information.

[our translation] (Loveluck, 2015, 256)

Therefore, the use of filtering algorithms can be perceived as a form of automated censorship (Shepherd & Landry, 2013). Although not all forms of censorship is in contravention of the ideals protected by freedom of information and expression, it remains necessary to consider the potential repercussions that such processes of information organization can have on public debates. The role of search engines and digital media newsfeeds can be problematic, inasmuch as it can be dissimulated in commercial activities that are not aimed at appropriately informing the citizen but rather presenting content to stimulate engagement (hits, page views, etc.). Here, engagement refers to a sustained interaction with the platform in order to maximize its presence and the intensity of the use of the services offered.

The discussion above leads to two findings. First, digital infrastructure has a daily impact on the usage and practices of media, and incidentally, on the fundamental rights of persons who mobilize technical devices. Second, economic interests structure the governance of these infrastructures. This informational and economic power is evidently exacerbated in light of the context of convergence and concentration of digital corporations.

Conclusion

Be it by defining it as praxis or through the constant technological mutations that are changing the media landscape and the tools available to citizens, media education is called upon to constantly renew itself both as a disciplinary field and as a set of pedagogical practices. Theoretical foundations based in communication studies, cultural studies and critical pedagogy enable us to address and analyze the media through the institutions that produce its messages, contents and the governance methods framing its practices. They also situate the learners in a proactive position by offering them the tools to produce and create media messages. Consequently, we have argued for the integration of another dimension to this approach in the form of technology education. Considering that the latter is typically associated with the development of a set of skills related to the ethical use of digital technology, our contribution positions technology education as a way to systematically deploy knowledge and skills related to technological infrastructure and associated governance practices.

Technology education thus becomes increasingly central to socio-economic issues especially considering the increasing number of jobs in the field of technology. Indeed, studies demonstrate that decision makers in the information technology field "fall into a relatively homogenous social class: highly educated, altruistic, liberal-minded science professionals from modernized societies around the globe" (Galloway, 2004, 122). As such, the inclusion of technology education in school curriculums constitutes an essential aspect of critical media education, because it fosters a more egalitarian training for learners while reducing the social, economic and cultural disparities that are currently characterizing the technology field.

Finally, the issues raised in relation to media and technology education highlight the need to update the locations where the development of this type of skill unfolds. Indeed, an important segment of the literature identifies schools as the preferred location to address the themes inherent to media education, specifically technology literacy. However, we argue that it is important to build bridges between initiatives that take place in school environments and the transformative and militant ones that are studied in research areas such as social movements, digital and media technologies and militant practices that are dedicated to them (Milan, 2013; Landry et al., 2016). The enrichment of media education in light of the issues raised by technology education results in broadening its scope and diversifying its resulting contributions. Media education participates in a collective process of appropriation of spaces where the interests of economic, political and social actors related to mediatized communication are brought in relation and confronted. In doing so, it highlights the power and domination relations that are part of and contribute to its transformation.

Notes

1 This chapter translates, rearticulates, and regroups content published in French. It includes a novel discussion on technology literacy. It mobilizes and adapts excerpts found in previous publications. The reader is invited to refer to this material. The publications mobilized are the following: Landry, N. & Letellier, A-S. (eds.). (2016). Introduction. *L'éducation aux médias à l'ère numérique: entre fondations et renouvellement.* Montreal: Presses de l'Université de Montréal, 253 p.; Landry, N., Pilote, A-M, & Brunelle, A-M. (2017). « L'éducation aux médias en tant que pratique militante: luttes et résistances au sein des espaces médiatiques et de gouvernance. » In Bonenfant, M., Dumais, F., & Trépanier-Jobin, G. *Les cahiers du Gerse,* vol. 12. (pp. 119–139). Quebec: Presses de l'Université du Québec; Landry, N., Basque, J. & Agbobli, C. (2016). « Éducation aux médias au Canada: état des savoirs et perspective de recherche en communication. » In Kiyindou, A. Barbey, F. & Corroy-Labardens, L. (eds.), *De l'éducation par les médias à l'éducation aux médias.* (pp. 37–51) Paris: L'Harmattan; Landry, N. (2017). « Articuler les dimensions constitutives de l'éducation aux médias. » TIC & Société, vol (11): 7–45; Landry, N. & Basque, J. (2015). « L'éducation aux médias: contributions, pratiques et perspectives de recherche en sciences de la communication. » *Communiquer, revue de communication sociale et publique* 15: 47–63; Landry, N. & Basque, J. (2015). "L'éducation aux médias dans le Programme de formation de l'école québécoise: intégration, pratiques et problématiques." *Canadian Journal of Education* 38(2): 1–33; Letellier, A-S (2016). *La littératie technologique.* In "L'éducation aux médias à l'ère numérique: entre fondations et renouvellement," ed. Landry, N. & Letellier, A-S, Presses de l'Université de Montréal; Letellier, A-S (2017). « Le Data Haven: repenser les politiques publiques et la régulation de l'information. » In Mondoux & Ménard, *Big Data et société: industrialisation des médiations symboliques.* Montreal: Presses de l'Université du Québec, pp. 105–124.

2 See http://ec.europa.eu/culture/policy/audiovisual-policies/literacy_en.

3 See http://www.unesco.org/new/en/communication-and-information/capacity-building-tools/media-and-information-literacy/.

4 See http://www.csem.be.
5 See https://www.ofcom.org.uk/research-and-data/media-literacy.
6 See http://www.clemi.fr/fr.html.
7 See https://namle.net.
8 See http://habilomedias.ca.
9 See http://www.aml.ca.
10 See https://www.ctf-fce.ca/en/Pages/Issues/Media-Literacy.aspx.
11 Metadata is data about data. For example, metadata associated with a Google search would be the following: the IP address of the computer used to do this search, the type of browser as well as its version (e.g., Mozilla, version 46.0.1), the location of the computer in question, etc.

References

Addison, C. & Meyers, E. (2013). Perspectives on information literacy: A framework for conceptual understanding. *Information Research: An International Electronic Journal*, 18(3), pp. 1–14.

Aharony, N. (2010). Information literacy in the professional literature: An exploratory analysis. *Aslib Proceedings*, 62(3), pp. 261–282.

Anderson, N., Duncan, B. & Pungente, J. (1999). Media education in Canada – The second spring. In C. Von Feilitzen & U. Carisson (eds.), *Children and media: Image, education, participation. Children and media violence. Yearbook from the UNESCO international clearinghouse on children and violence on the screen* (pp. 139–162). Göteborg, Sweden: UNESCO, Nordicom.

Angwin, J. (2014). *Dragnet nation: A quest for privacy, security, and freedom in a world of relentless surveillance* (304 p.). New York, NY: Times Books.

Bawden, D. (2001). Information and digital literacies: A review of concepts. *Journal of Documentation*, 57(2), pp. 218–259.

boyd, D. & Marwick, A. E. (2011). Social privacy in networked publics: Teen's attitudes, practices, and strategies. In Taylor & Francis (ed.), *A decade in Internet time: OII symposium on the dynamics of the Internet and society* (pp. 1–29). Oxford, UK: University of Oxford.

Bragg, S. (2002). Wrestling in woolly gloves: Not just being critically media literate. *Journal of Popular Film and Television*, 30(1), pp. 41–51.

Buckingham, D. (2003). *Media education: Literacy, learning, and contemporary culture*. Cambridge, Angleterre: Polity Press.

Buckingham, D. (2008). Defining digital literacy. What do young people need to know about digital media? In C. Lankshear & M. Knobel (eds.), *Digital literacies: Concepts, policies, and practices* (pp. 73–91). Oxford, UK: Peter Lang.

Buckingham, D. & Domaille, K. (2002). Where are we going and how can we get there? General findings from the UNESCO youth media education survey 2001. University of Southampton: Centre for Language in Education, UNESCO.

Buckingham, D. & Sefton-Green, L. (1994). *Cultural studies goes to school: Reading and teaching popular media* (260 p.). London, UK/Bristol, PA: Taylor & Francis.

Cardon, D. (2010). Confiner le clair-obscur: Réflexions sur la protection de la vie personnelle sur le Web 2.0. In F. Millerand, et al. (ed.), *Web social. Mutation de la communication* (pp. 315–328). Montréal, Canada: Presses de l'Université du Québecpp.

Carrington, V. & Robinson, M. (eds.). (2009). *Digital literacies: Social learning and classroom practices* (184 p). Los Angeles, CA: Sage Publications.

Chomsky, N. (2012). *Pour une éducation humaniste* (87 p.). Paris, France: Éditions de L'Herne.

Conseil de l'Europe. (1989, octobre). Résolution sur la Société de l'information: un défi pour les politiques de l'éducation? (no 1). Présenté à la Conférence permanente des ministres de l'Éducation du Conseil de l'Europe, Istanbul, Turquie. Repéré à <http://www.coe.int/t/f/coop%E9ration_culturelle/education/conf%E9rences_permanentes/j.16esessionl_istanbul1989.asp>

Corriveau, R. (2016). L'éducation aux médias: Entre nécessité et contradiction. In N. Landry & A. S. Letellier (eds.), *Éducation aux médias: Fondations, enjeux et politiques* (pp. 133–146). Montréal, Canada: Presses de l'Université de Montréal.

Crypto.Québec. (2018). *On vous voit* (240 p.). Montreal, Canada: Éditions Trécarré.

Culler, J. (2006). *Literary theory: A very short introduction* (184 p.). Oxford, UK: Oxford University Press.

Deibert, R. (2013). *Black code: Surveillance, privacy, and the dark side of the Internet* (336 p.). Toronto, Canada: McClelland & Stewart Limited.

DeNardis, L. & Musiani, F. (2016). Governance by infrastructure. In L. De Nardis & F. Musiani (eds.), *The turn to infrastructure in Internet governance* (pp. 3–21). London, UK: Palgrave Macmillan.

During, S. (ed.). (1999). *The cultural studies reader* (576 p.). New York, NY: Routledge.

Erstad, O. & Amdam, S. (2013). From protection to public participation: A review of research literature on media literacy. *Javnost*, 20(2), pp. 83–98.

Eshet-Alkalai, Y. (2004). Digital literacy: A conceptual framework for survival skills in the digital era. *Journal of Educational Multimedia and Hypermedia*, 13, pp. 93–106.

Eubanks, V. (2018). *Automating inequality: How high-tech tools profile, police, and punish the poor* (272 p.). New York, NY: St. Martin's Press.

European Audiovisual Observatory. (2016). *Mapping of media literacy practices and actions in EU-28*. Strasbourg, UK: Council of Europe.

Fahser-Herro, D. & Steinkuehler, C. (2010). Web 2.0 literacy and secondary teacher education. *Journal of Computing in Teacher Education*, 26(2), pp. 55–62.

Feenberg, A. (2012). *Questioning technology* (266 p.). New York, NY: Routledge.

Ferguson, A. G. (2017). *The rise of big data policing: Surveillance, race, and the future of law enforcement* (272 p.). New York, NY: NYU Press.

Fish, A. (2017). *Technoliberalism and the end of participatory culture in the United States* (217 p.). London, UK: Palgrave Macmillan.

Frau-Meigs, D. & Torrent, J. (2009). *Mapping media education policies in the world: Visions, programs, and challenges* (259 p.). New York, NY: The United Nations-Alliance of Civilizations & Grupo Comunicar.

Frau-Meigs, D., Velez, I. & Flores Michel, J. (ed.). (2017). *Public policies in media and information literacy in Europe. Cross-country comparisons* (304 p.). New York, NY: Routledge.

Freire, P. (2000). *Pedagogy of the oppressed* (183 p.). New York, NY: Continuum.

Galloway, A. R. (2004). *Protocol: How control exists after decentralization* (260 p.). Cambridge, MA: MIT Press.

Galloway, A. R. & Thacker, E. (2013). *The exploit: A theory of networks* (256 p.). Minneapolis, MN: U of Minnesota Press.

Gingras, A. M. (2009). *Médias et démocratie: Le grand malentendu* (290 p.). Montréal, Canada: Presses de l'Université du Québec.

Gonnet, J. (2001). *Éducation aux médias: Les controverses fécondes* (142 p.). Paris, France: Hachette.

Greenhow, C. & Robelia, B. (2009). Old communication, new literacies: Social network sites as social learning resources. *Journal of Computer-Mediated Communication*, 14(4), pp. 1130–1161.

Greenwald, G. (2014). *No place to hide Edward Snowden, the NSA, and the US surveillance state* (304 p.). London, UK: Palgrave Macmillan.

Gurak, L. J. (2008). *Cyberliteracy: Navigating the internet with awareness* (208 p.). New Haven, CT: Yale University Press.

Hall, S (1999), "Cultural studies and its theoretical legacies" in During, S. (ed.), *The cultural studies reader* (pp. 97–109, 576 p.). New York, NY: Routledge.

Hammer, R. & Kellner, D. (eds.). (2009). *Media/cultural studies: Critical approaches* (644 p.). New York, NY: Peter Lang.

Hands, J. (2011). *@ Is for activism: Dissent, resistance and rebellion in a digital culture* (224 p). London, UK/ New York, NY: Pluto.

Hartai, L. (2014). Report on formal media education in Europe 2014. Hungarian Institute for Education Research and Development.

Hobbs, R. (2010). Digital and media literacy: A plan of action. A White Paper on the Digital and Media Literacy Recommendations of the Knight Commission on the Information Needs of Communities in a Democracy. Aspen Institute.

Hobbs, R. (2011). The state of media literacy: A response to Potter. *Journal of Broadcasting & Electronic Media*, 55(3), pp. 419–430.

Hobbs, R. (2016). *Exploring the roots of digital and media literacy through personal narrative* (226 p.). Philadelphia, PA: Temple University Press.

Hoeschmann, M. & Poyntz, S. (2012). *Media literacies: A critical introduction* (246 p.). Malden, MA/Hichester, UK: Wiley-Blackwell.

Hogan, M. (2015). Data flows and water woes: The Utah data center. *Big Data & Society*, 2(2). doi:10.1177/2053951715592429

Hogan, M. & Shepherd, T. (2015). Information ownership and materiality in an age of big data surveillance. *Journal of Information Policy*, 5, pp. 6–31.

Horton, F. W. J. (1983). Information literacy vs. Computer literacy. *American Society for Information Science Bulletin*, 9(4), pp. 14–16.

Hu, T. H. (2015). *A prehistory of the cloud* (240 p.). Cambridge, MA: MIT Press.

Jasanoff, S. (ed.). (2004). *States of knowledge: The co-production of science and the social order* (332 p.). New York, NY: Routledge.

Jenkins, H. (2009). *Confronting the challenges of participatory culture: Media education for the 21ˢᵗ century* (145 p.). Cambridge, MA: The MIT Press.

Kellner, D. & Hammer, R. (2009). *Media/cultural studies: Critical approaches* (644 p.). New York, NY: Peter Lang.

Kellner, D. & Share, J. (2005). Toward critical media literacy: Core concepts, debates, organizations, and policy. *Discourse: Studies in the Cultural Politics of Education*, 26(3), pp. 369–386.

Kellner, D. & Share, J. (2007). Critical media literacy, democracy, and the reconstruction of education. In D. Macedo & S. R. Steinberg (eds.), *Media literacy: A reader* (pp. 3–23). New York, NY: Peter Lang.

Knight Abowitz, K. (2000). A pragmatist revisioning of resistance theory. *American Educational Research Journal*, 37(4), pp. 877–907.

Laborderie, P. (2012). Les offices du cinéma scolaire et éducateur à l'épreuve des publics. *Conserveries Mémorielles*, 12(1) <https://journals.openedition.org/cm/1230>, last consulted on February 17th 2020.

Landry, N. (2013). *Droits et enjeux de la communication* (290 p.). Quebec, Canada: Presses de l'Université du Québec.

Landry, N. (2017). « Articuler les dimensions constitutives de l'éducation aux médias. » *TIC & Société*, vol (11): 7–45.

Landry, N. & Basque, J. (2015). « L'éducation aux médias: contributions, pratiques et perspectives de recherche en sciences de la communication. » *Communiquer, revue de communication sociale et publique 15*, 47–63.

Landry, N., Basque, J. & Agbobli, C. (2016). Éducation aux médias au Canada: État des savoirs et perspective de recherche en communication. In A. Kiyindou, F. Barbey & L. Corroy-Labardens (eds.), *De l'éducation par les médias à l'éducation aux médias* (pp. 37–51). Paris, France: L'Harmattan.

Landry, N. & Letellier, A. S. (2016). *L'éducation aux médias à l'ère numérique: Entre fondations et renouvellement* (262 p.). Montreal, QC: Les Presses de l'Université de Montréal.

Landry, N., Pilotte, A.-M. & Brunelle, A.-M. (2017). L'éducation aux médias en tant que pratique militante: Luttes et résistances au sein des espaces médiatiques et de gouvernance. In M. Bonenfant, F. Dumais & G. Trépanier-Jobin (eds.), *Les pratiques transformatrices de l'espace socionumérique: Appropriation, résistance et expérimentation* (pp. 119–139). Quebec, Canada: Presses de l'Université du Québec.

Landry, N. & Roussel, C. (2018). Élèves, éducation aux médias et citoyenneté: Analyse du programme de formation de l'école québécoise. *Lien Social Et Politiques*, 80(1), p. 38.

Lebrun, M., Lacelle, N. & Boutin, J. F. (2012). *La littératie médiatique multimodale: De nouvelles approches en lecture-écriture à l'école et hors de l'école* (274 p.). Montreal, Canada: Presses de l'Université du Québec.

Lee, A. et al. (2013). *Conceptual relationship of information literacy and media literacy in knowledge societies*. Paris, France: UNESCO.

Lessig, L. (2008). *Remix: Making art and commerce thrive in the hybrid economy* (352 p.). New York, NY: NYL Penguin Press (*Internet: Research and policy challenges in comparative perspective* (382 p.). Bristol, UK/Portland, OR: Policy Press).

Lessig, L. (2009). *Code: And other laws of cyberspace* (297 p.). New York, NY: Penguin Press.

Livingstone, S. & Haddon, L. (eds.) (2009). *Kids online: Opportunities and risks for children* (296 p.). Bristol, UK/Portland, OR: Policy Press.

Livingstone, S. et al. (eds.). (2012). *Children, risk and safety on the Internet: Research and policy challenges in comparative perspective* (382 p.). Bristol, UK/Portland, OR: Policy Press.

Loveluck, B. (2015). *Réseaux, libertés et contrôle: Une généalogie politique d'internet* (368 p.). Paris, France: Armand Colin.

Macdonald, M. F. (2008). Media literacy in action: An exploration of teaching and using media literacy constructs in daily classroom practice. Doctoral thesis, University of California, United States.

Mackey, T. P. & Jacobson, T. E. (2011). Reframing information literacy as a metaliteracy. *College and Research Libraries*, 72(1), pp. 62–78.

Markauskaite, L. (2006). Towards an integrated analytical framework of information and communications technology literacy: From intended to implemented and achieved dimensions. *Information Research: An International Electronic Journal*, 11(3), pp. 252–277.

Martens, H. (2010). Evaluating media literacy education: Concepts, theories and future directions. *Journal of Media Literacy Education*, 2(1), pp. 1–22.

Masterman, L. (1985), *Teaching the media* (364 p.). London/New York, NY: Routledge.

Masterman, L. & Council of Europe (1988). *Le développement de l'éducation aux médias dans l'Europe des années 80: En particulier dans le domaine de la télévision et des médias électroniques.* Strasbourg, France: Council for Cultural Co-operation, Council of Europe.

McChesney, R. W. (1995). *Telecommunications, mass media, and democracy: The battle for the control of US broadcasting, 1928–1935* (416 p.). Oxford, UK: Oxford University Press.

McChesney, R. W. (2000). *Rich media, poor democracy: Communication politics in dubious times* (427 p). New York, NY: The New Press.

McLuhan, M. (1994). *Understanding media: The extensions of man* (392 p.). Cambridge, MA: MIT Press.

Milan, S. (2013). *Social movements and their technologies: Wiring social change* (233 p.). New York, NY: Palgrave Macmillan.

Musiani, F. (2015). Practice, plurality, performativity, and plumbing: Internet governance research meets science and technology studies. *Science, Technology & Human Values*, 40 (2), pp. 272–286.

Musiani, F., Cogburn, D. L., DeNardis, L. & Levinson, N. S. (eds.). (2016). *The turn to infrastructure in Internet governance* (268 p.). London, UK: Palgrave Macmillan.

Nissenbaum, H. (2011). A contextual approach to privacy online. *Daedalus*, 140(4), pp. 32–48.

Parks, L. (2015). "Stuff you can kick": Conceptualizing media infrastructure. In D. T. Golberg & P. Svensson (eds.), *Humanities and the digital* (p. 335). Cambridge, MA: MIT Press.

Piette, J. (1996). *Éducation aux médias et fonction critique* (362 p.). Paris, France/Montreal, Canada: L'Harmattan.

Plante, T. (2016). Les médias numériques: Un défi pour les éducateurs ». In N. Landry & A.-S. Letellier (eds.), *L'éducation aux medias à l'ère numérique: Entre fondations et renouvellements* (pp. 115–130). Montréal, Canada: Presses de l'Université de Montréal.

Potter, J. W. (2013). *Media literacy* (7th ed., 576 p.). London, UK: Sage Publications.

Potter, W. J. (2012). *Media effects* (400 p.). London, UK: Sage Publications.

Provenzo, E. F. et al., (2011). *Multiliteracies: Beyond text and the written word* (226 p.). Charlotte, NC: Information Age Pub.

Schneier, B. (2018). *Click here to kill everybody: Security and survival in a hyper-connected world* (336 p.). New York, NY and London, UK: W. W. Norton & Company.

Schwarz, G. (2005). Overview: What is media literacy, who cares, and why? *Yearbook of the National Society for the Study of Education*, 104(1), pp. 5–17.

Share, J. (2009a). *Media literacy is elementary: Teaching youth to critically read and create media.* New York, NY: Peter Lang.

Share, J. (2009b). Young children and critical media literacy. In D. D. Kellner et R. Hammer (dir.), *Media/cultural studies: Critical approaches* (p. 126–151), New York, NY: Peter Lang.

Shariff, S. (2008). *Cyber-bullying: Issues and solutions for the school, the classroom and the home* (320 p.). London, UK/New York, NY: Routledge.

Shepherd, T. & Landry, N. (2013). Technology design and power: Freedom and control in communication networks. *International Journal of Media & Cultural Politics*, 9(3), pp. 259–275.

Shepherd, T. & Shade, L. R. (2016). «La vie privée et les jeunes au Québec et au Canada. In N. Landry & A.-S. Letellier (eds.), *L'éducation aux medias à l'ère numérique: Entre fondations et renouvellements* (pp. 191–208). Montreal, Canada: Presses de l'Université de Montréal.

Silver, A. (2009). Forewords. A European approach to media literacy: Moving toward an inclusive knowledge society. In D. Frau-Meigs & J. Torrent (eds.), *Mapping media education policies in the world: Visions, programs, and challenges* (pp. 11–13). New York, NY: The United Nations-Alliance of Civilizations & Grupo Comunicar.

Solove, D. J. (2008). *Understanding privacy* (272 p.). Cambridge, UK: Harvard University Press.

Stordy, P. (2015). Taxonomy of literacies. *Journal of Documentation Journal of Documentation*, 71(3), pp. 456–476.

Taylor, A. (2014). *The people's platform: Taking back power and culture in the digital age* (288 p.). New York, NY: Palgrave Macmillan.

Thoman, E. et Jolls, T. (2004). Media literacy: A national priority for a changing world. *American Behavioral Scientist*, 48(1), 18–29.

UNESCO (1982). Déclaration de Grunwald sur l'éducation aux médias <http://aeema.net/2011/01/unesco-la-declaration-de-grunwald-sur-leducation-aux-medias-22-janvier-1982/> last consulted on 7 March 2017.

UNESCO et al. (2013). *Global media and information literacy (MIL). Assessment framework: Country readiness and competencies.* Paris, France: UNESCO.

Van Deursen, A. & Van Dijk, J. (2011). Internet skills and the digital divide. *New Media and Society*, 13(6), pp. 893–911.

Mayers-West, S. M. (2017). Data capitalism: Redefining the logics of surveillance and privacy. *Business & Society*, 58(1), pp. 20–41. doi:10.1177/0007650317718185

Winner, L. (1980). Do artifacts have politics? *Daedalus*, 19(1), pp. 121–136.

Wu, T. (2018). Is the first amendment obsolete. *Michigan Law Review*, 117, 548–571.

Yousman, B. (2008). Media literacy: Creating better citizens or better consumers? In R. Andersen & J. Gray (eds.), *Battleground: The media* (pp. 238–247). Westport, CT: Greenwood Press.

15

FROM MEDIA ECOLOGY TO MEDIA EVOLUTION

Toward a Long-Term Theory of Media Change

Carlos A. Scolari

A New Paradigm

The web is coming up to its 10,000-day birthday. It was in August of 1991 that Tim Berners-Lee made the first web page and started the story. These 10,000 days have changed everything: there is no area of our individual or social life that has not been transformed by digital networks. Although just over half the world's population is online, the effects and consequences of these technologies also reach those who are offline. Education, finances, politics, economics, social and couple relationships have been transformed in little more than two decades. In the specific field of media, the arrival of the world wide web meant a profound change of paradigm: the transformation from a media ecology centered on broadcasting to one centered on networking. The ways of producing, distributing and consuming media content are changing. The traditional roles of the sender and receiver have also been transformed and have become blurred (consider the concept of "prosumer", which is the combination of "producer" and "consumer").

Does this mutation mean the end of broadcasting? No, but broadcasting is no longer central in the media diet of citizens. Although there will continue to be live broadcasts to be consumed "in real-time" (for example sports events), more and more citizens spend their time producing, distributing and consuming media content in networked environments.

In parallel to this paradigm shift, we are witnessing an explosion of new media and communication platforms. As in the biological evolution of the planet, which has had explosive appearances of new species, like the Cambrian explosion,[1] we are now witnessing a similar burst of new "media species." The world wide web is a metamedium capable of giving birth and hosting new forms of communication (Facebook was born on the world wide web, like YouTube and Twitter).

Have there ever been other Cambrian media explosions? Yes. At the end of the nineteenth century, there was the first explosion (burst) of new communication devices and environments. In just a couple of decades the mass press, photography, telegraphy, cinema, comics and the telephone, were all consolidated, and this is without counting the patents that the laboratories of Thomas A. Edison registered every week. However, the current Cambrian media expansion has no parallel in the history of humanity: we are talking about media or platforms that in just ten years have reached 25% of the world's population (i.e., Facebook).

Another aspect to take into account is the emergence of new media players. Any map of new communications should include YouTubers and the different digital celebrities, not to mention the increasing weight of the content generated by the fans of transmedia narrative worlds.

All these changes have led, on the one hand, to the progressive atomization of the audiences: families no longer watch television sitting in front of the screen like in *The Simpsons*. Now, each member of the family consumes content asynchronously on a personal screen. This atomization has led to a radical change in the audiences' media diet (by the way: can we continue talking about "audiences"?): before, they used to spend a lot of time using a few media, now, they spend a small amount of time in many media.

These transformations are blasting away the old advertising-centered business models of broadcasting. The picture is confusing and there is no manual for managing the transition from broadcasting to networking. Trial and error seem to be the only possible strategy. In this context, the old actors who cannot adapt to this new media ecology run the serious risk of "extinction".

Considering the current situation of the media "environment", it is worth asking: Is there any theoretical proposals capable of capturing and integrating all these transformations? The previous paragraphs clearly evoked the eco-evolutionary metaphor; indeed, concepts like "media ecology", "media species", "mutations" or "extinction" direct us to that biological analogy. However, could we move from simply using the eco-evolutionary metaphor to a more solid theoretical construction? Is it possible to go beyond the analogy and develop an eco-evolutionary theory of media change? The answers to these questions are in the next sections of this chapter.

Media Ecology[2]

Origins and Development

The consolidation of an ecological vision for media and communication ran parallel with the diffusion of ecologist ideas from the 1960s. Although the concept of media ecology was officially introduced by Neil Postman in a talk for the National Council of Teachers of English in 1968, Postman himself recognized that Marshall McLuhan had used it at the beginning of that decade, when the Canadian's brilliance was at its brightest (*The Gutenberg Galaxy* is from 1962 and *Understanding Media* is from 1964). However, other researchers prefer to award the distinction of this semantic coining to Postman (Lum, 2006, 9). Whatever the case, during his talk, Postman (1970) defined media ecology as "the study of media as environments". It can be said that Postman brought about the shift from metaphor to theory or, better, the journey from a purely metaphoric use of the term media ecology to the start of the delimitation of a specific scientific field. Postman fought hard for the new concept: in 1971 he created the first degree in media ecology at New York University, thereby providing media ecology with its first step toward academic institutionalization.

Beyond the semantic origins of media ecology, it's clear that this conception, which aims to integrate different components and processes from the techno-social-communicational sphere, did not arise from spontaneous generation nor from a stroke of genius from McLuhan or Postman. As Jorge L. Borges maintained about Franz Kafka and his predecessors (how many writers were unwittingly Kafkaesque before Kafka was born?), similarly we can identify a series of researchers who were "McLuhanesque" before McLuhan himself.

Media Ecology: Theoretical Contributions

Media Ecology is a recognized theoretical approach that has become consolidated in the last decades. For reasons of space, in this section, I will describe the main contributions of Marshal McLuhan and Neil Postman.

McLuhan had a double effect on media ecology: on the one hand, he was one of the pioneers of introducing an ecological approach to media; on the other hand, his fame eclipsed other researchers who were also constructing the media ecological approach. Regarding his main influences, McLuhan updated and integrated within one single approach the ideas of scholars such as Harold Innis (1950, 1951) and Eric Havelock (1963, 1981, 1986) among many others. McLuhan insisted that media create an environment (a "medium") in which we all move; like a fish in water, we do not realize it exists until we stop perceiving it for some reason. In this context, it could be said that humans shape communication instruments but, at the same time, communication instruments shape humans.

Neil Postman was one of the great thinkers of media studies. In works like *Amusing Ourselves to Death: Public Discourse in the Age of Show Business* (1985), *Technopoly: The Surrender of Culture to Technology* (1992) and *The End of Education: Redefining the Value of School* (1995) Postman developed an ecological, critical and ethical view of the American media system (Gencarelli, 2006). According to Postman,

> Technological change is not additive; it is ecological. I can explain this best by an analogy. What happens if we place a drop of red dye into a beaker of clear water? Do we have clear water plus a spot of red dye? Obviously not. We have a new coloration to every molecule of water. That is what I mean by ecological change. A new medium does not add something; it changes everything. In the year 1500, after the printing press was invented, you did not have old Europe plus the printing press. You had a different Europe. After television, America was not America plus television. Television gave a new coloration to every political campaign, to every home, to every school, to every church, to every industry, and so on.
>
> *(Postman, 1998)*

The figure of Postman is fundamental to media ecology, not only for his theoretical ideas but also for the establishment, in 1971, of the first degree in media ecology at the Steinhardt School of Education (New York University). In addition, Postman trained, inspired and worked with distinguished researchers such as Paul Levinson (1979, 1997, 1999), Joshua Meyrowitz (1985) and Lance Strate (2004, 2008, 2017). If *Wired* magazine "resurrected" Marshall McLuhan from a digital perspective in the early 1990s (they chose McLuhan as "Patron Saint" of the publication in the first issue), the explosion of fake news and the discussions about post-truth in recent years have generated a renewed interest in the works of Neil Postman.

From Media Ecology to Media Evolution

In the last two decades the texts written by McLuhan, Postman and other pioneers of media ecology have undergone a digital reinterpretation. In a world marked by profound changes in the ways humans produce, distribute and consume information, the comparison with past processes such as the discovery of writing or the invention of the press has much to offer. Within this context, the works of Eric Havelock, Marshall McLuhan, Neil Postman, Walter Ong (2002) and other media ecologists become compulsory reading for researchers interested in the new forms that digital interactive communication is taking on. Specifically, the re-reading of McLuhan in the digital age has generated works of great value such as *Digital McLuhan: A Guide to the Information Millennium* (Levinson, 1999), *The Sixth Language: Learning a Living in the Internet Age* (Logan, 2000), and *Understanding New Media: Extending Marshall McLuhan* (Logan, 2010).

In a scenario marked by the consolidation of global information networks, convergence processes and the explosion of new media and communication platforms, the appearance of transmedia narratives and the emergence of many-to-many communications that break the traditional model of broadcasting, media ecology becomes an almost essential reference for understanding these processes. It proposes themes, concepts and questions that enrich scientific conversations about interactive digital communication. Re-reading McLuhan without the academic prejudices that isolated him from some colleagues in the 1960s, rediscovering Postman's analysis of education and communication in the midst of a crisis in schooling, or going back to the astute reflections of Ong or Havelock on the transition from orality to literacy, can offer us new key interpretations for understanding the shape that the media ecosystem is adopting in the twenty-first century.

The contributions of researchers like Robert K. Logan (1986, 2000, 2007, 2010), who inaugurated an incredibly rich conversation between media ecology, complexity theories and biological models, and Paul Levinson, who defended the first PhD Thesis on Media Evolution in 1979 under the supervision of Neil Postman, are the best possible interfaces between the second and third sections of this chapter.

Media Evolution

The Darwin Machine

A good part of the intellectuals, artists and scientists of the second half of the nineteenth century fell victim to Charles Darwin's theory of evolution. The evolutionary model was so simple, economical (in the sense of William de Ockham's razor), rational and consistent for modern minds that it was almost impossible not to adopt it. The influence of the dual variation/natural selection not only transformed the natural sciences but also left a significant mark on economics, sociology, anthropology and even literature. Novels like *Erewhon* (1872) by Samuel Butler, *The War of the Worlds* (1898) by H. G. Wells and *The Lost World* (1912) by A. C. Doyle could not have been written before Darwin's theory of evolution.

If Darwin theorized the processes of natural selection and showed that biological diversity arose as a consequence of a branched model of common evolution and descent, Karl Marx was the first to extend these ideas to technology. Marx published the first of the three volumes of *Das Kapital* in 1867, eight years after *On the Origin of Species*. In that volume Marx wrote:

> Darwin has directed attention to the history of natural technology, i.e. the formation of the organs of plants and animals, which serve as the instruments of production for sustaining their life. Does not the history of the productive organs of man in society, of organs that are the material basis of every particular organization of society, deserve equal

attention? And would not such a history be easier to compile, since, as Vico says, human history differs from natural history in that we have made the former, but not the latter?

(Marx, 1976, 493)

Just as the diversity of new life forms in the ecosystem of the Galapagos Islands caught Darwin's attention, Marx was surprised by the existence of more than 500 different models of hammers in Birmingham in the mid-nineteenth century. Researchers from different disciplines would soon apply the Darwinian model to technological evolution.

In the specific field of economics, Karl Marx's first attempts to understand the diversity of hammers were followed by Joseph Schumpeter's theory of innovation (see Fagerberg, 2003) and contemporary research on the evolution of technology, the life cycle of goods and innovation processes. In *Technological Evolution, Variety and the Economy* Paolo Saviotti argued that the evolutionary theory could provide a "better generalization of the phenomenon of technological innovation or, more generally, of any type of innovation that leads to a qualitative change in the economic system" (Saviotti, 1996, 7).

Fossils do not lie: evolution has been a continuous process of variation, bifurcation, adaptation, survival and extinction of living entities. Darwin's theory integrated in the same picture the appearance of new species due to variation and the extinction of those that did not adapt. The basic principle behind this process can be reduced to two words: natural selection. It is also known by another name: the "Darwin machine". George Basalla, author of the classic *The Evolution of Technology* (1996), considers that industrial society produces a much greater variety of artifacts than what it really requires to meet its needs. According to Basalla, this diversity is the result of technological evolution. As in biological systems, a variation/selection process would choose the best artifacts to be replicated. The Darwin machine applied to machines.

The Evolution of Media: Theoretical Contributions

Never in the long history of Homo sapiens has our social-technological environment gone through such an accelerated and unpredictable paradigm shift. Different scientific disciplines have addressed the transformation of media over time, some with a long history (such as media history) and more recent ones (such as media archaeology). Media history, at a general level, is characterized by proposing a perspective based on the succession of different communication technologies (for example the series *clay tablet > roll > codex > printed book > press > cinema > radio > television > internet*) (i.e., Briggs and Burke, 2009). Media archaeology, on the other hand, focuses on specific devices or technologies and analyses them from a perspective inspired by theorists such as Michel Foucault (Huhtamo and Parikka, 2011; Parikka, 2012). Media archaeologists do not agree with historical storytelling and focus instead, as Siegfried Zielinski put it, on the "deep time of media" (2008). In this chapter, I will introduce a third perspective that complements media history and media archaeology: media evolution. Before describing its main characteristics, I will briefly review a series of media researchers who have applied, to a greater or lesser extent, evolutionary models in their studies.

Media ecology has always included a strong interest in the evolution of media. Beyond the classic works of scholars such as Innis (1950, 1951), researchers like Levinson (1979, 1997, 1999) and Logan (1986, 2000, 2007, 2010, 2014) have made useful contributions to the understanding of media ecosystem transformation. In recent years many scholars have been using the metaphor of media evolution in their theoretical and analytical discourses (i.e., Lehman-Wilzig and Cohen-Avigdor, 2004; Manovich, 2013; Natale, 2016; Neuman, 2010; Olesen, 2016; Stöber, 2004; van Dijck, 2013).

Media evolution has an ambitious research agenda. Issues like the life cycle of media (emergence, domination, survival/extinction) (Lehman-Wilzig and Cohen-Avigdor, 2004; Scolari, 2013, 2018a, 2018b), the recapture of elements from natural life transformed by "new" media extensions (i.e., the telephone replaces the telegraph due to a human evolutionary pressure for retrieving the lost element of voice) (Levinson, 1979, 1997, 1999), the rise of new media as an emergent phenomenon (Logan, 2007, 2010), or the combinations of different media as if they were biological entities (Manovich, 2013) should all be considered at the center of media evolution interest (for a deeper analysis of these contributions, see Scolari (2018a)).

Beyond media ecology, media evolution dialogs with different fields and theoretical approaches, from Actor-Network Theory (Latour, 2005) to evolutionary epistemology (Ziman, 2000), social construction of technology (Bijker et al., 1987), and social-ecological systems (Glaser et al., 2012). Although media evolution's general approach to media change retrieves and analyses examples of media transformation in different historical moments (from the first mediated communication experiences thousands of years ago to the latest transformations of software interfaces), it takes a wide-angle/long-term perspective. In this sense, the media evolution macro-approach is the best complement and scientific partner for the media archaeology micro-perspective (see next section).

Toward a Theory of Media Evolution

A scientific discipline, if it wants to become consolidated, must: 1) delimit a field or series of problems to be researched (which can take the form of questions), 2) define a series of concepts and 3) propose methodologies to answer these questions. Let's look at an example. According to Wikipedia, sociology is

the scientific study of society, patterns of social relations, social interaction, and culture of everyday life. It is a social science that uses various
methods of empirical investigation and critical analysis to develop a body
of knowledge about social order, acceptance, and change or social evolution ... The different traditional focuses of sociology include social stratification, social class, social mobility, religion, secularization, law,
sexuality, gender, and deviance.

(Wikipedia)

As can be seen, this definition delimits a field of action – "the scientific study of
society, patterns of social relationships, social interaction, and culture of everyday
life" – and proposes a first series of key concepts that researchers must define
precisely (from "society" to "social relationships", "social interactions", "social
order", "stratification", "social class", "mobility", "secularization", etc.). In
other words, researchers must share a common dictionary if they want to
"speak" the sociological language. Regarding the methodology, as every scholar
knows sociological research methods are divided into two broad categories:
quantitative and qualitative methods. Quantitative methods rely on quantifiable
evidence, based on both statistical analyses and, in recent years, on big data
(complete universes). Qualitative methods, on the other hand, work with direct
observation, exchanges with participants or textual analysis.

From this perspective, it could be said that *media evolution is a proto-discipline
that studies media and social change from a long-term, holistic and complex point of
view.* Let's look at the components of this definition:

Proto-Discipline

Media evolution is far from being a consolidated discipline. Even this definition should be considered as just an operative characterization of a theoretical
work-in-progress. The first steps of media evolution should be aimed at identifying continuities, emergent and repetitive phenomena within the processes of
media and social change. At the same time, this "under construction" discipline should refine its own dictionary of concepts and strengthen exchanges
with the closest disciplines and theories (media history, media archaeology,
media ecology, ANT, social construction of technology, etc.)

Long-Term

As we have seen, media archaeology privileges a close-up perspective while
media history develops a long-term linear approach. In this context, media
evolution proposes activating a wide-angle view of media change, understanding "wide" as a broad spatial and temporal perspective.

Scholars like José L. Fernández (2018) have proposed a set of three "observation distances": the macro-perspective (the "society", the "culture", etc. and their respective "key conflict elements": "social classes", "lifestyles", etc.), the meso-perspective (closer to the social phenomena, this perspective observes the "conflict and exchange scenes"), and the micro-perspective (focused on "products and their processes") (Fernández, 2018, 35–36). In this framework, media archaeology should be included in the micro-perspective while media evolution aspires to a macro-perspective of media change. However, in the same way that Charles Darwin needed fossils to build his theory, the expert in media evolution must often work with "media fossils". In this sense, the dialog between media archaeology and media evolution could take on a strategic character: if media archaeology proposes a micro-approach often focused on a single device, media evolution aims to reconstruct the change from the perspective of large networks of actors, relationships and processes.

Holistic

This conception proposes that any kind of system (i.e., physical, biological, chemical, social, economic, mental or linguistic system) and its properties should be considered as a whole, and not as just a collection of parts. According to Jan C. Sumuts (1927), who coined the concept, "wholes are not artificial constructions of thought: they actually exist; they point to something real in the universe, and Holism is a real operative factor, a vera causa" (88). For Smuts

> The idea of wholes and wholeness should therefore not be confined to the biological domain: it covers both inorganic substances and mental structures as well as the highest manifestations of the human spirit … As holism is a processes of creative synthesis, the resulting wholes are not static but dynamic, evolutionary creative. Hence Evolution has an ever-deepening inward spiritual character; and the wholes of evolution and the evolutionary process itself can only be understood in reference to this fundamental character of wholeness.
>
> *(Sumuts, 1927, 88–89)*

In the second half of the twentieth-century holism led to system thinking (von Bertalanffy, 1968). From this perspective, biological or sociological systems are frequently so complex that their behavior shows emergent phenomena. In other words, these emergent configurations cannot be deduced from the properties of the individual elements. In the context of a theoretical reflection on media evolution, holism should be understood as a synonym for the "systemic" or "ecological" approach.

Complex

Despite its success, many researchers suspected that Darwin's model was not enough: is natural variation/selection the only possible model for interpreting the exuberant diversity of species? Without contradicting Darwin, researchers like Stuart Kauffman (1995) of the Santa Fe Institute (New Mexico) tried to answer this question by developing a broader theoretical framework: the sciences of complexity, chaos, emergency and self-organization. Any sociologist, biologist, semiotician or economist of the twenty-first century must take note of this fascinating paradigm.

Talking about "complexity" is not at all simple. A system is "complex" when it is composed of interrelated elements that exhibit general properties not evident in the sum of the individual parts. The intensification of the interactions between individual elements increases the complexity of a system. In some complex systems, Kauffman explains in *At Home in the Universe* (1995), a minor change can cause catastrophic transformations in the behavior of the whole. A self-organized system has emerging properties that, as Ricard Solé explains in *Redes Complejas* (*Complex Networks*) (2009), cannot be explained by the properties of the components. In these systems

> The whole is more than the sum of the parts or, perhaps more appropriately, the whole is different from the sum of the parts … The complex has much more to do with the nature of the interactions than with the nature of the objects that interact, although the latter impose some limitations on what can happen at the next level.
>
> *(Solé, 2009, 19–25)*

Kauffman and other scientists at the Santa Fe Institute, like Brian Arthur, author of *The Nature of Technology* (2009), believe that technological evolution is based on laws that are similar to those that govern the biological domain.

Regarding the relationships between media ecology and media evolution, as I stated in a previous text,

> If the ecological approach studies the network of relations between organisms at the same time, then the evolutionary approach investigates the diversification of these organisms into new species, the extinction of species (macroevolution), and the smaller changes such as adaptations (microevolution). In other words, while the ecologist reconstructs webs of organisms, the evolutionary scholar draws trees of life. Or, in another sense, ecology thinks in space and evolution thinks in time. Both conceptions – ecology and evolution – are complementary and can be reorganized following the traditional linguistic opposition between diachronic/synchronic levels.
>
> *(Scolari, 2013) (see Figure 15.1)*

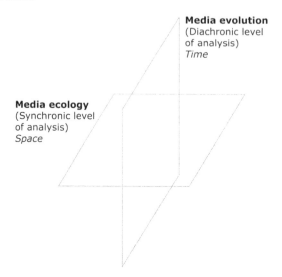

Media evolution
(Diachronic level
of analysis)
Time

Media ecology
(Synchronic level
of analysis)
Space

FIGURE 15.1 Media ecology and media evolution as complementary approaches
(Scolari, 2013)

Work-In-Progress

Media evolution has to complete two tasks if it wants to consolidate itself as
a discipline. It needs a dictionary of concepts and categories, and it needs to
define its research methodologies.

In the first task, concepts such as "coevolution", "extinction", "adaptation",
etc. from the evolutionary theory allow a reinterpretation from a media perspective (Scolari, 2013, 2018a, 2018b, 2019). Concepts like "interface" could also be
very useful in the development of this new discipline (Scolari, 2018b).

For the second task, media evolution is open to both qualitative and quantitative methods (Scolari, 2013, 2018a). From the perspective of qualitative
research, both media history (i.e., Briggs and Burke, 2009) and media archaeology (i.e., Huhtamo and Parikka, 2011; Parikka, 2012) offer an interesting set
of techniques that, conveniently adapted, could be adopted by media evolution
researchers. The experience of researchers working in fields such as the Actor-
Network Theory (Latour, 2005) and Social Construction of Technology (Bijker
et al., 1987), or the morphological approaches of researchers like Basalla (1988),
should also be considered as indispensable references for media evolution.

Regarding quantitative methods, media evolution should recover the analytical experience of evolutionary economics (Nelson and Winter, 1982), evolutionary epistemology (Ziman, 2000), literary criticism (Moretti, 2005) and
cultural analytics (Manovich, 2009, 2013). For example, the analysis of 44
genres in British fiction between 1740 and 1900 allowed Franco Moretti
(2005) to identify patterns, isolate major bursts of creativity (genre emergence)

and describe genre extinction. The same approach could be adapted to media evolution research. Pattern recognition was one of McLuhan's favorite analytical tools (Moretti talks about "distant reading"), and now it is possible to recover this approach working with data sets coming from media content, media devices and digital traces left when people discuss, create, publish, consume, share, edit and remix these media. Researchers need to recognize the common traits of the evolution of media species, improving and deepening the field drafted in the *Laws of Media* (McLuhan and McLuhan, 1992).

Conclusions

Why are these biological metaphors so useful for understanding the system of interfaces? The great variety of artifacts that humans have created in the last 2.5 million years and the increasing complexity acquired by the technological sphere can only be compared to the variety and complexity of the biological world. A path that encompasses both the idea of ecosystem and the theory of evolution is a path that deserves to be explored. Aristotle wrote memorable texts on nature and technology (his was one of the first reflections on the camera obscura) but never established an analogy between the two universes. It was in the Renaissance that intellectuals began to discover parallels between the organic and the technological domains. Initially, technological models were used to describe living entities but in the nineteenth century, after Darwin, the analogies were reversed and went from biology to technology. Metaphors are interchangeable: we can see the biological universe from the perspective of technology, for example when a scientist considers the human body as a machine or the brain as a computer, or the technological universe from the perspective of biology, for example when we look at the mutations in media (and, more generally, technological change) from an eco-evolutionary perspective.

The Limits of the Eco-Evolutionary Metaphor

Working with analogies can be problematic. Some laws and biological principles do not support a technological translation. Human bodies are not machines, in the same way that interfaces are not living entities. Biological analogies can provide new concepts ("variety", "selection", "coevolution", etc.) and suggest promising lines of research (Can media become extinct? How do media coevolve with other media or with their users?), but they do not always provide good answers. These analogies should be approached with caution due to the huge differences between the biological and technological domains. For example, the transmission of acquired characters is central in the technological domain, whereas, in the biological realm, it is only verified very exceptionally at the level of microorganisms. Another large difference is found in the different evolutionary speeds: rapid in the technological domain, slow in the biological one.

Another point of disagreement between the biological and technological domains is the distance between the actors that establish a relationship. Two biological species separated by thousands of kilometers will probably never be able to establish a relationship; however, two technological artifacts located in different domains can join and create a new interface at any time. For example, the telephone system was far from related to computers until the creation of ARPANET in the late 1960s. In the socio-technological world, the distance between two "media species" is almost non-existent.

Returning to the question of metaphor, it is possible that at some point the analogy between the technological and biological domains has more limitations than advantages. It would then be time to look for new analogies to interpret technological (and media) change. In any case, the eco-evolutionary metaphor has just begun to be explored and there is a long way to go.

Notes

1 The Cambrian explosion was the sudden appearance (from a geological point of view) and rapid diversification of complex multicellular macroscopic organisms in the early Cambrian period, 542/530 million years ago.
2 This section is based on Scolari (2010, 2015).

References

Arthur, B. (2009). *The nature of technology. What it is and how it evolves*. New York, NY: Penguin.

Basalla, G. (1988). *The evolution of technology*. Cambridge: Cambridge University Press.

Bijker, W. E., Hughes, T. P. and Pinch, T. (eds.) (1987). *The social construction of technological systems*. Cambridge, MA: The MIT Press.

Briggs, A. and Burke, P. (2009). *Social history of the media: From Gutenberg to the internet* (3rd ed.). Cambridge: Polity Press.

Fagerberg, J. (2003). Schumpeter and the revival of evolutionary economics: An appraisal of the literature. *Journal of Evolutionary Economics* 13(2): 125–159. doi: 10.1007/s00191-003-0144-1.

Fernández, J. L. (2018). *Plataformas mediáticas*. Buenos Aires: La Crujía.

Gencarelli, T. (2006). Neil Postman and the rise of media ecology. In: C. M. K. Lum (ed.) *Perspectives on culture, technology and communication. The media ecology tradition*. Cresskill, NJ: Hampton Press, pp. 201–254.

Glaser, M., Krause, G., Ratter, B. M. W. and Welp, M. (eds.) (2012). *Human-nature interaction in the anthropocene: Potentials of social-ecological systems analysis*. London: Routledge.

Havelock, E. (1963). *Preface to plato*. Cambridge: Harvard University Press.

Havelock, E. (1981). *The literate revolution in Greece and its cultural consequences*. Princeton, NJ: Princeton University Press.

Havelock, E. (1986). *The muse learns to write: Reflections on orality and literacy from antiquity to the present*. New Haven, CT: Yale University Press.

Huhtamo, E. and Parikka, J. (2011). Introduction: An archaeology of media archaeology. In: E. Huhtamo and J. Parikka (eds.) *Media archaeology: Approaches, applications, and implications*. Berkeley, CA: University of California Press, pp. 1–26.

Innis, H. (1950). *Empire and communications*. Oxford: Oxford University Press.

Innis, H. (1951). *The bias of communication*. Toronto: University of Toronto Press.

Kauffman, S. (1995). *At home in the universe: The search for laws of self-organization and complexity*. New York, NY: Oxford University Press.

Latour, B. (2005). *Reassembling the social: An introduction to actor-network-theory*. Oxford: Oxford University Press.

Lehman-Wilzig, S. and Cohen-Avigdor, N. (2004). The natural life cycle of new media evolution: Inter-media struggle for survival in the internet age. *New Media & Society* 6 (6): 707–730.

Levinson, P. (1979). *Human replay: A theory of the evolution of media*. New York, NY: New York University dissertation.

Levinson, P. (1997). *The soft edge: A natural history and future of the information revolution*. New York, NY: Routledge.

Levinson, P. (1999). *Digital McLuhan: A guide to the information millennium*. New York, NY: Routledge.

Logan, R. K. (1986). *The alphabet effect*. Cresskill, NJ: Hampton.

Logan, R. K. (2000). *The sixth language: learning a living in the internet age*. Caldwell NJ: Blackburn Press.

Logan, R. (2007). The biological foundation of media ecology. *Explorations in Media Ecology* 6: 19–34.

Logan, R. K. (2010). *Understanding new media: Extending Marshall McLuhan*. New York, NY: Peter Lang.

Logan, R. K. (2014). *What is information? Propagating organization in the biosphere, the symbolosphere, the technosphere and the econosphere*. Toronto: Demo Publishing.

Lum, C. M. K. (2006). Notes toward an intellectual history of media ecology. In: C. M. K. Lum (ed.) *Perspectives on culture, technology and communication. The media ecology tradition*. Cresskill, NJ: Hampton Press, pp. 1–60.

Manovich, L. (2009). Cultural analytics: Visualizing cultural patterns in the era of "more media". Retrieved December 7, 2018, from http://manovich.net/content/04-projects/063-cultural-analytics-visualizing-cultural-patterns/60_article_2009.pdf

Manovich, L. (2013). *Software takes command*. New York, NY: Bloomsbury Academic.

Marx, K. (1976). *Capital. A critique of political economy*. Vol. 1. Harmondsworth: Penguin Books.

McLuhan, M. (1962). *The Gutenberg galaxy: The making of typographic man*. Toronto: University of Toronto Press.

McLuhan, M. (1964). *Understanding media: The extensions of man*. New York, NY: New American Library.

McLuhan, M. and McLuhan, E. (1992). *Laws of media: The new science*. Toronto: University of Toronto Press.

Meyrowitz, J. (1985). *No sense of place: The impact of electronic media on social behavior*. New York, NY: Oxford University Press.

Moretti, F. (2005). *Graphs, maps, trees: Abstract models for a literary history*. London: Verso.

Natale, S. (2016). There are no old media. *Journal of Communication* 66(4): 585–603.

Nelson, R. R. and Winter, S. G. (1982). *An evolutionary theory of economic change*. Cambridge, MA: Harvard University Press.

Neuman, W. R. (2010). *Media, technology, and society: Theories of media evolution*. Ann Arbor, MI: University of Michigan Press.

Olesen, M. (2016). Media evolution and "epi-technic" digital media: Media as cultural selection mechanisms. *Explorations in Media Ecology* 14(1 & 2): 141–160.

Ong, W. (2002). *Orality and literacy. The technologizing of the world* (1st ed., 1982). New York, NY: Routledge.

Parikka, J. (2012). *What is media archaeology?* Cambridge: Polity Press.

Postman, N. (1970). The reformed English curriculum. In: A. C. Eurich (ed.) *High school 1980: The shape of the future in american secondary education*. New York, NY: Pitman Pub. Corp, pp. 160–168.

Postman, N. (1985). *Amusing ourselves to death: Public discourse in the age of show business*. New York, NY: Viking Penguin.

Postman, N. (1992). *Technopoly: The surrender of culture to technology*. New York, NY: A. Knopf.

Postman, N. (1995). *The end of education: Redefining the value of school*. New York, NY: A. Knopf.

Postman, N. (1998). *Five things we need to know about technological change*. Conference in Denver, Colorado, 27 March 1998. Retrieved December 7, 2018, from www.mat. upm.es/~jcm/neil-postman-five-things.html

Saviotti, P. P. (1996). *Technological evolution, variety and the economy*. Cheltenham: Edward Elgar.

Scolari, C. A. (2010). Media ecology. Map of a theoretical niche. *Quaderns Del CAC 34* XIII(1): 17–25.

Scolari, C. A. (2013). Media evolution: Emergence, dominance, survival, and extinction in the media ecology. *International Journal of Communication* 7(2013): 1418–1441. doi: 1932–8036/20130005.

Scolari, C. A. (2015). *Ecología de los medios*. Barcelona: Gedisa.

Scolari, C. A. (2018a). Media evolution. In: P. Napoli *Mediated communication*. Berlin: de Gruyter Mouton. pp. 149–168.

Scolari, C. A. (2018b). *Las leyes de la interfaz*. Barcelona: Gedisa.

Scolari, C. A. (2019). *Media evolution. Sobre el origen de las especies mediáticas*. Buenos Aires: La Marca.

Sociology (n.d.). In Wikipedia. Retrieved December 7, 2018, from https://en.wikipedia. org/wiki/Sociology.

Solé, R. (2009). *Redes complejas. Del genoma a Internet*. Barcelona: Tusquets.

Stöber, R. (2004). What media evolution is: A theoretical approach to the history of new media. *European Journal of Communication* 19(4): 483–505.

Strate, L. (2004). A media ecology review. *Communication Research Trends* 23(2): 3–48.

Strate, L. (2008). Studying media as media: McLuhan and the media ecology approach. *MediaTropes eJournal* 1: 127–142.

Strate, L. (2017). *Media ecology: An approach to understanding the human condition*. New York, NY: Peter Lang.

Sumuts, J. C. (1927). *Holism and evolution*. London: Macmillan.

van Dijck, J. (2013). *The culture of connectivity: A critical history of social media*. New York, NY: Oxford University Press.

von Bertalanffy, L. (1968). *General system theory: Foundations, development, applications.* New York, NY: George Braziller.

Zielinski, S. (2008). *Deep time of the media: Toward an archaeology of hearing and seeing by technical means.* Cambridge, MA: The MIT Press.

Ziman, J. (ed.) (2000). *Technological innovation as an evolutionary process.* Cambridge: Cambridge University Press.

16

MEDIA PSYCHOLOGY

Emma Rodero

Definition of Media Psychology

Media psychology is a scientific discipline that examines how people interact with media technology (media uses) and the influence of this interaction (media effects). This discipline applies psychological theories and methods to understand how individuals and groups perceive, process and respond to the exposure to the media. Therefore, the subject matter is the cognitive processes, emotion and behavior underlying this media interaction (Dill, 2013).

Media interactions play a crucial role in our lives. We live surrounded and in permanent contact with media messages and inputs the entire day. We interact with our cellphone, we read and write social media messages, we listen to the news, we watch videos, to mention just some examples. This proliferation of different media technologies in the twenty-first century has led its users to be named "over-whelmed consumers" (Bryant & Davies, 2006; Okdie et al., 2014). According to eMarketer (2018), adults in the United States spend 12 hours and 8 minutes per day in direct interaction with media. New media have added to the traditional press, radio and television. At the same time, people not only have become an active audience, but they are also producers and distributors of media content, such as blogs or podcasts (Jenkins, Ford & Green, 2013). Consequently, research about the impact of these growing and influential technologies, especially about both their benefits and threats, is crucial to improving the communication process in media technology.

Background of Media Psychology

Media psychology is a recent science very linked to the evolution of technology, as is the development of communication studies. This discipline was born as a result of a paradigm shift in psychology: from behaviorism to cognitivism in the 1950s (Miller, 2003; Potter, 2013). Behaviorism focused on the study of behavior and their determinants. Researchers studied the association between an environmental stimulus and the behavioral input that this stimulus produced, so the focus was behavior. The mind was considered a "black box" that could not be explored. Cognitivism was a reaction to this approach. This movement understood the brain as a complex and dynamic system of information processing that influences behavior. With this new paradigm, psychologists studied the mental processes produced when individuals interacted with media messages and technology. Consequently, the origin of the discipline was in the field of psychology, but in later years this conception expanded to communication. The interaction between human–media technology was the subject area where psychology and communication converged. *Mass communication research* had a strong influence in this evolution, as it was the seed of media effects research (McQuail & Windahl, 1993). The influential theories of the *mass communication research* can be divided into those that conceived the effects and influence of mass media as very powerful (until the 1940s), as having only limited effects (1950–1970), and as causing an indirect impact (from 1970 onwards).

During the first years, due to the historical context (Nazi propaganda, radio popularization and impact of Hollywood productions), these theories considered that the media had a substantial direct effect on people's behavior. The audience was conceived as "mass" – as passive individuals easily moldable and influenceable. The most representative theory about the powerful effects of media was the *Magic Bullet Theory* or the *Hypodermic Needle Model*. As the name indicates, the approach suggested a powerful influence of media. Graphically, we can imagine a bullet directly inserted in the audience's brain to get the desired goal. Cantril described this powerful effect applied to a medium, the radio, in 1940 in the book *The Invasion from Mars. A Study in the Psychology of Panic*, now re-edited (2017). The author analyzed the impact of the broadcast of *The War of the Worlds* on Halloween night in 1938. Later, some studies reviewed this initial conception.

From 1950 to 1970, the media effects theories suggested that the influence of media was determined by the personal characteristics of the individuals, their social status and education. Therefore, the media's impact was limited and conditioned by these factors. One of the most representative theories of this period was the *Two-step Flow of Mass Communication* (Lazarsfeld, Berelson & Gaudet, 1944). This theory posited that media information flows in two steps. The first step of this media information process is the opinion leaders.

These leaders are the individuals who process and receive much information from the media. The second step occurs when these leaders spread their opinions to other people. As the influence of the leaders is big, they can change the attitudes or behaviors of the general audience. Another approach of this period is the *Uses and Gratifications Theory* (Rosengren, 1974). This theory was born as a reaction to the traditional conception of mass communication studies. It focuses on the analysis of the uses and gratifications that individuals and society achieve from the media: how they use media to satisfy their needs; what are the reasons why they expose themselves to the media; and what are the consequences.

From 1970 on, the influence of media is considered strong again, although now with indirect effects. The theories in the decades since assume that there is a powerful impact of media on social perception and interaction, but this impact is indirect. One of the main representative theories is the *Agenda-Setting Theory* (McCombs & Shaw, 1972). The underlying assumption of this theory is that the media select news based on an agenda, highlighting some topics over others. Therefore, the audience assumes these issues are the most important thanks to this prominence. Linked to this model is the *Framing Theory* (Goffman, 1974). The basic idea of this approach is that the media highlights some information over other events and, by doing so, place them in an outstanding frame. A frame can be defined as an abstraction that serves to structure the meaning of messages. These events, presented to media users as frames, influence how people process this information. Therefore, the media not only establishes the agenda of the most important events, but these events are selected in frames that influence information processing. The *Agenda-Setting Theory* is also related to the *Priming Theory* (Iyengar, Peters & Kinder, 1982). Priming is an essential model in media effects, although it is a cognitive psychology theory, rooted in the associative network model of human memory. By underlying some events, the media places them on the agenda. The model posits that memories are stored as nodes in networks. Priming is the activation of a node and helps people to understand and interpret the new information, as nodes act as a previous reference. Therefore, certain nodes are activated due to the prominence of some events shown in the media. The other representative theory in the last period of the study of media effects is the *Spiral of Silence Theory* (Noelle-Neumann, 1984). This assumption suggested that people remain silent when they feel that they are in the minority, as they are afraid to be isolated. Therefore, the media can influence the opinion and behavior of the individuals. Lastly, *Cultivation Theory* (Gerbner & Gross, 1976) postulated that the more you are exposed to a media, the more you will think that its messages are valid and real. High frequent viewers of TV believe that the world described in their messages is a picture of the real world. If these messages contain violent actions, then heavy media users will think that the world is a violent and dangerous place.

Evolution of Media Psychology

There are some steps in the evolution of media psychology that were influential to its development. Concerned about the growing impact of technology on society, the *American Psychological Association* (APA) founded Division 46 of Media Psychology in 1986. This act is considered the official emergence of media psychology as a field (Fischoff, 2005). As Giles explains (2003), many members of this division were clinicians. Due to this origin, for some years there was some confusion about the nature of the discipline. The initial goal in those years was to disseminate psychological information in media programs. Many psychologists hosted or participated in radio or television programs to inform people about psychological problems and treatments. The popularity of radio call-in psychology programs in the 1980s can be considered part of the evolution (Bouhoutsos, Goodchilds & Huddy, 1986). The first publication in Division 46 was *Perspectives on Psychology and the Media*, edited by Kirschner and Kirschner (1997). The objective was to understand the effect of technology on human behavior or how the brain interacted with an environment mediated by technology. In 1998, Division 46 added new occupations and roles to media psychology and included digital media.

Another essential step in the evolution of this emerging discipline was the foundation of different journals devoted to media psychology. The *Journal of Media Psychology* (Hogrefe & Huber) was born as *Zeitschrift für Medienpsychologie* in Germany in 1989. The journal *Media Psychology* was born in 1999 in the United States (Taylor & Francis/Routledge). Both are indexed in communication and psychology in the *Journal Citation Reports* (Thomson Reuters). *Media Psychology* was founded with the purpose of understanding and explaining "the roles, uses, processes, and effects of mediated communication in complex information societies" (Bryant & Roskos-Ewoldsen, 1999). *The Amplifier Magazine* is the official newsletter of Division 46 and showcases the discussions and activities of the members. Also, *Media Psychology Review* is an academic online journal devoted to studying media psychology and emerging technologies. The *Media Psychology Research Center* publishes this journal, a non-profit aimed at examining the interaction between human experience and media technologies.

Along with the journals, some master programs about media psychology emerged at the beginning of the twenty-first century. The *Fielding Graduate University* in the United States launched the first doctoral program in 2003. Since then, some communication faculties set up laboratories devoted to studying the cognitive processing of media messages, first in the United States and more recently in Europe, especially in Germany. The study by Okdie et al. (2014) reported 16 media psychology courses (12 undergraduate and 4 graduate) in the top 53 psychology programs in the United States. In Europe, there is a longer tradition – especially in England, The Netherlands, and Germany – of studying the psychological effects of media exposure (Roskos-Ewoldsen, 2001).

The evolution of the discipline was fast. In the first years (2003–2004), according to Google Trends, the term media psychology attained considerable popularity (66 over 100). Division 46 also extended the interests and research of the field to social and mobile technologies (Rutledge, 2012). Despite this growth, a content analysis of four journals in psychology showed a low number of papers devoted to media analysis (Roskos-Ewoldsen, 2004). Since then, the interest has remained, but with moderate popularity in Google Trends (45). The study by Okdie et al. (2014) examined five top journals in psychology from 2003 to 2014 including two from Rosskos-Ewolsen's study. The results were very similar, and only 1.6% of the papers focused on media. A search of the term in Google during 2018 produced more than 550,000 results between articles, web pages and other resources. However, this interest was concentrated in a few countries in this order: Pakistan; Philippines; England; Australia; the United States; Canada; India; The Netherlands; and Germany. On these grounds, Okdie et al. (2014) consider the interest of psychologists in the development of the study of media as "negligible." Part of the problem has the origin in the interdisciplinarity that it has to combine interests from two very different disciplines: psychology and communication. For example, one interesting question is if the media psychology research should be in psychology or a communication department (Giles, 2003; Igartua & Moral, 2012). In any case, this field can represent an opportunity for media researchers, as there is still a long way to extend the discipline in other countries.

Current Situation and Applications of Media Psychology

Today, in psychology, this discipline is considered as one of the most innovative fields. As Reeves and Anderson (1991) state, the study of media enriches the psychology discipline, as it provides a new framework to study the cognitive processing in complex, dynamic and real situations. From the perspective of communication, media psychology is conceived in a broad sense and not focused exclusively on technology but the human being conceived as a processor interacting with many forms of communication. The discipline embraces not only the traditional mass communication media (press, radio and television) but also all the mediated forms of communication (video games, internet, social networks, etc.). Consequently, media psychology is a productive field that benefits two main areas. On the one hand, psychology provides the great theories and research about the human cognitive system and the study of how the audience use and process mediated messages. On the other hand, communication enriches psychology with the knowledge and studies about complex mediated messages in a real context.

This productive combination of research has evolved gradually. In recent years, media psychology has benefited the emergence of two important fields related to technology and psychology – human–computer interaction and neuroscience. Regarding technology, human–computer interaction or computer-mediated communication specifically studies the interaction mediated through a computer. In 2017, the journal with the most impact factor in communication in the JCR was the *Journal of Computer-Mediated Communication*, which gives us an idea about the importance of this area in communication research. Technology and its effects on people continue to be very present as an area of study. In 2017, four journals about technology were indexed in communication. Moreover, the change of the name of Division 46 in 2012 from *Media Psychology* to *The Society for Media Psychology and Technology* is a clear sign of technology's importance. Regarding neuroscience, there is a new field named *Media Neuroscience* in the intersection of media psychology and cognitive neuroscience (Weber, Eden, Huskey, Mangus & Falk, 2015).

Media psychology covers a wide variety of topics. Baker Derwin and De Merode (2013) identified the most discussed topics in the journal *Media Psychology* between 1999 and 2010: ethics; advertising; marketing; art; news; attitude; gender; identity; politics; education; narrative; and health. By the number of mentions, the medium of television was the most cited, although the analysis of video games and social networks is increasing. A recent study by Reeves, Yeykelis and Cummings (2016) identified the most common stimuli measured in studies published in *Media Psychology* from 2004 to 2013. These stimuli included video games, texts, television content, movies, public service messages, images, websites, VR, print messages, music messages and non-music messages. These analyses confirm that there are many studies about the impact of violent content in games and television and their influence on children and adults (Anderson & Bushman, 2002). In conclusion, the applications can be varied in: the field (health and well-being, gender, moral, education, politics, etc.); genre (advertising, entertainment, narrative, marketing, non-verbal communication, business, interactive communication, news, games, etc.); outlet or technology (television, radio, mobile communication, cinema, digital media, social media, VR, etc.); and the analysis (behavior, cognition, addictions, violence or emotions) (Gahlowt, 2015).

Main Theories and Approaches of Media Psychology

As a discipline, media psychology represents a paradigm that integrates diverse theories, conceptions and methods. We will briefly review the essential theoretical approaches in the field.

Cognitive Dissonance Theory (Festinger, 1957) assumes that an individual can experience a cognitive dissonance or mental discomfort when holding two or more contradictory ideas. In this situation, people will try to reduce this mental discomfort by solving the contradiction between the new ideas and the pre-existing beliefs so that they can recover the cognitive balance. Therefore, people always look for information that supports their initial positions and, in this manner, to achieve psychological consistency to survive in the real world. *Selective Exposure Theory* is rooted in *Cognitive Dissonance* and argues that people tend to select aspects of the information that reinforce their pre-existing perspectives, perceptions, decisions or beliefs. As with the *Uses and Gratifications Theory*, it postulates that people use media to satisfy their needs, but, in this case, according to previous beliefs. Therefore, when people select information, there is confirmation bias toward favorable messages. People tend to select media content that confirms their thoughts and beliefs. As media users are not fully aware of the reasons that motivate this selection of information, this selective exposure can profoundly affect people's decisions and behavior. Research based on this theory usually uses observational methods (Knobloch-Westerwick, 2015).

Social Learning Theory (Bandura, 1971) posits that we learn by observing and imitating the social behavior of other people. Learning is a cognitive process developed in a social context by looking at social behavior and observing the rewards and punishments. If the behavior is continuously rewarded, then it will be reinforced. Conversely, if the behavior is constantly punished, the individual will desist. In recent years, this approach has applied to how we learn from new media especially the internet and social networks. Social interaction is also crucial in *Social Information Processing* (Walther, 1992). The model considers that forms of computer-mediated communication provide an opportunity to connect and build interpersonal relationships with others. These relationships are based on the impressions that people create on media. People need to build an identity and use media to create favorable impressions on the others, for example, in social networks like Facebook. This theory has given rise to different applications in interpersonal communication, education and marketing. Dodge and Crick (1990) examined aggressive behavior in children. In the same vein, the *general aggression model* is a theory that analyzes aggression and violence from a cognitive-social perspective (Allen, Anderson & Bushman, 2018). The model integrated the *Priming* and *Social Learning* theories. This approach posited that that human aggression was influenced by a wide variety of factors such as beliefs, attitudes, perceptions, cognition or feelings.

Another theory about behavior is the *Theory of Planned Behavior* (Ajzen, 1991), which represents the development of the *Theory of Reasoned Action* (Ajzen, 1992). This approach assumes that the individual's behavior is conditioned by subjective norms (normative beliefs), attitudes (behavioral beliefs) and perceived behavioral control (control beliefs). This last element, the perceived behavioral control, is the part that was incorporated into the final model. This control determines the intentions of enacting certain behaviors. If people have a more favorable attitude toward the behavior and subjective norm and more control over a given behavior, then they are more likely to do it. Therefore, behavioral intention grows when people assume that they can control this behavior. This model has been used in the study of media users' behavior and the adoption of new technologies.

Regarding technology, one important model is *Media Equation* (Reeves & Nass, 1996). This communication theory conceives that humans often treat and respond to computer-mediated communication, as if they were receiving information face-to-face from a real person. The reason for this behavior is that the human brain is not adapted to new technologies and, therefore, people behave like they would in a human interaction. People's social answers then do not come from media but from themselves. Consequently, the response can be the same for television as for a computer. This approach assumes that these natural and social responses are automatic, with no effort; thus, the media equation can happen even in passive use of media.

Concerning processing, one representative theory is the *Elaboration Likelihood Model of Persuasion* (Petty & Cacioppo, 1986). This model is a dual process theory that focuses on the effectiveness of persuasion by describing how the change of attitudes is produced. It proposes two primary routes to persuasion – central and peripheral. These are two different ways of processing messages. The central route is based on the cognitive evaluation of the arguments (elaboration) while the peripheral depends on the affective associations linked to cues in persuasion messages. In other words, the central route is related to arguments, and cognitive evaluation and the peripheral is connected to attitudes and an affective evaluation. As the central route involves deep processing for elaborating strong arguments, the result is a resistant and durable attitude change. There are two influence factors of processing messages centrally: motivation, based on the personal interest of the individual in the topic; and the capacity to elaborate. If these factors are high, then the message will be processed mainly via the central route. If these factors are low, then the peripheral route becomes more important. In the peripheral route, people do not evaluate the message thoroughly. They are guided more by general impressions and moods, as they do not have an interest and look to reduce mental effort. Therefore, in this context, the elaboration is low, and the factors that affect the peripheral route are more related to the form of the message: credibility; the attractiveness of the sources; or the production quality. In this case, the attitude change, if produced, is not so consistent and robust as in the central route.

The most representative model for the study of cognitive processing of mediated messages is the *Limited Capacity Model Motivated Mediated of Message Processing* (LC4MP) (Lang, 2011). The LC4MP information-processing model of cognition is based on the idea that the capacity of the human cognitive system to process all the information contained in a message is limited. This information processing depends on the type and the number of cognitive resources that are allocated to process messages. Regarding the type, some cognitive processes demand automatic resources and others need controlled resources. The automatic resources do not require an excessive amount of attention and effort (i.e., driving). The controlled resources demand attention and a considerable effort to process (i.e., reading a book). Concerning the number, the amount of resources that an individual needs to process a message depends on the two characteristics of the stimuli – structural complexity and information density. Structural features of a message allocate automatic resources for processing the message while content features require controlled resources. This means that the structural features of the message can attract the listener's attention, automatically eliciting an orientating response (OR). An OR represents a temporal increase in attention to process a novel stimulus in the environment. A large body of research framed in this theory has been conducted for identifying structural features that can evoke ORs – in television (camera changes, sudden movement, etc.), in radio (music onsets, voice changes, intonation variations, sound effects, etc.) or in the internet (emotional pictures, etc.) (Lang, Park, Sanders-Jackson, Wilson & Wang, 2007; Potter, 2000; Potter, Lang & Bolls, 2008; Rodero, Potter & Prieto, 2017). Lang revisited the LC4MP later and named it *Dynamic human-centered communication systems theory* (2014). This new approach understands the mediated process between the human mind and media content as a dynamic interaction which varies depending on the level of motivation of the human being. To support their assumptions, this model uses psychophysiological measures (Lang, Potter & Bolls, 2009).

Methods and Techniques of Media Psychology

Due to the diversity of theories, media psychology employs a wide variety of qualitative and quantitative methods. Qualitative techniques are the minority in the discipline. In-depth interviews, surveys and focus groups are the most employed. An interview is a technique to obtain information about an individual's media uses and experiences. While interviews are individual, focus groups differentiate from interviews in that they are group discussions with a moderator. However, the topics for discussion can be the same based on examining media interactions. Information about media can also be collected using surveys about specific issues by responding to different questions. Surveys and psychological tests can also be quantitative methods. Researchers also can use standardized psychological tests to collect information about people's personality traits, emotional states, attitudes, or behaviors.

The most common research methods in media psychology are the quantitative techniques, due to the psychology influence. There are two primary methods when analyzing media interaction – behavioral and psychophysiological analysis.

Behavioral analysis is central to some theories, such as *Selective Exposure* (Knobloch-Westerwick, 2015). Observation is a method of collecting data for the study of media users' behavior. Researchers are immersed in a situation where the behavior to analyze is produced. There are different observation methods – naturalistic and controlled. First, naturalistic observation is based on the study of spontaneous behavior of participants in natural environments without interfering with them. Researchers record the events that they observe and the relationships among them. The main advantage of this method is that the studies have greater ecological validity, as the researchers observe behavior in the natural environment with no interference. However, a limitation is that these observations can be conducted on a reduced scale, in specific situations, and it is difficult to extrapolate the results to wide samples. Controlling the different variables acting at the same time also complicated. Second, controlled observation is a method where the observation is in a controlled environment as, for example, a laboratory. The researcher decides where the observation is performed and all the conditions of the experiment. With this technique, the researcher has total control over the environment and can use different equipment to measure and record the participants' behavior. Behavior is measured with different scales, which are structured in various categories. The strength of this technique is that the observation is controlled and can be replicated in other studies. The limitation has less ecological validity.

Along with behavior analysis, researchers studying media effects conduct experiments in laboratories using a biological perspective. Many studies in the field have started to adopt this emerging approach (see Weber et al., 2015). The biological perspective includes psychophysiology and neuroscience. The psychophysiological paradigm or *media psychophysiology* is central to the LC4MP (Lang et al., 2009), and according to Giles (2003), is one of the main approaches to the study of information processing. *Media psychophysiology* applies psychophysiological measurements to the analysis of cognitive processing (Bartholow & Bolls, 2013; Potter & Bolls, 2012). The main idea underlying this framework is that all that happens in our mind when processing messages have a direct effect on our body. This phenomenon is called "embodiment." Therefore, studying the physiological responses, we can understand the psychological processes involved in a media interaction. In this method, we can see how psychology has influenced communication.

Psychophysiology is a well-established discipline that stands between psychology and physiology. This discipline was founded in the 1950s in the United States and materialized in the creation of the *Society for Psychophysiological Research* in 1960. Psychophysiology aims at understanding the relationship between what people think, feel and do, and how our body reacts biologically. The measurement of the body's physiological response allows the understanding of the psychological processes (Cacioppo, Tassinary & Berntson, 2000). To do so, psychophysiology studies the human nervous system – central and peripheral. The central nervous system (CNS) is composed of the brain and the spinal cord. The main techniques to measure the activity of the CNS are: electroencephalography (EEG), to register the electrical activity of the brain; the event-related potentials (ERPs) which use EEG to measure the brain response to a specific cognitive, sensory or motor event; and the functional magnetic resonance imaging (fMRI), to analyzes brain activity by detecting changes in the blood flow. The peripheral nervous system (PNS), formed by the nerves and ganglia outside of the CNS, connects it to the organs of the body. It performs sensory and motor functions and is divided into the autonomic nervous system and the somatic-sensory nervous system. The autonomic nervous system, in turn, is divided into the sympathetic and parasympathetic nervous systems. The sympathetic nervous system activates the body and predisposes it to the action. It is responsible for the fight-or-flight response, the physiological reaction to threats. Conversely, the parasympathetic nervous system is the antagonist and coordinates the body at rest with a function of energy conservation. The most common measurements to record the PNS are the skin conductance or electrodermal activity by registering the galvanic response (EDA), the cardiovascular activity using an electrocardiogram (EKG or ECG), the salivary gland activity by analyzing samples or the pupil movement by using an Eye-Tracker. Finally, the somatic nervous system is responsible for sending sensory information to the CNS and control muscle activity. Some techniques to register the activity of the somatic nervous system are electromyography (EMG), respiratory activity, laryngeal activity or eye movements. The most used of all these techniques in media are mainly EEG for brain activity, EKG for heart rate, EDA for skin conductance, eye-tracking and facial muscle movements measured through EMG (Potter & Bolls, 2012).

What is now new in the communication field is the application of psychophysiology methods to study media interactions. The main benefit of this measurement is that it allows for registering responses to stimuli in real-time during the media exposure. Therefore, participants are not asked afterward how they have perceived or processed the message. In self-reported data, individuals can be conditioned by what is socially accepted and can exaggerate their answers or lie for this reason. Also, sometimes it is complicated for participants to measure certain cognitive variables, as, for example, attention. Using psychophysiological methods can solve these problems, as the data are being collected during the exposure to the stimulus and unconscious processing is also measured. Moreover, the answers can be analyzed in specific moments, which is very useful in media analysis. For example, we could analyze the influence of music in commercials and to collect the physiological response at the exact moment in which the melody appears. Therefore, temporal accuracy is an excellent benefit of these techniques (Laaksonen, Salminen, Falco, Aula & Ravaja, 2013). As this measurement can be complemented with other methods (for example, scales or post-hoc surveys), the technique provides a very complete radiography of the individual's information processing based on reliable data.

Conclusion

In conclusion, media psychology was born as a discipline to understand how people use media and how they are influenced by it. As technology constantly evolves and people spend a great amount of time exposed to media, we can conclude that it is an emerging and thriving discipline. Two fields fuel media psychology – psychology (tradition, theories and methods), and communication (application in real context, knowledge and theories about media). For communication researchers, media psychology can be a discipline that reinforces research into the vital role of media studies in our society, especially as communication has been considered a discipline in crisis (Lang, 2013; Mas & Rodero, 2018). Media psychology can be an interesting option to overcome this crisis by applying the psychological concepts, theories and methods to the study of cognitive processes, emotion and behavior underlying the numerous media interactions that people have each day (Rodero, Mas & Larrea, 2018). Media psychology can be a promising field to strengthen the scientific study of these communication processes.

References

Ajzen, I. (1991). The theory of planned behavior. *Organizational Behavior and Human Decision Processes*, *50*(2), 179–211.

Ajzen, I. (1992). A comparison of the theory of planned behavior and the theory of reasoned action. *Personality and Social Psychology Bulletin*, *18*, 3–9.

Allen, J. J., Anderson, C. A. & Bushman, B. J. (2018). The general aggression model. *Current Opinion in Psychology*, *19*, 75–80.

Anderson, C. A. & Bushman, B. J. (2002). Human aggression. *Annual Review of Psychology*, *53*, 27–51.

Baker Derwin, E. & De Merode, J. (2013). Inside media psychology: The story of an emerging discipline as told by a leading journal. In E. Baker Derwin (Ed.), *The Oxford handbook of media psychology* (pp. 75–95). Oxford: Oxford University Press.

Bandura, A. (1971). *Social learning theory*. New York: General Learning Press.

Bartholow, B. D. & Bolls, P. (2013). Media psychophysiology: The brain and beyond. In K. Dill (Ed.), *The Oxford handbook of media psychology* (pp. 474–495). New York: Oxford University Press.

Bouhoutsos, J. C., Goodchilds, J. D. & Huddy, L. (1986). Media psychology: An empirical study of radio call-in psychology programs. *Professional Psychology: Research and Practice*, *17*(5), 408.

Bryant, J. & Davies, J. (2006). Selective exposure processes. In J. Bryant & P. Vorderer (Eds.), *Psychology of entertainment* (pp. 19–33). Mahwah, NJ: Erlbaum.

Bryant, J. & Roskos-Ewoldsen, D. (1999). Raison d'Etre. *Media Psychology*, *1*, 1–2.

Cacioppo, J. T., Tassinary, L. G. & Berntson, G. G. (Eds.). (2000). *Handbook of psychophysiology*. Cambridge: Cambridge University Press.

Cantril, H. (2017). *The invasion from Mars: A study in the psychology of panic*. New York: Routledge.

Dill, K. (ed.). (2013). *The Oxford handbook of media psychology* (pp. 474–495). New York: Oxford University Press.

Dodge, K. A. & Crick, N. R. (1990). Social information-processing bases of aggressive behavior in children. *Personality and Social Psychology Bulletin, 16*(1), 8–22.

eMarketer (2018). US time spent with media 2018. Retrieved from www.emarketer. com/content/us-time-spent-with-media-2018.

Festinger, L. (1957). *A theory of cognitive dissonance.* California: Stanford University Press.

Fischoff, S. (2005). Media psychology: A personal essay in definition and purview. *Journal of Media Psychology, 10*(1), 1–21.

Gahlowt, R. (2015). *Introduction to media psychology.* Mumbai: Himalaya.

Gerbner, G. & Gross, L. (1976). Living with television: The violence profile. *Journal of Communication, 26*(2), 172–199.

Giles, D. (2003). *Media psychology.* USA: Routledge.

Goffman, E. (1974). *Frame analysis: An essay on the organization of experience.* New York, NY et al.: Harper & Row.

Igartua, J. J. & Moral, F. (2012). Media psychology: An overview and perspectives. *Escritos De Psicologia, 5*(3), 1–3.

Iyengar, S., Peters, M. D. & Kinder, D. R. (1982). Experimental demonstration of the 'not-so-minimal' consequences of television news programs. *American Political Science Review, 76,* 848–858.

Jenkins, H., Ford, S. & Green, J. (2013). *Spreadable media: Creating value and meaning in a networked culture.* New York, NY: New York University Press.

Kirschner, S. & Kirschner, D. A. (1997). *Perspectives on psychology and the media.* Washington, DC: American Psychological Association.

Knobloch-Westerwick, S. (2015). *Choice and preference in media use: Advances in selective exposure theory and research.* USA: Routledge.

Laaksonen, S. M., Salminen, M., Falco, A., Aula, P. & Ravaja, N. (2013). Use of psychophysiological measurements in communication research: Teachings from two studies of corporate reputation. *Journal for Communication Studies, 6*(1), 245–255.

Lang, A. (2011). The limited capacity model of motivated mediated message processing. In R. Nabi & M. B. Oliver (Eds.), *The SAGE handbook of mass media processes and effects* (pp. 193–204). Thousand Oaks, CA: Sage.

Lang, A. (2013). Discipline in crisis? The shifting paradigm of mass communication research. *Communication Theory, 23*(1), 10–24.

Lang, A. (2014). Dynamic human-centered communication systems theory. *The Information Society, 30*(1), 60–70.

Lang, A., Park, B., Sanders-Jackson, A. N., Wilson, B. D. & Wang, Z. (2007). Cognition and emotion in TV message processing: How valence, arousing content, structural complexity, and information density affect the availability of cognitive resources. *Media Psychology, 10*(3), 317–338.

Lang, A., Potter, R. F. & Bolls, P. D. (2009). Where psychophysiology meets the media: Taking the effects out of media research. In J. Bryant & M. B. Oliver (Eds.), *Media effects: Advances in theory and research* (pp. 185–206). New York, NY: Routledge.

Lazarsfeld, P. F., Berelson, B. & Gaudet, H. (1944). *The people's choice: How the voter makes up his mind in a presidential campaign.* New York: Columbia University Press.

Mas, L. & Rodero, E. (2018). The Interactive Communication Process (ICP): A model for integrating science, academia, and profession. *Communications, 43*(2), 173–207.

McCombs, M. E. & Shaw, D. (1972). The agenda-setting function of mass media. *POQ, 36,* 176–187.

McQuail, D. & Windahl, S. (1993). *Communication models for the study of mass communication*. London: Longman.

Miller, A. (2003). The cognitive revolution: A historical perspective. *Trends in Cognitive Sciences*, 7(3), 141–144.

Noelle-Neumann, E. (1984). *The spiral of silence: Public opinion – Our social skin*. Chicago: University of Chicago.

Okdie, B. M., Ewoldsen, D. R., Muscanell, N. L., Guadagno, R. E., Eno, C. A., Velez, J. A. … Smith, L. R. (2014). Missed programs (you can't TiVo this one) why psychologists should study media. *Perspectives on Psychological Science*, 9(2), 180–195.

Petty, R. E. & Cacioppo, J. T. (1986). The elaboration likelihood model of persuasion. In R. E. Petty & J. T. Cacioppo (Eds.,)*Communication and persuasion* (pp. 1–24). New York, NY: Springer.

Potter, R. (2000). The effects of voice changes on orienting and immediate cognitive overload in radio listeners. *Media Psychology*, 2(2), 147–177.

Potter, R. F. & Bolls, P. D. (2012). *Psychophysiological measurement and meaning: Cognitive and emotional processing of media*. Routledge Communication series. CONT.

Potter, R. F., Lang, A. & Bolls, P. D. (2008). Identifying structural features of audio: Orienting responses during radio messages and their impact on recognition. *Journal of Media Psychology: theories, methods, and Applications*, 20, 168–177.

Potter, W. J. (2013). A general framework for media psychology scholarship. In E. K. Dill (Ed.), *The Oxford handbook of media psychology* (pp. 474–495). Nueva York: Oxford University Press.

Reeves, B. & Anderson, D. R. (1991). Media studies and psychology. *Communication Research*, 18(5), 597–600.

Reeves, B. & Nass, C. (1996). *The media equation: How people treat computers, television, and new media like real people and places*. Cambridge: Cambridge University Press.

Reeves, B., Yeykelis, L. & Cummings, J. J. (2016). The use of media in media psychology. *Media Psychology*, 19, 49–71.

Rodero, E., Mas, L. & Larrea, O. (2018). Media psychology: A new discipline to understand the mediated processes. In C. Caffarel, et al. (Ed.), *Methodological tendencies in communication research* (pp. 189–201). Salamanca, Spain: Comunicación Social.

Rodero, E., Potter, R. F. & Prieto, P. (2017). Pitch range variations improve cognitive processing of audio messages. *Human Communication Research*, 43(3), 397–413.

Rosengren, K. E. (1974). Uses and gratifications: A paradigm outlined. In J. G. Blumler & E. Katz (Eds.), *The uses of mass communications: Current perspectives on gratifications research* (pp. 269–286). Beverly Hills, CA: Sage.

Roskos-Ewoldsen, D. R. (2001). Possible relationships between psychology and the media. *The American Journal of Psychology*, 114(4), 641.

Roskos-Ewoldsen, D. R. (2004). The psychology of the media: A missing program. *Contemporary Psychology*, 49, 560–562.

Rutledge, P. B. (2012). Arguing for a distinct field of media psychology. In K. Dill (Ed.), *Oxford handbook of media psychology* (pp. 43–58). New York: Oxford University Press.

Walther, J. B. (1992). Interpersonal effects in computer-mediated interaction: A relational perspective. *Communication Research*, 19(1), 52–90.

Weber, R., Eden, A., Huskey, R., Mangus, J. M. & Falk, E. (2015). Bridging media psychology and cognitive neuroscience. *Journal of Media Psychology*, 27(3), 146–156.

CONTRIBUTORS

Paul Cobley is Professor in Language and Media at Middlesex University. He is the author of a number of books, most recently *Cultural Implications of Semiotics* (2016) and *Narrative* 2nd edn (2014). He is co-series editor (with Kalevi Kull) of *Semiotics, Communication and Cognition* (de Gruyter Mouton), co-editor (with Peter J. Schulz) of the multi-volume *Handbooks of Communication Sciences* (de Gruyter), co-edits the journal *Social Semiotics* and is associate editor of *Cybernetics and Human Knowing*. He is the Ninth Thomas A. Sebeok Fellow of the Semiotic Society of America and is President of the International Association for Semiotic Studies.

Johan Fornäs is Professor Emeritus of Media and Communication Studies and a member of Academia Europaea. He lives in Stockholm in Sweden and Haut de Cagnes in France. He has studied and worked at the universities of Lund, Göteborg, Stockholm, Linköping and Södertörn. His publications include *Cultural Theory and Late Modernity* (1995); *In Garageland: Rock, Youth and Modernity* (1995); *Digital Borderlands: Cultural Studies of Identity and Interactivity on the Internet* (2002); *Consuming Media: Communication, Shopping and Everyday Life* (2007); *Signifying Europe* (2012); *Capitalism: A Companion to Marx's Economy Critique* (2013); *Europe Faces Europe: Narratives from Its Eastern Half* (2017); and *Defending Culture: Conceptual Foundations and Contemporary Debate* (2017).

Usha S. Harris is an academic in the Department of Media, Music, Communication and Cultural Studies at Macquarie University, Sydney, Australia. She uses information and communication technologies as tools for social inclusion, a philosophy that has guided her research and academic practice. Usha has provided participatory media training to NGO workers, media trainers and Pacific island communities on a broad range of environmental and social change issues.

Emma A. Jane is Associate Professor at UNSW Sydney. She researches the social implications of emerging technologies using transdisciplinary methods to interrogate the issues and consider proposed interventions. Having previously led a major study on gendered cyberhate, her current project involves using complex systems theory to explain high rates of irreplicability in science. Prior to her career in academia, Jane spent nearly 25 years working in the print, broadcast and electronic media during which time she won multiple awards for her writing and investigative reporting. Her tenth book – *Misogyny Online: A Short (and Brutish) History* – was published by Sage in 2017.

Kim Ebensgaard Jensen is Associate Professor in English Linguistics at the University of Copenhagen, Denmark. his research focuses on cognitive linguistics, corpus linguistics, and construction grammar.

Derek Johnson is an Associate Professor in Media and Cultural Studies at the University of Wisconsin-Madison. His research investigates the media industries and specifically the ways in which production operates as a site of cultural and creative struggle.

Gooyong Kim (PhD Cultural/Media Studies, UCLA) is Assistant Professor of Communication Arts at the Cheyney University of Pennsylvania. His research includes critical studies, media aesthetics/criticism, media literacy, political economy of the media, cultural politics of neoliberalism, and social movements. He has published a book on K-pop's broader economic, cultural and social implications, *From Factory Girls to K-Pop Idol Girls: Cultural Politics of Developmentalism, Patriarchy, and Neoliberalism in Korean Popular Music Industry* with Lexington Books.

Normand Landry is Canada Research Chair in Media Education and Human Rights and Professor at TÉLUQ University. He holds a PhD from McGill University and has conducted postdoctoral research projects at Concordia University. Normand's work focuses on communication rights, media education, social movement theory, law and democratic communications.

Anne-Sophie Letellier is Doctoral Student and Lecturer at the *École des Médias* of the University of Quebec in Montreal (UQAM). She is Research Assistant at the Canada Research Chair in Media Education and Human Rights at TÉLUQ University and in the "Groupe de recherche sur l'information et la surveillance au quotidian" (GRISQ) at UQAM. Her research focuses on critical infrastructure studies, public policy, digital security, social justice and hacktivism.

Christophe Magis is Associate Professor of Communication and Media Studies at the Université Paris 8, France, and a member of the Cemti laboratory. His research focuses on the critical political economy of the cultural and communication industries and the epistemology of critical theories in communication studies. He recently co-edited (in French) the third volume of the series *Matérialismes, culture et communication* on the figures of the political economy of communications: Fabien Granon, Jacques Guyot et Christophe Magis (dir.), *Matérliasmes, culture et communication. T.3 Économie politique de la culture, des medias et de la communication*, Paris, Presses des Mines, 2019.

Maria Murumaa-Mengel (PhD in Media and Communication) is Social Media Lecturer and the Program Director of Journalism and Communication at the Institute of Social Studies, University of Tartu. She is involved in research focusing mainly on young people's use (and non-use) of social media, digital literacies (e.g., social media literacies and porn literacies) and the transformation of private and public in online spaces. More specifically, her most recent research has looked into how online risks (cyberbullying, e-bile, online shaming) and opportunities (online-participation and creation, intimacy) are changing everyday practices of youth.

Tanner Mirrlees is Associate Professor of Communications and Digital Media Studies in the Faculty of Social Sciences and Humanities, Ontario Tech University. He is also Vice President of the Canadian Communication Association. Mirrlees is a critical political economist of the ICT and cultural industries, and the author of books such as *Global Entertainment Media: Between Cultural Imperialism and Cultural Globalization* (Routledge) and *Hearts and Mines: The US Empire's Culture Industry* (University of British Columbia Press); he is also the co-editor of *The Television Reader: Critical Perspectives in Canadian and US Television Studies* (Oxford University Press) and *Media Imperialism: Continuity and Change* (Rowman & Littlefield).

Sara Monaci is Associate Professor in Media and Communication at Politecnico di Torino (Italy). She teaches Technologies, Communication and Society and Future Storytelling in the Cinema and Media Engineering Degree and she's a member of R3/Responsible Risk Resilience Center, an interdepartmental research center on natural and anthropic risk and on processes related to resilience. Among her recent publications: "Designing a Social Media Strategy against Violent Extremism Propaganda: The #heartofdarkness Campaign," In: *Digital Medien und politisch-weltanschaulicher Extremismus im Jugendalter* (2018) and; "Explaining the Islamic State's Online Media Strategy: A Transmedia Approach," *International Journal Of Communication*, University of Southern California's Annenberg Center for Communication (2017).

Emma Rodero is the Director of the Media Psychology Lab in the Department of Communication at Pompeu Fabra University, Spain, PhD in Communication, PhD in Psychology, Master in Pathology of Voice and Master in Psychology of Cognition. She obtained a Marie Curie fellowship (European Union) to conduct research in the U.S.A. about cognitive processing of sound messages using psychophysiological techniques. She is author of 12 books and 70 scientific papers about voice, radio and sound. Rodero usually teaches public speaking and advertising on the radio at UPF. She has over a decade of experience in the radio industry. She is currently a voice-over artist and has received awards for several radio documentaries and dramas.

Michael Schandorf is the author of *A Gesture Theory of Communication: Media(tion), Meaning, and Movement* (Emerald Publishing Group, 2019). He has studied and taught communication and the making of meaning at institutions including the University of Illinois Chicago and the Massachusetts Institute of Technology, and is currently Lecturer at the University of British Columbia.

Adina Schneeweis is Associate Professor in the Department of Communication and Journalism at Oakland University. She specializes in representations of race and ethnicity in popular, institutional and political discourse, international communication and advocacy communication. Her research examines the Roma/Gypsy communities in the press, within politics and the activist movement for Roma rights, and in worldwide popular culture.

Carlos A. Scolari is Professor in the Department of Communication, University Pompeu Fabra, Barcelona, Spain. His research interests include interfaces, digital media, transmedia and media ecology/evolution. Incorporating mass media theories, he focuses on the emerging forms of communication facilitated by the rise of the internet.

Andra Siibak (PhD in Media and Communication) is Professor of Media Studies And Director of the doctoral program on Media and Communication at the Institute of Social Studies, University of Tartu. Her main field of research has to do with the opportunities and risks surrounding internet use, social media usage practices, intergenerational relationships on social media, new media audiences and privacy. In her most recent projects she has explored the topics related to the datafication of childhood, sharenting, teacher–student interactions on social media and digital device use of toddlers.

Nicole A Vincent is Senior Lecturer in the Faculty of Transdisciplinary Innovation at the University of Technology Sydney. She has taught, written and delivered talks on a wide range of topics including transgender-related public policy, gendered cyberhate and cybercrime, smart drugs (aka cognitive enhancement), biomedical moral enhancement, free will and determinism and how emerging technologies such as gene editing, blockchain and autonomous vehicles can foster human flourishing. Her edited volume *Neuroscience and Criminal Responsibility* (OUP) is a core reference work on the topic of neuro-law, and she has another forthcoming volume in the same series entitled *Neurointerventions and the Law: Regulating Human Mental Capacity*.

Olivier Voirol is Senior Lecturer at the University of Lausanne and associated member of the Institut für Sozialforschung (IfS) in Frankfurt am Main. After studies in social sciences and philosophy in Switzerland, France and Germany, he worked as Researcher at the IfS and the EHESS, Paris. His research interests are in the transformations of the public sphere, recognition and value theory, affects and technologies in digital capitalism. He has devoted numerous texts to the topics of media, culture, communication and recognition in Critical Theory. He is currently researching the "pathologies of the public sphere."

INDEX